Recent Titles in This Series

163 **Zhong-Ci Shi and Chung-Chun Yang, Editors,** Computational mathematics in China, 1994

162 **Ciro Ciliberto, E. Laura Livorni, and Andrew J. Sommese, Editors,** Classification of algebraic varieties, 1994

161 **Paul A. Schweitzer, S. J., Steven Hurder, Nathan Moreira dos Santos, and José Luis Arraut, Editors,** Differential topology, foliations, and group actions, 1994

160 **Niky Kamran and Peter J. Olver, Editors,** Lie algebras, cohomology, and new applications to quantum mechanics, 1994

159 **William J. Heinzer, Craig L. Huneke, and Judith D. Sally, Editors,** Commutative algebra: Syzygies, multiplicities, and birational algebra, 1994

158 **Eric M. Friedlander and Mark E. Mahowald, Editors,** Topology and representation theory, 1994

157 **Alfio Quarteroni, Jacques Periaux, Yuri A. Kuznetsov, and Olof B. Widlund, Editors,** Domain decomposition methods in science and engineering, 1994

156 **Steven R. Givant,** The structure of relation algebras generated by relativizations, 1994

155 **William B. Jacob, Tsit-Yuen Lam, and Robert O. Robson, Editors,** Recent advances in real algebraic geometry and quadratic forms, 1994

154 **Michael Eastwood, Joseph Wolf, and Roger Zierau, Editors,** The Penrose transform and analytic cohomology in representation theory, 1993

153 **Richard S. Elman, Murray M. Schacher, and V. S. Varadarajan, Editors,** Linear algebraic groups and their representations, 1993

152 **Christopher K. McCord, Editor,** Nielsen theory and dynamical systems, 1993

151 **Matatyahu Rubin,** The reconstruction of trees from their automorphism groups, 1993

150 **Carl-Friedrich Bödigheimer and Richard M. Hain, Editors,** Mapping class groups and moduli spaces of Riemann surfaces, 1993

149 **Harry Cohn, Editor,** Doeblin and modern probability, 1993

148 **Jeffrey Fox and Peter Haskell, Editors,** Index theory and operator algebras, 1993

147 **Neil Robertson and Paul Seymour, Editors,** Graph structure theory, 1993

146 **Martin C. Tangora, Editor,** Algebraic topology, 1993

145 **Jeffrey Adams, Rebecca Herb, Stephen Kudla, Jian-Shu Li, Ron Lipsman, and Jonathan Rosenberg, Editors,** Representation theory of groups and algebras, 1993

144 **Bor-Luh Lin and William B. Johnson, Editors,** Banach spaces, 1993

143 **Marvin Knopp and Mark Sheingorn, Editors,** A tribute to Emil Grosswald: Number theory and related analysis, 1993

142 **Chung-Chun Yang and Sheng Gong, Editors,** Several complex variables in China, 1993

141 **A. Y. Cheer and C. P. van Dam, Editors,** Fluid dynamics in biology, 1993

140 **Eric L. Grinberg, Editor,** Geometric analysis, 1992

139 **Vinay Deodhar, Editor,** Kazhdan-Lusztig theory and related topics, 1992

138 **Donald St. P. Richards, Editor,** Hypergeometric functions on domains of positivity, Jack polynomials, and applications, 1992

137 **Alexander Nagel and Edgar Lee Stout, Editors,** The Madison symposium on complex analysis, 1992

136 **Ron Donagi, Editor,** Curves, Jacobians, and Abelian varieties, 1992

135 **Peter Walters, Editor,** Symbolic dynamics and its applications, 1992

134 **Murray Gerstenhaber and Jim Stasheff, Editors,** Deformation theory and quantum groups with applications to mathematical physics, 1992

133 **Alan Adolphson, Steven Sperber, and Marvin Tretkoff, Editors,** *p*-Adic methods in number theory and algebraic geometry, 1992

(*Continued in the back of this publication*)

CONTEMPORARY MATHEMATICS

163

Computational Mathematics in China

Zhong-Ci Shi
Chung-Chun Yang
Editors

American Mathematical Society
Providence, Rhode Island

1991 *Mathematics Subject Classification*. Primary 62–02; Secondary 01A25.

Papers in this volume are final versions, which will not be published elsewhere.

Library of Congress Cataloging-in-Publication Data

Computational mathematics in China/Zhong-Ci Shi and Chung-Chun Yang, editors.
p. cm. — (Contemporary mathematics; v. 163)
Includes bibliographical references.
ISBN 0-8218-5163-2
1. Numerical calculations. 2. Mathematics—China—Data processing. I. Shih, Chung-tz'u. II. Yang, Chung-Chun, 1942– . III. Series: Contemporary mathematics (American Mathematical Society); v. 163.
QA297.C634 1994 — 93-45006
519.4—dc20 — CIP

∞ The paper used in this book is acid-free and falls within the guidelines
established to ensure permanence and durability.
♻ Printed on recycled paper.
This volume was printed from copy prepared by the authors.
This volume was typeset by the authors using AMS-LaTeX,
the American Mathematical Society's TeX macro system.

10 9 8 7 6 5 4 3 2 1 — 99 98 97 96 95 94

IN MEMORY OF FENG KANG

— The founder and pioneer of China's computational mathematics

Contents

Preface

Until two to three hundred years ago Chinese mathematics was flourishing, especially in the subject of arithmetic. Examples of its achievements include the calculation of π in the third century and the development of the Chinese Remainder Theorem in the twelfth century. Unfortunately, from this time until the 1950s, Chinese mathematics entered a period of stagnation. A re-awakening came in 1958 when, with the installation of the first Chinese made computer in the Chinese Academy of Sciences, China began to study modern computational mathematics.

Professor Feng Kang played a key role in this development. In the early 1960s his team and he, working in almost complete isolation from the Western World, established the finite element method. In 1965 he wrote a Chinese mathematical paper in which the first rigorous proof of the convergence of the method was given. This was a historically significant contribution to the field of numerical analysis and it was made by a Chinese mathematician. It did not, however, receive timely notice from Western researchers because Professor Feng's paper appeared only in a computational mathematical journal published in China, which ceased publication right after eruption in 1966 of the disastrous 10 year-long Cultural Revolution.

Since 1976, however, the finite element method has regained its momentum in China and impressive progress has been made both in theoretical research of the subject and in its applications. Computational mathematics,too, has rapidly gained a foothold in universities and in government industrial departments all over China. When the great enthusiasm among young students for computational mathematics and the great number of researchers and their abundant results in the subject is compared to the bleak and ignorant situation before 1960, we can proudly say that computational mathematics is one mathematical area that has really taken off in China.

In this collection, twelve key Chinese numerical analysts have presented their important and representative results obtained since the 1960s. The emphasis is on results gained in the last ten years and the topics covered fall into the following fields: finite element method, computational fluid mechanics, numerical solutions of differential equations, computational methods in dynamical systems, numerical algebra, approximation and optimization. We regret the coverage here may inevitably omit some important works due to the limitations of our energy and time.

We sincerely hope that the publications of this volume will promote further the academic exchange between East and West and allow the Western reader to have a systematic and updated comprehension of the contributions made by Chinese numerical analysts. We believe strongly that through the mutual understanding, common research problems will be identified and further cooperation can be expected. Undoubtedly both East and West will benefit from developments of this kind.

Although the research conducted by Chinese numerical analysts has its own merits, it has also suffered from certain drawbacks. The key problem is that China lags behind in her computing facilities. This has greatly hampered the testing of new algorithms and the implementation of theorems by applying them to practical problems. We are trying to overcome this situation.

When the collection was near to its completion, we were shocked by the news that Professor Feng–the founder and pioneer of China's computational mathematics research passed away suddenly in this August. Before the tragedy, he had been very concerned with the progress of the volume and even contributed an article himself. This is his last article which summarizes his most recent researches and contains many new and stimulating ideas. We regret that he would not witness the publication of this volume.

As the editors of the volume, we want to thank the contributors for their support and cooperation throughout the editing process. We also want to thank American Mathematical Society for its endorsement of the project. Finally, we are grateful to Drs. Wang Dao-Liu and Yu Xi-Jun for their work in preparing the typescript for this volume.

Zhong-Ci Shi 石钟慈
Chung-Chun Yang 杨重骏

Contributors

CHAN, Raymod H. 陈汉夫
Department of Mathematics, Chinese University, Hong Kong

FENG, Kang 冯康
Computing Center, Academia Sinica, Beijing 100080, China

GUO, Ben-yu 郭本瑜
Shanghai University of Science and Technology, Shanghai 201800, China

HAN, Hou-de 韩厚德
Department of Applied Mathematics, Tsinghua University, Beijing 100084, China

HUANG, Hong-ci 黄鸿慈
Department of Mathematics, Hong Kong Baptist College, Hong Kong

JIANG, Er-xiong 蒋尔雄
Department of Mathematics, Fudan University, Shanghai 200433, China

KWOK, Yue-kuen 郭宇权
Department of Mathematics, Hong Kong University of Science and Technology, Hong Kong

LIN, Qun 林群
Institute of System Science, Academia Sinica, Beijing 100080, China

SHI, Zhong-ci 石钟慈
Computing Center, Academia Sinica, Beijing 100080, China

SUN, Jia-chang 孙家昶
Computing Center, Academia Sinica, Beijing 100080, China

WANG, Dao-liu 汪道柳
Computing Center, Academia Sinica, Beijing 100080, China

WANG, Ming 王鸣
Department of Mathematics, Peking University, Beijing 100871, China

WANG, Ren-hong 王仁宏
Institute of Mathematical Sciences, Dalian University of Technology, Dalian 116024, China

WANG, Xing-hua 王兴华
Department of Mathematics, Hangzhou University, Hangzhou 310028, China

YING, Long-an 应隆安
Department of Mathematics, Peking University, Beijing 100871, China

YU, De-hao 余德浩
Computing Center, Academia Sinica, Beijing 100080, China

YUAN, Ya-xiang 袁亚湘
Computing Center, Academia Sinica, Beijing 100080, China

Contemporary Mathematics
Volume **163**, 1994

Dynamical Systems and Geometric Construction of Algorithms*

FENG KANG AND WANG DAO-LIU

ABSTRACT. We give a brief survey on the research, undertaken by the senior author and his group, on the construction and analysis of numerical algorithms for ordinary differential equations, including specifically Hamiltonian systems and other systems with algebraico-geometric structures from the view-point of dynamical systems and ∞-dimensional Lie algebras of vector fields on manifolds and the associated ∞-dimensional Lie transformation groups.

1. Introduction

Consider the autonomous system on $\mathbf{R}^m$

$$\dot{z} = a(z), \quad z = (z_1, \cdots, z_m)^T, \quad a(z) = (a_1(z), \cdots, a_m(z))^T \tag{1.1}$$

defined by its "right hand function", i.e., a smooth vector field $z \to a(z)$ on $\mathbf{R}^m$. The phase flow operator of dynamical system (1.1) defined by the vector field a is determined by the one parameter family $e_a^t = e^t$ of near-identity diffeomorphisms (or synonymously, transformations, operators) on $\mathbf{R}^m$ satisfying

$$\begin{cases} \dfrac{d}{dt} e_a^t = a \circ e_a^t \\ e_a^0 = \text{ identity } := 1_m. \end{cases} \tag{1.2}$$

It completely determines the dynamical evolution of the given continuous-time dynamical system. The phase flow operator e_a^t satisfies the one parameter group property

$$\begin{cases} e_a^{t+s} = e_a^t \circ e_a^s, \quad \forall t, s \in \mathbf{R}, \\ e_a^0 = 1_m \end{cases} \tag{1.3}$$

1991 *Mathematics Subject Classification.* **70H05, 58F13, 58F27, 65L20.**
*This work was supported by National Natural Science Foundation of China and State Commission of Science and Technology.

and depends analytically on t, expressible as a convergent, at least locally in space and time, power series

$$e_a^t = \sum_{k=0}^{\infty} t^k e_k, \tag{1.4}$$

where the coefficients are smooth vector functions from $\mathbf{R}^m$ to $\mathbf{R}^m$ and can be determined recursively

$$e_0 = e^0 = 1_m, \; e_1 = a, \; e_k = \frac{1}{k}(e_{k-1})_* e_1, \quad k = 1, 2, \cdots, \tag{1.5}$$

here Jacobian matrix of $u : \mathbf{R}^m \to \mathbf{R}^m$ is denoted by u_*.

Now to any difference scheme solving the system (1.1) there corresponds a step-transition operator $g^s = g_a^s$ for step size s which completely (for single-step methods) or essentially (for multi-step methods) determines the dynamical evolution of the discrete-time dynamical system. Operator g^s is a one parameter family in step size s of near-identity diffeomorphisms

$$g_a^s = \sum_{k=0}^{\infty} s^k g_k, \quad g_0 = 1_m. \tag{1.6}$$

The algorithm g_a^s is said to be consistent with system (1.1) if $g_1 = a$; it is said to be of (accuracy) order $p(\geq 1)$ if

$$g_k = e_k, \quad k = 0, 1, \cdots, p; \quad g_{p+1} \neq e_{p+1}. \tag{1.7}$$

If g_a^s is consistent with system (1.1) and is a one-parameter group in s, then $g_a^s = e_a^s$, i.e., the difference scheme is exact. So, apart from some exceptional cases, g_a^s does not satisfy the group property $g_a^{s+t} = g_a^s \circ g_a^t$. Therefore, in the most cases, g_a^s is an approximation of e_a^s. The problem is to construct the algorithmic approximation g^s to the phase flow e_a^s in a *proper* way.

As is well known, all (smooth) vector fields on $\mathbf{R}^m$ form an ∞-dim Lie algebra, denoted by $\mathbf{V}(\mathbf{R}^m) = \mathbf{V}_m$, under the Lie bracket $[a, b] = a_* b - b_* a$, where a and b are two vector fields on $\mathbf{R}^m$. The Lie algebra $\mathbf{V}_m$ is associated with the ∞-dimensional Lie group $\mathbf{D}(\mathbf{R}^m) = \mathbf{D}_m$ of near-identity diffeomorphisms on $\mathbf{R}^m$. Every Lie subalgebra $\mathbf{L}$ of $\mathbf{V}_m$ is associated with a Lie subgroup $\mathbf{G}$ of $\mathbf{D}_m$. For any vector field a in $\mathbf{L}$, the phase flow e_a^t is a one-parameter subgroup in $\mathbf{G}$.

The three classical cases of dynamical systems with Lie-algebraic structures are: 1. Hamiltonian systems on $\mathbf{R}^{2n}$ — with the symplectic structure. $\mathbf{L}$ is the Lie algebra $\mathbf{SpV}_{2n}$ of Hamiltonian vector fields, $\mathbf{G}$ is the Lie group $\mathbf{SpD}_{2n}$ of symplectic transformations. 2. Contact systems on $\mathbf{R}^{2n+1}$ — with the contact structure. $\mathbf{L}$ is the Lie algebra $\mathbf{KV}_{2n+1}$ of contact vector fields, $\mathbf{G}$ is the Lie group $\mathbf{KD}_{2n+1}$ of contact transformations. 3. Source-free systems on $\mathbf{R}^m$ — with the volume structure. $\mathbf{L}$ is the Lie algebra $\mathbf{SV}_m$ of source-free vector fields, $\mathbf{G}$ is the Lie group $\mathbf{SD}_m$ of volume-preserving transformations. It is natural and even mandatory to require that the numerical algorithms for solving dynamical systems with structures to be structure-preserving, i.e., g^s belongs to the associated group $\mathbf{G}$ for small $|s|$. This is the main theme of this survey.

In section 2, we discuss the revertibility property of a one-parameter family of operators. Since the phase flows of all autonomous systems are revertible, their algorithms should be revertible too. In section 3, we give a kind of explicit structure-preserving algorithms for systems with Lie algebraic structures. In section 4, based on explicit Euler schemes, a class of explicit symplectic schemes for separable Hamiltonian systems are given. Symplectic schemes are implicit in general. But for specific Hamiltonians explicit Euler schemes may be symplectic and even exact. Consequently, if a Hamiltonian can be decomposed as a sum of specific Hamiltonians, then explicit symplectic schemes can be constructed by composing explicit Euler schemes for these specific Hamiltonians and higher order explicit symplectic schemes can be obtained as well. This kind of Hamiltonians is not too rare but can cover many important cases. In section 5 we give simple families of algorithms which are symplectic for all Hamiltonian systems. The general and systematic method to construct unconditional symplectic schemes is the generating function method which is discussed in section 6. Section 7 is to construct Hamiltonian algorithms for Hamiltonian systems with a perturbation parameter. In sections 8 and 9 we consider contact algorithms for contact systems and volume preserving algorithms for source-free systems respectively. In section 10 we consider the general theory of numerical approximations for ordinary differential equations based on the apparatus of formal power series and their exponential and logarithmic transforms.

2. Revertibility of operators

Let g^t be a one-parameter family of near-1 diffeomorphisms expansible as a convergent power series in t,

$$g^t = \sum_0^\infty t^k g_k, \quad g_0 = 1_m. \tag{2.1}$$

g^t is called *revertible* if $g^t \circ g^{-t} = 1_m$, $\forall t$. We define the reversion of g^t as

$$\check{g}^t = \left(g^{-t}\right)^{-1}. \tag{2.2}$$

g^t is revertible if and only if $g^t = \check{g}^t$. We have

(1) $\check{g}^t \circ g^t$ and $g^t \circ \check{g}^t$ are always revertible for all g^t;
(2) $\check{\check{g}}^t = g^t$, $(f^t \circ g^t)^\vee = \check{g}^t \circ \check{f}^t$;
(3) f^t and g^t are revertible $\Longrightarrow$ $f^{\alpha t} \circ g^{\beta t} \circ f^{\alpha t}$ and $f^{\alpha t} \circ g^{\beta t} \circ g^{\beta t} \circ f^{\alpha t}$ are revertible $\forall \alpha, \beta \in \mathbf{R}$;
(4) f_i^t are revertible, $i = 1, \cdots, k \Longrightarrow f_1^{\alpha_1 t} \circ \cdots \circ f_k^{\alpha_k t} \circ f_k^{\alpha_k t} \circ \cdots \circ f_1^{\alpha_1 t}$ are revertible for $\forall \alpha_i \in \mathbf{R}$, $i = 1, \cdots, k$;

(5) When f^t and g^t are revertible then $f^t \circ g^t$ is revertible if and only if $f^t \circ g^t = g^t \circ f^t$.

Since revertibility is a weaker form of the group property (1.3) all one-parameter groups of diffeomorphisms are always revertible.

For example, consider the system (1.1). The explicit and implicit Euler schemes are

$$z \to \hat{z} = E_a^s z: \quad \hat{z} = z + sa(z), \quad E_a^s = 1_m + sa; \tag{2.3}$$

$$z \to \hat{z} = I_a^s z: \quad \hat{z} = z + sa(\hat{z}), \quad I_a^s = (1_m - sa)^{-1}, \tag{2.4}$$

where E_a^s and I_a^s are step transition operators of the explicit and implicit Euler schemes respectively. So $I_a^s = (E_a^{-s})^{-1} = \check{E}_a^s$, i.e., the reversion of the explicit Euler operator is the implicit Euler operator.

Two-leg and one-leg weighted Euler schemes are

$$\begin{aligned} z \to \hat{z} = T_{a,c}^s z: \ \hat{z} &= z + s(ca(\hat{z}) + (1-c)a(z)), \\ T_{a,c}^s &= (1_m - sca)^{-1} \circ (1_m + s(1-c)a); \end{aligned} \tag{2.5}$$

$$\begin{aligned} z \to \hat{z} = E_{a,c}^s z: \ \hat{z} &= z + sa(c\hat{z} + (1-c)z), \\ E_{a,c}^s &= (1_m + s(1-c)a) \circ (1_m - sca)^{-1}. \end{aligned} \tag{2.6}$$

So

$$\check{T}_{a,c}^s = T_{a,1-c}^s, \qquad \check{E}_{a,c}^s = E_{a,1-c}^s.$$

Consequently, $T_{a,c}^s$ and $E_{a,c}^s$ are revertible for any vector field a if and only if $c = \frac{1}{2}$, i.e., they are trapezoidal and centered Euler schemes respectively.

We write $g^s \approx e_a^s$ if $g^s = e_a^s + O(s^2)$, $g^s \approx e_a^s$, ord p if $g^s = e_a^s + O(s^{p+1})$, $g^s \approx e_a^s$, ord ∞ if $g^s = e_a^s$. A consistent revertible step transition operator is always of even order $2l (l \geq 1)$. For $g^s \approx e_a^s$, ord 1, $g^{s/2} \circ \check{g}^{s/2}$ and $\check{g}^{s/2} \circ g^{s/2} \approx e_a^s$ are revertible and of order 2. If $g^s \approx e_a^s$ is revertible and of order 2, then the revertible composite of g^s:

$$g^{\alpha s} \circ g^{\beta s} \circ g^{\alpha s} \approx e_a^s \tag{2.7}$$

is of order 4 when

$$2\alpha + \beta = 1, \quad 2\alpha^3 + \beta^3 = 0,$$

i.e.,

$$\alpha = \frac{1}{2 - 2^{1/3}} > 0, \qquad \beta = 1 - 2\alpha < 0. \tag{2.7'}$$

Generally, if $g^s \approx e_a^s$ is revertible and of order $2l$, then the revertible composite (2.7) of g^s is of order $2(l+1)$, when

$$2\alpha + \beta = 1, \quad 2\alpha^{2l+1} + \beta^{2l+1} = 0, \tag{2.8}$$

i.e.,

$$\alpha = \frac{1}{2 - 2^{1/(2l+1)}} > 0, \qquad \beta = 1 - 2\alpha < 0. \tag{2.8'}$$

For more details, see [19, 28, 31, 44].

We give some elementary algorithms, i.e., the explicit Euler method E, the implicit Euler method I, the centered Euler method C. E, I, C, especially the oldest and simplest E, will be the basic components for the composition of structure preserving algorithms. Then transition maps for step-size s for $x \longrightarrow$ next step $\hat{x}$ will be denoted by E^s, I^s, C^s.

$$E: \quad \hat{x} = x + sa(x), \quad E_a^s = 1_m + sa \approx e_a^s,$$
$$\text{ord 1, non-revertible.} \tag{2.9}$$

$$I: \quad \hat{x} = x + sa(\hat{x}), \quad I_a^s = (1_m - sa)^{-1} = \check{E}_a^s \approx e_a^s,$$
$$\text{ord 1, non-revertible.} \tag{2.10}$$

$$C: \quad \hat{x} = x + sa\Big(\frac{1}{2}\hat{x} + \frac{1}{2}x\Big), \quad \text{or in equivalent 2-stage form}$$
$$\bar{x} = x + \frac{s}{2}a(\bar{x}), \quad \hat{x} = 2\bar{x} - x. \tag{2.11}$$
$$C_a^s = E_a^{s/2} \circ I_a^{s/2} = E_a^{s/2} \circ \check{E}_a^{s/2} = 2I_a^{s/2} - 1_m \approx e_a^s,$$
$$\text{ord 2, revertible.}$$

A vector field a is called *nilpotent of degree 2* if

$$a^2 = a_* a = a^* a = 0 \Longleftrightarrow e_a^s = 1_m + sa = E_a^s, \tag{2.12}$$

so with such "simple" vector fields, the method E gives exact solutions.

3. Explicit structure-preserving algorithms

Consider a dynamical system

$$\frac{dz}{dt} = a(z), \quad a \in \mathbf{L} \tag{3.1}$$

where $\mathbf{L} \subset \mathbf{V}_m$, a Lie algebra of vector fields under the Lie bracket. Then the phase flow e_a^t is a one parameter subgroup of $\mathbf{G} \subset \mathbf{D}_m$, the Lie group associated with the Lie algebra $\mathbf{L}$.

We want to construct algorithms g_a^s approximating to e_a^s in the same group $\mathbf{G}$, i.e., $g_a^s \in \mathbf{G}$, for $|s|$ small. Such algorithm g_a^s is called *structure-preserving*, or $\mathbf{L}$-preserving. We have the following general propositions.

Let

$$a = \sum_{i=1}^{k} a^i, \qquad a^i \in \mathbf{L} \tag{3.2}$$

and let e_i^s be the phase flow of a^i, $e_i^s \in \mathbf{G}$, and $f_i^s \approx e_i^s$, ord ≥ 1, $f_i^s \in \mathbf{G}$,

$\forall i = 1, \cdots, k$. Then

$$f^s := f_1^s \circ \cdots \circ f_k^s \approx e_a^s, \quad \text{ord 1, } \mathbf{L}\text{-preserving,} \tag{3.3}$$

$$g^s := f^{s/2} \circ \check{f}^{s/2} = f_1^{s/2} \circ \cdots \circ f_k^{s/2} \circ \check{f}_k^{s/2} \circ \cdots \circ \check{f}_1^{s/2} \approx e_a^s,$$
$$\text{ord 2, revertible } \mathbf{L}\text{-preserving,} \tag{3.4}$$

$$h^s := g^{\alpha s} \circ g^{\beta s} \circ g^{\alpha s} \approx e_a^s, \quad \text{with the } \alpha, \beta \text{ in (2.7')}$$
$$\text{ord 4, revertible } \mathbf{L}\text{-preserving; etc.} \tag{3.5}$$

A case with $a = a^1 + a^2 + a^3$ for construction of 3D volume preserving algorithms will be shown in §9. In particular, as an ideal case, take $f_i^s = e_{a^i}^s = e_i^s$, $i = 1, \cdots k$, then

$$f^s := e_1^s \circ \cdots \circ e_k^s \approx e_a^s, \quad \text{ord 1, } \mathbf{L}\text{-preserving,} \tag{3.6}$$

$$g^s := e_1^{s/2} \circ \cdots \circ e_{k-1}^{s/2} \circ e_k^s \circ e_{k-1}^{s/2} \circ \cdots \circ e_1^s \approx e_a^s,$$
$$\text{ord 2, revertible } \mathbf{L}\text{-preserving.} \tag{3.7}$$

This construction gives actually a *numerical algorithm* only when the component phase flows e_i^s are available and algorithmically implementable, this is quite often the case.

We call a vector field $a \in \mathbf{L}$ to be $\mathbf{L}$-separable if it admits a decomposition

$$a = \sum_{i=1}^{k} a^i, \quad a^i \in \mathbf{L} \text{ and nilpotent, i.e., } a_*^i a^i = 0. \tag{3.8}$$

In this case, by (2.12), $e_{a^i}^s = E_{a^i}^s =: E_i^s$, so f^s, g^s become

$$f^s := E_1^s \circ \cdots \circ E_k^s \approx e_a^s, \quad \text{ord 1, } \mathbf{L}\text{-preserving,} \tag{3.9}$$

$$g^s := E_1^{s/2} \circ \cdots \circ E_{k-1}^{s/2} \circ E_k^s \circ E_{k-1}^{s/2} \circ \cdots \circ E_1^{s/2} \approx e_a^s,$$
$$\text{ord 2, revertible } \mathbf{L}\text{-preserving,} \tag{3.10}$$

and all the related revertible $\mathbf{L}$-preserving algorithms are explicit. A simple $\mathbf{L}$-separable case with decomposition $a = a^1 + a^2$ is important for the construction of explicit symplectic method for Hamiltonian systems. See immediately the next section.

4. Explicit Euler-type Hamiltonian algorithms

We consider now Hamiltonian systems on $\mathbf{R}^{2n}$. The Hamiltonian vector field is of the form

$$a = J\nabla H, \quad J = J_{2n} = \begin{pmatrix} 0 & -I_n \\ I_n & 0 \end{pmatrix} \tag{4.1}$$

defined by an energy function $H(z)$ on $\mathbf{R}^{2n}$, which defines the Hamiltonian system

$$\frac{dz}{dt} = a(z) = J\nabla H(z), \quad z \in \mathbf{R}^{2n}, \tag{4.2}$$

where $\nabla H(z) = H_z = (H_{z_1}(z), \cdots, H_{z_{2n}}(z))^T$ is the gradient of H, or

$$\frac{dp}{dt} = -H_q, \quad \frac{dq}{dt} = H_p, \qquad \begin{pmatrix} p \\ q \end{pmatrix} = z \in \mathbf{R}^{2n}, \quad p = \begin{pmatrix} p_1 \\ \vdots \\ p_n \end{pmatrix}, \quad q = \begin{pmatrix} q_1 \\ \vdots \\ q_n \end{pmatrix}.$$

Intrinsic to all Hamiltonian systems is the symplectic structure defined by the differential 2-form

$$\omega = \sum_{i=1}^{n} dp_i \wedge dq_i.$$

A *symplectic* (or *canonical*) transformation is a diffeomorphism $g : z \to \hat{z}$ on $\mathbf{R}^{2n}$ preserving the symplectic structure

$$\sum d\hat{p}_i \wedge d\hat{q}_i = \sum dp_i \wedge dq_i, \qquad \text{i.e.} \qquad \big(g_*(z)\big)^T J_{2n}\big(g_*(z)\big) \equiv J_{2n}.$$

All Hamiltonian vector fields form a Lie algebra $\mathbf{L} = \mathbf{SpV}_{2n}$ under Lie bracket. The associated Lie group $\mathbf{G} = \mathbf{SpD}_{2n}$ consists of all near-identity symplectic transformations on $\mathbf{R}^{2n}$. The phase flows $e^t_{J\nabla H}$, abbreviated as e^t_H, of Hamiltonian systems are symplectic transformations. So it is natural and mandatory to require the algorithms with transition operator $g^s_{J\nabla H}$, abbreviated as g^s_H, to be symplectic too, such algorithms are called *symplectic* (or *Hamiltonian* or *canonical*). Non-symplectic algorithms inevitably bring in distortions alien to Hamiltonian dynamics, such as artificial dissipation and other parasitic effects. See [2–5, 14–15, 20].

The phase flow e^t_H of (4.2) is a one parameter group of symplectic maps

$$e^t_H = e_0 + te_1 + t^2 e_2 + \cdots, \tag{4.3}$$

where

$$e_0 = 1_{2n}, \; e_1 = J\nabla H, \; e_2 = \frac{1}{2}(J\nabla H)_* J\nabla H,$$
$$e_k = \frac{1}{k}(e_{k-1})_* J\nabla H, \; k \geq 3. \tag{4.4}$$

We need to construct *symplectic* algorithms.

By (2.12), a Hamiltonian vector field $J\nabla H$ is nilpotent of degree 2 if

$$(J\nabla H)_* J\nabla H = 0 \quad \text{or} \quad H_{zz} J H_z = 0, \tag{4.5}$$

for simplicity we say H is nilpotent of degree 2. For the Hamiltonian H nilpotent of degree 2, the explicit Euler scheme E^s_H is just the phase flow of the system (4.2) at $t = s$

$$e^s_H(z) = z + se_1(z) = z + sJ\nabla H(z) \equiv E^s_H(z).$$

So it must be symplectic.

If

$$H(z) = \sum_{i=1}^{k} H_i(z), \tag{4.6}$$

and every $H_i(z)$ is nilpotent of degree 2, then $H(z)$ is said to be *symplectically separable.* The following composite

$$z \to \hat{z} = f^s z, \quad f^s := E^s_{H_1} \circ E^s_{H_2} \circ \cdots \circ E^s_{H_k}, \tag{4.7}$$

is explicit, symplectic and of order 1. The composite

$$\begin{aligned} &z \to \hat{z} = g^s z, \\ &g^s := f^{s/2} \circ \check{f}^{s/2} = E^{s/2}_{H_1} \circ \cdots \circ E^{s/2}_{H_{k-1}} \circ E^s_{H_k} \circ E^{s/2}_{H_{k-1}} \circ \cdots \circ E^{s/2}_{H_1} \end{aligned} \tag{4.8}$$

is explicit, symplectic and revertible of order 2. High order revertible symplectic schemes can be constructed from these schemes by the procedure (2.7) and (2.8).

Examples. If

$$H(z) = H(p,q) = \phi(p) \quad \text{or} \quad \psi(q),$$

where ϕ and ψ are functions of n variables, then $H(z)$ is nilpotent of degree 2

$$\begin{pmatrix} H_p(p,q) \\ H_q(p,q) \end{pmatrix} = \begin{pmatrix} \phi_p(p) \\ 0 \end{pmatrix} \quad \text{or} \quad \begin{pmatrix} 0 \\ \psi_q(q) \end{pmatrix}.$$

The corresponding explicit Euler schemes

$$\begin{aligned} \hat{p} &= p, \\ \hat{q} &= q + s\phi_p(p), \end{aligned}$$

or

$$\begin{aligned} \hat{p} &= p - s\psi_q(q), \\ \hat{q} &= q, \end{aligned}$$

are symplectic. If

$$H(p,q) = \phi(p) + \psi(q), \tag{4.9}$$

which is symplectically separable, then the composite schemes

$$E^s_\psi \circ E^s_\phi : \quad \hat{p} = p - s\psi_q(\hat{q}), \quad \hat{q} = q + s\phi_p(p), \tag{4.10}$$

$$E^s_\phi \circ E^s_\psi : \quad \hat{p} = p - s\psi_q(q), \quad \hat{q} = q + s\phi_p(\hat{p}), \tag{4.11}$$

are explicit, symplectic and of order 1. The revertible scheme

$$\begin{aligned} &g^s := (E^{s/2}_\psi \circ E^{s/2}_\phi)^\vee \circ (E^{s/2}_\psi \circ E^{s/2}_\phi) = E^{s/2}_\phi \circ E^s_\psi \circ E^{s/2}_\phi : \\ &q^1 = q + \frac{s}{2}\phi_p(p), \quad \hat{p} = p - s\psi_q(q_1), \quad \hat{q} = q^1 + \frac{s}{2}\phi_p(\hat{p}) \end{aligned} \tag{4.12}$$

is symplectic and of order 2. Its revertible composite

$$h^s := g^{\alpha s} \circ g^{\beta s} \circ g^{\alpha s}$$

gives a symplectic scheme of order 4 with the parameters (2.7′), i.e.,

$$\begin{aligned} p^1 &= p - c_1 s\psi_q(q), & q^1 &= q + d_1 s\phi_p(p^1), \\ p^2 &= p^1 - c_2 s\psi_q(q^1), & q^2 &= q^1 + d_2 s\phi_p(p^2), \\ p^3 &= p^2 - c_3 s\psi_q(q^2), & q^3 &= q^2 + d_3 s\phi_p(p^3), \\ \hat{p} &= p^3 - c_4 s\psi_q(q^3), & \hat{q} &= q^3 + d_4 s\phi_p(\hat{p}), \end{aligned} \tag{4.13}$$

with the parameters $\alpha = (2 - 2^{1/3})^{-1}$, $\beta = 1 - 2\alpha$ and either

$$c_1 = 0, \quad c_2 = c_4 = \alpha, \quad c_3 = \beta, \quad d_1 = d_4 = \alpha/2, \quad d_2 = d_3 = (\alpha + \beta)/2,$$

or

$$c_1 = c_4 = \alpha/2, \quad c_2 = c_3 = (\alpha + \beta)/2, \quad d_1 = d_3 = \alpha, \quad d_2 = \beta, \quad d_4 = 0.$$

Symplectically separable Hamiltonians have wide coverage in applications. It is easy to see that, the Hamiltonian of the form

$$H(p, q) = \phi(Ap + Bq), \quad AB^T = BA^T, \tag{4.14}$$

where $\phi(x)$ is a function of n variables, A and B are $n \times n$ matrices, is also nilpotent of degree 2. Then the explicit Euler scheme $E_H^s = E^s(\phi) = e_H^s$ is

$$E^s(\phi): \quad \begin{aligned} \hat{p} &= p - sB^T \phi_x(Ap + Bq) \\ \hat{q} &= q + sA^T \phi_x(Ap + Bq) \end{aligned} \tag{4.15}$$

so we get a class of symplectically separable Hamiltonians [19]

$$H(p, q) = \sum_{i-1}^{k} H_i(p, q), \quad H_i(p, q) = \phi_i(A_i p + B_i q), \quad A_i B_i^T = B_i A_i^T, \tag{4.16}$$

where $\phi_i(x)$ arc functions of n variables, A_i, B_i are $n \times n$ matrices. Moreover, all polynomial Hamiltonians belong to this class; in fact, it has been proved in [19] that every polynomial $H(p, q)$ in $2n$ variables p, q can be decomposed in the form of (4.16), where $\phi_i(x)$ are polynomials of n variables x, $A_i = \operatorname{diag}(a_1^i, \cdots, a_n^i)$ and $B_i = \operatorname{diag}(b_1^i, \cdots, b_n^i)$.

For symplectically separable Hamiltonians of the above class, all explicit revertible symplectic schemes formed by compositions contain solely the Euler schemes $E^s(\phi_i)$ in the form (4.15) as basic components. A related conjecture is: this class already covers all the symplectically separable cases, i.e., H is nilpotent of degree 2 if and only if it is expressible in the form (4.14).

More generally, if we have a decomposition $H = H_1 + \cdots + H_k$, for which each H_i is integrable and its phase flow $e_{H_i}^s$ is algorithmically implementable, then by (3.4), (3.5) the symmetrical composites

$$g_H^s = e_{H_1}^{s/2} \circ \cdots \circ e_{H_{k-1}}^{s/2} \circ e_{H_k}^s \circ e_{H_{k-1}}^{s/2} \circ \cdots \circ e_{H_1}^{s/2} \tag{4.17}$$

$$h_H^s = g_H^{\alpha s} \circ g_H^{\beta s} \circ g_H^{\alpha s} \tag{4.18}$$

give explicit symplectic, revertible algorithms of order 2, 4, etc. This general approach is widely applicable to *many body problems* in different physical contexts for which the *2-body problem* is *solvable.* Since in such problems the Hamiltonian usually admits a natural decomposition $H = \sum_{i<j} H_{ij}$, each H_{ij} accounts for a 2-body problem.

An interesting result from this approach is a construction [47] of explicit symplectic method for computing the Hamiltonian system of N vortices $z_i = (x_i, y_i)$ with intensities k_i

$$k_i \frac{dx_i}{dt} = \frac{\partial H}{\partial y_i}, \qquad k_i \frac{dy_i}{dt} = -\frac{\partial H}{\partial x_i}, \qquad i = 1, \cdots, N \tag{4.19}$$

with symplectic structure $\omega_k = \sum \frac{1}{k_i} dx_i \wedge dy_i$ and

$$H = \sum_{i<j} H_{ij}, \qquad H_{ij} = -\frac{1}{2\pi} k_i k_j \ln r_{ij},$$
$$r_{ij} = ((x_i - x_j)^2 + (y_i - y_j)^2)^{1/2}. \tag{4.20}$$

Each H_{ij} accounts for a solvable 2-vortex motion in which both vortices z_i, z_j rotate about their center of vorticity $z = (k_i z_i + k_j z_j)(k_i + k_j)^{-1}$ with angular velocity $a = (k_i + k_j)/2\pi r_{ij}^2$ if $k_i + k_j \neq 0$ or translate with linear velocity $(b(y_i - y_j), -b(x_i - x_j))$, $b = k_i/2\pi r_{ij}^2$ if $k_i + k_j = 0$. These simple phase flows $e_{H_{ij}}^s$ serve as the basic algorithmic components for successive compositions, resulting in efficient explicit revertible methods, symplectic for this specific system. They are promising in application to incompressible ideal flows for tracking the vortex particles, which ought to be done in the structure-preserving way.

5. Unconditional Hamiltonian algorithms

Now we want to construct *unconditional* Hamiltonian algorithms, i.e., they are symplectic for *all* Hamiltonian systems.

It can be shown that

(a) The two-leg weighted Euler schemes (2.5) are generically non-symplectic for any real number c or real $2n \times 2n$ matrix c.

(b) The one-leg weighted Euler schemes (2.6), i.e.,

$$\hat{z} = g_H^s z : \quad \hat{z} = z + sJH_z(c\hat{z} + (1-c)z), \tag{5.1}$$

with real number c is unconditionally symplectic if and only if $c = \frac{1}{2}$, which corresponds to the *centered Euler scheme*

$$\hat{z} = z + sJH_z\left(\frac{\hat{z} + z}{2}\right). \tag{5.2}$$

These simple propositions illustrate a general situation: apart from some very rare exceptions, the vast majority of conventional schemes are non-symplectic.

However, if we allow c in (5.1) to be a real matrix of order $2n$, we get a far-reaching generalization: (5.1) is symplectic iff

$$c = \frac{1}{2}(I_{2n} + J_{2n}B), \quad B^T = B, \quad c^T J + Jc = J. \tag{5.3}$$

The simplest and important cases are

$$\begin{array}{lll} C: & c = \frac{1}{2}I_{2n}, & \hat{z} = z + sJH_z(\frac{\hat{z}+z}{2}), \\ P: & c = \begin{pmatrix} I & 0 \\ 0 & 0 \end{pmatrix}, & \begin{array}{l} \hat{p} = p - sH_q(\hat{p}, q), \\ \hat{q} = q + sH_p(\hat{p}, q), \end{array} \\ Q: & c = \begin{pmatrix} 0 & 0 \\ 0 & I \end{pmatrix}, & \begin{array}{l} \hat{p} = p - sH_q(p, \hat{q}), \\ \hat{q} = q + sH_p(p, \hat{q}). \end{array} \end{array} \tag{5.4}$$

For $H(p,q) = \phi(p) + \psi(q)$, the above schemes P and Q reduce to explicit schemes (4.11) and (4.10).

Note that we may split the Hamiltonian vector field as

$$J\nabla H = cJ\nabla H + (I-c)J\nabla H \tag{5.5}$$

and it is more convenient to implement (5.1) as the composite

$$\begin{aligned} &z \to \bar{z} = z + scJ\nabla H(\bar{z}) = I^s_{cJ\nabla H}(z), \\ &\bar{z} \to \hat{z} = \bar{z} + s(I-c)J\nabla H(\bar{z}) = E^s_{(I-c)J\nabla H}(\bar{z}), \\ &g^s_{H,c} = E^s_{(I-c)J\nabla H} \circ I^s_{cJ\nabla H}. \end{aligned} \tag{5.6}$$

The first step is the implicit Euler method for the vector field $cJ\nabla H$, the second step is the explicit Euler method for the vector field $(I-c)J\nabla H$, both maps are not symplectic, but their composite is symplectic.

Scheme (5.1) is revertible of order 2 for $c = \frac{1}{2}I$, this is (5.2). (5.1) is of order 1 for $c \neq \frac{1}{2}I$, from §2 it follows that $\check{g}^s_{H,c} = g^s_{H,I-c}$. Then the composites

$$g^{s/2}_{H,c} \circ g^{s/2}_{H,I-c} \quad \text{and} \quad g^{s/2}_{H,I-c} \circ g^{s/2}_{H,c}$$

are symplectic, revertible and of order 2 for step size s.

So we get a great variety of simple symplectic schemes of order 1 or 2, classified according to type matrices $B \in sm(2n) :=$ space of symmetric matrices of order $2n$, which is a linear space of dimension $2n^2 + n$.

The conservation properties of the above symplectic schemes are well-understood [4, 15, 18, 43]. In fact we have proved:

Let $\phi_A(z) = \frac{1}{2}z^T Az$, $A = A^T$ be a quadratic form and invariant under the system with Hamiltonian H. Then ϕ_A is invariant under symplectic difference scheme $g^s_{H,c}$ iff A commutes symplectically with the type matrix B in (5.3), i.e., $AJB = BJA$.

We list some of the most important weight matrices c, the type matrices B, together with the corresponding form of symmetric matrices A of the conserved quadratic invariants ϕ_A:

$$c = I - c = \frac{1}{2}I, \quad B = 0, \quad A \text{ arbitrary.}$$

$$\left.\begin{array}{l} c = \begin{pmatrix} I_n & 0 \\ 0 & 0 \end{pmatrix}, \quad B = \begin{pmatrix} 0 & -I_n \\ -I_n & 0 \end{pmatrix}, \\ c = \begin{pmatrix} 0 & 0 \\ 0 & I_n \end{pmatrix}, \quad B = \begin{pmatrix} 0 & I_n \\ I_n & 0 \end{pmatrix}, \end{array}\right. \quad A = \begin{pmatrix} 0 & b \\ b^T & 0 \end{pmatrix}, \quad \begin{array}{c} b \text{ arbitrary; angular} \\ \text{momentum type.} \end{array}$$

$$c = \frac{1}{2}\begin{pmatrix} I_n & \pm I_n \\ \mp I_n & I_n \end{pmatrix}, \quad B = \mp I_{2n}, \quad A = \begin{pmatrix} a & b \\ -b & a \end{pmatrix}, \quad \begin{array}{c} a^T = a,\ b^T = -b; \\ \text{Hermitian type.} \end{array}$$

$$c = \frac{1}{2}\begin{pmatrix} I & \pm I \\ \pm I & I \end{pmatrix}, \quad B = \pm\begin{pmatrix} I_n & 0 \\ 0 & -I_n \end{pmatrix}, \quad A = \begin{pmatrix} a & b \\ -b & -a \end{pmatrix}, \quad \begin{array}{c} a^T = a, \\ b^T = -b. \end{array}$$

Numerical evidences and theoretical analysis so far accumulated show unanimously and indisputably with the qualitative and the quantitative superiority of performance of symplectic methods over conventional non-symplectic ones for Hamiltonian systems. The contrasts are overwhelming and striking especially in crucial aspects of global behavior, orbital stability and long-time tracking capabilities. For the first time one now knows *consciously* a *proper* or descent way to compute *Newton's equations* of *motion*, i.e., first put them in Hamiltonian form, then compute by symplectic algorithms [4–5, 10, 15].

6. Generating function methods

The construction of unconditional symplectic algorithms in §5 presents a special case of a general methodology based on generating functions explained in this section.

A matrix α of order $4n$ is called a *Darboux matrix* if

$$\alpha^T J_{4n} \alpha = \tilde{J}_{4n}, \quad J_{4n} = \begin{pmatrix} 0 & -I_{2n} \\ I_{2n} & 0 \end{pmatrix}, \qquad \tilde{J}_{4n} = \begin{pmatrix} J_{2n} & 0 \\ 0 & -J_{2n} \end{pmatrix},$$

$$\alpha = \begin{pmatrix} a & b \\ c & d \end{pmatrix}, \quad \alpha^{-1} = \begin{pmatrix} a_1 & b_1 \\ c_1 & d_1 \end{pmatrix}.$$

Every Darboux matrix induces a (linear) *fractional transform* between symplectic and symmetric matrices

$$\sigma_\alpha : Sp(2n) \to sm(2n),$$
$$\sigma_\alpha(S) = (aS + b)(cS + d)^{-1} = A, \quad \text{for} \quad |cS + d| \neq 0$$

with the inverse transform $\sigma_\alpha^{-1} = \sigma_{\alpha^{-1}}$

$$\begin{aligned}\sigma_\alpha^{-1} :& sm(2n) \longrightarrow Sp(2n),\\ &\sigma_\alpha^{-1}(A) = (a_1 A + b_1)(c_1 A + d_1)^{-1} = S, \quad \text{for } |c_1 A + d_1| \neq 0,\end{aligned}$$

where $Sp(2n) = \{S \in GL(2n, \mathbf{R})|\ S^T J_{2n} S = J_{2n}\}$ is the group of symplectic matrices.

The above machinery can be extended to generally non-linear operators on $\mathbf{R}^{2n}$. Denote $\mathbf{SpD}_{2n}$ the totality of symplectic operators, and $symm(2n)$ the totality of symmetric operators (not necessary one-one). Every $f \in symm(2n)$ corresponds, at least locally, a real function ϕ (unique up to a constant) such that f is the gradient of ϕ: $f(w) = \nabla\phi(w)$, where $\nabla\phi(w) = (\phi_{w_1}(w), \cdots, \phi_{w_{2n}}(w))^T = \phi_w(w)$. Then we have

$$\begin{aligned}&\sigma_\alpha : \mathbf{SpD}_{2n} \longrightarrow symm(2n),\\ &\sigma_\alpha(g) = (a \circ g + b) \circ (c \circ g + d)^{-1} = \nabla\phi, \quad \text{for} \quad |cg_z + d| \neq 0,\end{aligned}$$

or alternatively

$$ag(z) + bz = (\nabla\phi)(cg(z) + dz),$$

where ϕ is called the *generating function* of Darboux type α for the symplectic operator g. Then

$$\begin{aligned}&\sigma_\alpha^{-1} : \quad symm(2n) \longrightarrow \mathbf{SpD}_{2n},\\ &\sigma_\alpha^{-1}(\nabla\phi) = (a_1 \circ \nabla\phi + b_1) \circ (c_1 \circ \nabla\phi + d_1)^{-1} = g,\\ &\qquad\qquad \text{for} \quad |c_1\phi_{ww} + d_1| \neq 0\end{aligned} \tag{6.1}$$

or alternatively

$$a_1\nabla\phi(w) + b_1(w) = g(c_1\nabla\phi(w) + d_1 w), \tag{6.2}$$

where g is called the symplectic operator of Darboux type α for the generating function ϕ.

For the study of symplectic difference scheme we may narrow down the class of Darboux matrices to the subclass of *normal Darboux matrices*, i.e., those satisfying $a + b = 0$, $c + d = I_{2n}$. The normal Darboux matrices α can be characterized as

$$\alpha = \begin{pmatrix} a & b \\ c & d \end{pmatrix} = \begin{pmatrix} J & -J \\ c & I - c \end{pmatrix}, \quad c = \frac{1}{2}(I + JB), \quad B^T = B, \tag{6.3}$$

$$\alpha^{-1} = \begin{pmatrix} a_1 & b_1 \\ c_1 & d_1 \end{pmatrix} = \begin{pmatrix} (c - I)J & I \\ cJ & I \end{pmatrix}. \tag{6.4}$$

The fractional transform induced by a normal Darboux matrix establishes a 1-1 correspondence between *symplectic operators near identity* and *symmetric operators near nullity.* Then the determinantal conditions could be taken for granted. Those B's listed in section 5 correspond to the most important normal Darboux matrices.

For every Hamiltonian H with its phase flow e_H^t and for every normal Darboux matrix α, we get the *generating function* $\phi(w,t) = \phi_H^t(w) = \phi_{H,\alpha}^t(w)$ of *normal Darboux type* α for the *phase flow* of H by

$$\nabla\phi_{H,\alpha}^t = (Je_H^t - J)\circ(ce_H^t + I - c)^{-1}, \quad \text{for small} \quad |t|. \tag{6.5}$$

$\phi_{H,\alpha}^t$ satisfies the *Hamilton-Jacobi* equation

$$\frac{\partial}{\partial t}\phi(w,t) = -H(w + a_1\nabla\phi(w,t)) = -H(w + c_1\nabla\phi(w,t)) \tag{6.6}$$

and can be expressed by Taylor series in t

$$\phi(w,t) = \sum_{k=1}^{\infty}\phi^{(k)}(w)t^k, \quad |t| \quad \text{small.} \tag{6.7}$$

The coefficients can be determined recursively

$$\phi^{(1)}(w) = -H(w), \quad \text{and for} \quad k \geq 0, \quad a_1 = (c-I)J:$$

$$\begin{aligned}\phi^{(k+1)}(w) = & \frac{-1}{k+1}\sum_{m=1}^{k}\frac{1}{m!}\sum_{\substack{j_1+\cdots+j_m=k\\ j_l\geq 1}} D^m H(w)\times \\ & \times(a_1\nabla\phi^{(j_1)}(w),\cdots,a_1\nabla\phi^{(j_m)}(w)),\end{aligned} \tag{6.8}$$

where we use the notation of the m-linear form

$$\begin{aligned}&D^m H(w)(a_1\nabla\phi^{(j_1)}(w),\cdots,a_1\nabla\phi^{(j_m)}(w)) \\ &\quad := \sum_{i_1,\cdots,i_m=1}^{2n} H_{z_{i_1}\cdots z_{i_m}}(w)(a_1\nabla\phi^{(j_1)}(w))_{i_1}\cdots(a_1\nabla\phi^{(j_m)}(w))_{i_m}.\end{aligned}$$

Let ψ^s be a truncation of $\phi_{H,\alpha}^s$ up to a certain power s^m, say. Using inverse transform σ_α^{-1} we get the symplectic operator

$$g^s = \sigma_\alpha^{-1}(\nabla\psi^s), \quad |s| \quad \text{small,} \tag{6.9}$$

which depends on s, H, α (or equivalently B) and the mode of truncation. It is a symplectic approximation to the phase flow e_H^s and can serve as the transition operator of a symplectic difference scheme (for the Hamiltonian system (4.2))

$$z \to \hat{z} = g^s z: \quad \hat{z} = z - J\nabla\psi^s(c\hat{z} + (I-c)z), \quad c = \frac{1}{2}(I + JB). \tag{6.10}$$

Thus, using the machinery of phase flow generating functions we have constructed, for every H and every normal Darboux matrix, a hierarchy of symplectic schemes by truncation. The simple symplectic schemes (5.1, 5.3) correspond to the lowest truncation. The conservation properties of all these higher order schemes are the same as stated in section 5.

7. Hamiltonian algorithms for Hamiltonian systems with a perturbation parameter

The machinery above can also be applied to construct symplectic algorithms for perturbed Hamiltonian systems defined by the perturbed Hamiltonian

$$H(z;\epsilon) = \sum_{k=0}^{\infty} \epsilon^k H_k(z) = H_0(z) + \sum_{k=1}^{\infty} \epsilon^k H_k(z), \tag{7.1}$$

where ϵ is the small perturbation parameter. $H_0(z)$ is usually an integrable Hamiltonian. The corresponding perturbed Hamiltonian system is

$$\frac{dz}{dt} = JH_z(z,\epsilon), \qquad z \in \mathbf{R}^{2n}. \tag{7.2}$$

Its phase flow, denoted by $e^t_\epsilon = e^t_{H,\epsilon}$, depends on the parameter ϵ. The Hamilton-Jacobi equation and generating function are also parameterized by ϵ. That means, the parameterized generating function $\phi^t_\epsilon(w) = \phi^t(w,\epsilon) = \phi(w,t,\epsilon)$ satisfies the parameterized Hamilton-Jacobi equation

$$\frac{\partial}{\partial t}\phi(w,t,\epsilon) = -H(w + a_1\nabla\phi(w,t,\epsilon),\epsilon). \tag{7.3}$$

$\phi(w,t,\epsilon)$ can be expanded as a power series in ϵ in stead of t

$$\phi(w,t;\epsilon) = \sum_{k=0}^{\infty} \epsilon^k \phi^{(k)}(w,t). \tag{7.4}$$

The coefficients $\phi^{(k)}(w,t)$ satisfy the following equations:

$$\phi^{(0)}_t(w,t) = -\, H_0(w + a_1\nabla\phi^{(0)}(w,t)), \tag{7.5}$$

$$k \geq 1: \ \phi^{(k)}_t(w,t) = -\, H_k(w^*) - \sum_{i-1}^{k}\sum_{m=1}^{i} \frac{1}{m!} \sum_{\substack{i_1+\cdots+i_m=i\\ i_j \geq 1}} D^m H_{k-i}(w^*)$$
$$\times\, (a_1\nabla\phi^{(i_1)}(w,t),\cdots, a_1\nabla\phi^{(i_m)}(w,t)) \tag{7.6}$$

with the initial values $\phi^{(i)}(w,0) = 0$, where $w^* = w + a_1\nabla\phi^{(0)}(w,t)$.

(7.5) is just the Hamilton-Jacobi equation for the unperturbed Hamiltonian system with Hamiltonian $H_0(z)$. The right hand side of (7.6) can be written as

$$-DH_0(w^*)\cdot a_1\nabla\phi^{(k)}(w,t) + R^{(k)}(w,t),$$

where the remainder $R^{(k)}(w,t)$ depends only on $H_i(w^*)$, $i = 0,1,\cdots,k$ and $\phi^{(i)}(w,t)$, $i = 0,1,\cdots,k-1$. Once $\phi^{(0)}(w,t),\cdots,\phi^{(k-1)}(w,t)$ are known, $R^{(k)}(w, t)$ is also known. Therefore, if $\phi^{(0)}(w,t)$ can be solved from (7.5), then for $k \geq 1$,

$$\phi^{(k)}_t(w,t) = -DH_0(w^*)\cdot a_1\nabla\phi^{(k)}(w,t) + R^{(k)}(w,t)$$

are the linear partial differential equations for $\phi^{(k)}(w,t)$. They have the same coefficients, only those for $R^{(k)}(w,t)$ are different.

In some cases, we can solve (7.5) easily, refer to [38]. In general, it is difficult to solve (7.5) by analytical method. Nevertheless we can always give an approximative solution, for example, using the methods discussed above.

Let now $\psi_\epsilon^s(w) = \psi(w, s, \epsilon)$ be a truncation of $\phi(w, s, \epsilon)$ up to a certain power ϵ^m. Using the inverse transform σ_α^{-1} we get the symplectic operator

$$g_\epsilon^s = \sigma_\alpha^{-1}(\nabla \psi_\epsilon^s), \quad |s| \quad \text{small}. \tag{7.7}$$

It is a symplectic approximation to the phase flow e_ϵ^s of order m in ϵ and can serve as the transition operator of a symplectic difference scheme for the perturbed Hamiltonian system (7.2)

$$z \to \hat{z} = g_\epsilon^s z: \quad \hat{z} = z - J\nabla\psi(c\hat{z} + (I - c)z, s, \epsilon), \quad c = \frac{1}{2}(I + JB). \tag{7.8}$$

For general perturbed Hamiltonians $H(z, \epsilon)$, these schemes are only consistent in the time stepsize s. But for the perturbed Hamiltonians with the form

$$H(z, \epsilon) = H_0(z) + \epsilon H_1(z), \tag{7.9}$$

the order of the schemes (7.8) in the time stepsize s is the same as in ϵ. More precisely, for the m-th order scheme g_ϵ^s for the perturbed Hamiltonian (7.9),

$$g_\epsilon^s = e_\epsilon^s + O\big((s\epsilon)^{m+1}\big). \tag{7.10}$$

Therefore, for small ϵ the time stepsize s can be taken quite large. For more details, refer to [38].

8. Contact algorithms for contact systems

Contact systems occur only on space $\mathbf{R}^{2n+1}$ of odd dimensions. We use 3-symbol notation to denote the coordinates and vectors on $\mathbf{R}^{2n+1}$

$$\begin{pmatrix} x \\ y \\ z \end{pmatrix}, \quad x = \begin{pmatrix} x_1 \\ \vdots \\ x_n \end{pmatrix}, \quad y = \begin{pmatrix} y_1 \\ \vdots \\ y_n \end{pmatrix}, \quad z = (z),$$

$$\begin{pmatrix} a(x,y,z) \\ b(x,y,z) \\ c(x,y,z) \end{pmatrix}, \quad a = \begin{pmatrix} a_1 \\ \vdots \\ a_n \end{pmatrix}, \quad b = \begin{pmatrix} b_1 \\ \vdots \\ b_n \end{pmatrix}, \quad c = (c).$$

A *contact* system is defined by a function $K(x, y, z)$, called contact Hamiltonian, on $\mathbf{R}^{2n+1}$ as follows

$$\begin{aligned} \frac{dx}{dt} &= -K_y + K_z x = a, \\ \frac{dy}{dt} &= K_x = b, \\ \frac{dz}{dt} &= K_e = c, \quad K_e(x, y, z) := K(x, y, z) - \langle x, K_x(x, y, z)\rangle. \end{aligned} \tag{8.1}$$

The right hand side is the general form of a *contact* vector field.

Intrinsic to all contact systems is the *contact* structure on $\mathbf{R}^{2n+1}$ defined by the differential 1-form

$$\alpha = \sum_{i=1}^{n} x_i dy_i + dz = (0 \ x^T \ 1) \begin{pmatrix} dx \\ dy \\ dz \end{pmatrix} \tag{8.2}$$

up to an everywhere non-vanishing multiplier function. A *contact* transformation f is a diffeomorphism on $\mathbf{R}^{2n+1}$

$$f: \quad \begin{pmatrix} x \\ y \\ z \end{pmatrix} \to \begin{pmatrix} \hat{x}(x,y,z) \\ \hat{y}(x,y,z) \\ \hat{z}(x,y,z) \end{pmatrix}$$

preserving the contact structure, that means

$$\sum_{i=1}^{n} \hat{x}_i d\hat{y}_i + d\hat{z} = \mu_f \left(\sum_{i=1}^{n} x_i dy_i + dz \right) \tag{8.3}$$

for some function $\mu_f \neq 0$ everywhere on $\mathbf{R}^{2n+1}$, called the multiplier of f. The explicit expression of (8.3) is

$$(0 \ \hat{x}^T \ 1) \begin{pmatrix} \hat{x}_x & \hat{x}_y & \hat{x}_z \\ \hat{y}_x & \hat{y}_y & \hat{y}_z \\ \hat{z}_x & \hat{z}_y & \hat{z}_z \end{pmatrix} = \mu_f (0 \ x^T \ 1).$$

All contact vector fields on $\mathbf{R}^{2n+1}$ form a Lie algebra $\mathbf{L} = \mathbf{KV}_{2n+1}$ under Lie bracket. The associated Lie group $\mathbf{G} = \mathbf{KD}_{2n+1}$ consists of all near-identity contact transformations on $\mathbf{R}^{2n+1}$. For $f \in \mathbf{G}$, $\mu_f > 0$ everywhere. The phase flows e_K^t of contact systems defined by contact Hamiltonians K on $\mathbf{R}^{2n+1}$ are contact transformations. So the proper algorithms for contact systems should be *contact* too. For elements of contact geometry, see [1].

All the existing numerical methods including the symplectic ones are generically non-contact even for linear contact systems. A linear contact system is necessarily of the form,

$$\begin{aligned} \frac{dx}{dt} &= (\lambda I - L^T)x, \qquad L \in gl(n), \quad \lambda \in \mathbf{R}, \\ \frac{dy}{dt} &= Ly, \\ \frac{dz}{dt} &= \lambda z, \end{aligned} \tag{8.4}$$

the 3 equations are independent. The matrix of a linear contact transformation is necessarily of the block-diagonal form

$$\begin{pmatrix} \mu M^{-T} & 0 & 0 \\ 0 & M & 0 \\ 0 & 0 & \mu \end{pmatrix}, \qquad M \in \mathbf{GL}(n), \quad \mu > 0. \tag{8.5}$$

Using the trapezoidal method (identical with the centered Euler method in linear cases), one gets

$$\hat{x} = \left(\left(1 - \frac{s\lambda}{2}\right)I + \frac{s}{2}L^T\right)^{-1}\left(\left(1 + \frac{s\lambda}{2}\right)I - \frac{s}{2}L^T\right)x =: Nx,$$
$$\hat{y} = \left(I - \frac{s}{2}L\right)^{-1}\left(I + \frac{s}{2}L\right) =: My,$$
$$\hat{z} = \left(1 - \frac{s\lambda}{2}\right)^{-1}\left(1 + \frac{s\lambda}{2}\right)z =: \mu z.$$

The map is contact $\iff N = \mu M^{-T}$, $\forall s \iff \lambda = 0$ or $L^2 = \lambda L$; but this is highly exceptional.

The construction of explicit Euler-type structure preserving algorithms in §3 is applicable to contact systems when the vector field is contact-separable, i.e., decomposable into

$$a = \sum_{i=1}^{n} a^i, \qquad a^i \in \mathbf{KL}_{2n+1}, \quad a^i_* a^i = 0, \quad i = 1, \cdots, n. \tag{8.6}$$

Our main concern will be the construction of unconditional contact algorithms for general contact systems. The method is based on the well-known correspondence between contact geometry on $\mathbf{R}^{2n+1}$ and homogeneous (or conic) symplectic geometry on $\mathbf{R}^{2n+2}$.

We use 4-symbol notation for the coordinates on $\mathbf{R}^{2n+2}$

$$\begin{pmatrix} p_0 \\ p_1 \\ q_0 \\ q_1 \end{pmatrix} \in \mathbf{R}^{2n+2}, \quad p_0 = (p_0), \; q_0 = (q_0), \quad p_1 = \begin{pmatrix} p_{11} \\ \vdots \\ p_{1n} \end{pmatrix}, \quad q_1 = \begin{pmatrix} q_{11} \\ \vdots \\ q_{1n} \end{pmatrix}.$$

Consider *homogeneous* near-identity transformation $g : (p_0, p_1, q_0, q_1) \to (\hat{p}_0, \hat{p}_1, \hat{q}_0, \hat{q}_1)$ and *homogeneous* function $H(p_0, p_1, q_0, q_1)$ on $\mathbf{R}^{2n+2}$, they are defined for $p_0 \neq 0$ and satisfy the conditions

$$\begin{aligned} \forall \lambda \neq 0 : \quad & \hat{p}_i(\lambda p_0, \lambda p_1, q_0, q_1) = \lambda \hat{p}_i(p_0, p_1, q_0, q_1), \\ & \hat{q}_i(\lambda p_0, \lambda p_1, q_0, q_1) = \hat{q}_i(p_0, p_1, q_0, q_1), \qquad i = 1, 2, \\ & H(\lambda p_0, \lambda p_1, q_0, q_1) = \lambda H(p_0, p_1, q_0, q_1). \end{aligned}$$

They depend essentially only on $2n + 1$ variables

$$\begin{aligned} p_0 \neq 0 : \quad & \hat{p}_i(p_0, p_1, q_0, q_1) = p_0 \hat{p}_i(1, \frac{p_1}{p_0}, q_0, q_1), \\ & \hat{q}_i(p_0, p_1, q_0, q_1) = \hat{q}_i(1, \frac{p_1}{p_0}, q_0, q_1), \\ & H(p_0, p_1, q_0, q_1) = p_0 H(1, \frac{p_1}{p_0}, q_0, q_1). \end{aligned}$$

The phase flow e_H^t for Hamiltonian system defined by homogeneous function H on $\mathbf{R}^{2n+2}$ is homogeneous and symplectic.

Near-identity homogeneous symplectic transformation g and homogeneous function H on $\mathbf{R}^{2n+2}$ can be put in correspondence with near-identity contact transformation f and function K on $\mathbf{R}^{2n+1}$ as follows:

$$\begin{pmatrix} x \\ y \\ z \end{pmatrix} \to \begin{pmatrix} p_0 \\ p_1 \\ q_0 \\ q_1 \end{pmatrix} = \begin{pmatrix} p_0 \\ p_0 x \\ z \\ y \end{pmatrix} \xrightarrow{g} \begin{pmatrix} \hat{p}_0 \\ \hat{p}_1 \\ \hat{q}_0 \\ \hat{q}_1 \end{pmatrix} = \begin{pmatrix} \hat{p}_0 \\ \hat{p}_0 \hat{x} \\ \hat{z} \\ \hat{y} \end{pmatrix}, \tag{8.7}$$

where $p_0 > 0$ is an arbitrary chosen parameter. Then $\hat{p}_0 > 0$ and the map

$$\begin{pmatrix} x \\ y \\ z \end{pmatrix} \xrightarrow{f} \begin{pmatrix} \hat{x} \\ \hat{y} \\ \hat{z} \end{pmatrix} = \begin{pmatrix} \hat{p}_1/\hat{p}_0 \\ \hat{q}_1 \\ \hat{q}_0 \end{pmatrix}, \quad \mu_f := \hat{p}_0/p_0 > 0 \tag{8.8}$$

is a near-identity contact transformation independent of p_0 and μ_f is the multiplier of f.

$$\begin{aligned}
&(x, y, z) \longrightarrow K(x, y, z) := H(1, x, z, y), \\
&H(p_0, p_1, q_0, q_1) = p_0 K(\frac{p_1}{p_0}, q_1, q_0). \\
&H_{p_0}(p_0, p_1, q_0, q_1) = K(x, y, z) - \langle x, K_x(x, y, z)\rangle = K_e(x, y, z) \\
&H_{p_1}(p_0, p_1, q_0, q_1) = K_x(x, y, z), \\
&H_{q_0}(p_0, p_1, q_0, q_1) = p_0 K_z(x, y, z), \\
&H_{q_1}(p_0, p_1, q_0, q_1) = p_0 K_y(x, y, z), \\
&x = p_1/p_0, \quad y = q_1, \quad z = q_0.
\end{aligned}$$

The above process and its inverse are called *contactization* and *symplectization* respectively. Under this correspondence we have

1. $K(x, y, z) \longleftrightarrow H(p_0, p_1, q_0, q_1)$.
2. Contact system (8.1) $\longleftrightarrow$ homogeneous Hamiltonian system

$$\begin{aligned}
\frac{dp_0}{dt} &= -H_{q_0}, & \frac{dq_0}{dt} &= H_{p_0}, \\
\frac{dp_1}{dt} &= -H_{q_1}, & \frac{dq_1}{dt} &= H_{p_1}.
\end{aligned} \tag{8.9}$$

3. Contact phase flow e_K^t on $\mathbf{R}^{2n+1}$ $\longleftrightarrow$ homogeneous symplectic phase flow e_H^t on $\mathbf{R}^{2n+2}$

Now symplectic phase flow e_H^t can be approximated by various unconditional symplectic algorithms, e.g., $g_{H,c}^s$ generated by (5.4) and (6.10) on $\mathbf{R}^{2n+2}$. For weight matrices c on $\mathbf{R}^{2n+2}$ compatible with homogeneity conditions of functions and mappings, $g_{H,c}^s$ are homogeneous symplectic, the generating relations (5.4) and (6.10) are contactized to give the contactization $f_{K,c}^s$ on $\mathbf{R}^{2n+2}$ of $g_{H,c}^s$. They provide unconditional contact algorithms for contact systems. We give some basic contact algorithms in the following. For more details and the theory of contact generating functions see [12–13].

$\widetilde{C}$. $c = \frac{1}{2}I_{2n+2}$, contact version of method C, two stage form (write $\bar{K}_x = K_x(\bar{x}, \bar{y}, \bar{z})$, etc.)

$$\begin{aligned} &\bar{x} = x + \frac{s}{2}(-\overline{K}_y + x\overline{K}_z), \quad \bar{y} = y + \frac{s}{2}\overline{K}_x, \quad \bar{z} = z + \frac{s}{2}\overline{K}_e, \\ &\hat{x} = (\bar{x} - \frac{s}{2}\overline{K}_y)/(1 - \frac{s}{2}\overline{K}_z), \quad \hat{y} = 2\bar{y} - y, \quad \hat{z} = 2\bar{x} - z. \end{aligned} \tag{8.10}$$

$\widetilde{P}$. $c = \begin{pmatrix} I_{n+1} & 0 \\ 0 & 0_{n+1} \end{pmatrix}$, contact analog of symplectic method P:

$$\text{1-stage form :} \quad \begin{aligned} &\hat{x} = x + s(-K_y(\hat{x}, y, z) + xK_z(\hat{x}, y, z)), \\ &\hat{y} = y + sK_x(\hat{x}, y, z), \\ &\hat{z} = z + sK_e(\hat{x}, y, z). \end{aligned} \tag{8.11}$$

$$\text{2-stage form :} \quad \begin{aligned} &\bar{x} = x + s(-\overline{K}_y + x\overline{K}_z), \quad \bar{y} = y, \quad \bar{z} = z, \\ &\hat{x} = \bar{x}, \quad \hat{y} = \bar{y} + s\overline{K}_x, \quad \hat{z} = \bar{z} + s\overline{K}_e. \end{aligned}$$

$\widetilde{Q}$. $c = \begin{pmatrix} 0_{n+1} & 0 \\ 0 & I_{n+1} \end{pmatrix}$, contact analog of symplectic method Q:

$$\text{1-stage form :} \quad \begin{aligned} &\hat{x} = x + s(-K_y(x, \hat{y}, \hat{z}) + \hat{x}K_z(x, \hat{y}, \hat{z})), \\ &\hat{y} = y + sK_x(x, \hat{y}, \hat{z}), \\ &\hat{z} = z + sK_e(x, \hat{y}, \hat{z}). \end{aligned} \tag{8.12}$$

$$\text{2-stage form :} \quad \begin{aligned} &\bar{x} = x, \quad \bar{y} = y + s\overline{K}_x, \quad \bar{z} = z + s\overline{K}_e, \\ &\hat{x} = \bar{x} + s(-\overline{K}_y + \hat{x}\overline{K}_z), \quad \hat{y} = \bar{y}, \quad \hat{z} = \bar{z}. \end{aligned}$$

One might suggest, by analogy with (5.4), for example, the following scheme for (8.1):

$$\hat{x} = x + sa(\hat{x}, y, z), \quad \hat{y} = y + sb(\hat{x}, y, z), \quad \hat{z} = z + sc(\hat{x}, y, z).$$

It differs from (8.11) only in one term for $\hat{x}$, i.e., $\hat{x}K(\hat{x}, y, z)$ instead of $xK(\hat{x}, y, z)$. This minute but delicate difference makes (8.11) contact and other non-contact! For further developments of contact algorithms, see [10, 12–13]. For numerical comparisons of 3D contact vs non-contact computations, see [10].

9. Volume-preserving algorithms for source-free systems

The source-free systems on $\mathbf{R}^m$ are governed by the m-D volume structure, the underlying Lie algebra $\mathbf{L} = \mathbf{SV}_m$ consists of all *source-free* vector fields a, $\operatorname{div} a = \operatorname{tr} a_* = 0$, the underlying Lie group $\mathbf{G} = \mathbf{SD}_m$ consists of all near-identity *volume-preserving* transformations g, $\det g_* = 1$.

The phase flow e_a^t of source-free field a is always volume preserving. Hence the proper algorithms for source-free systems should be also *volume-preserving*, since otherwise the dynamics would be polluted by artificial sources and sinks.

For $m = 2$, source-free fields = Hamiltonian fields, area-preserving maps = symplectic maps; so the problem for area-preserving algorithms has been solved in principle.

For $m \geq 3$, the problem is new, since all the conventional methods plus even the symplectic methods are generically not volume-preserving, even for linear source-free systems. As an illustration, solve on $\mathbf{R}^3$

$$\frac{dx}{dt} = Ax, \qquad \operatorname{tr} A = 0$$

by the method C, we get

$$x \to \hat{x} = G^s x, \quad G^s = \left(I - \frac{s}{2}A\right)^{-1}\left(I + \frac{s}{2}A\right).$$

$\det G^s = 1 \Longleftrightarrow \det A = 0$, which is exceptional.

The construction of explicit Euler type structure-preserving algorithms by composition is applicable here when the vector field is source-free separable, i.e., decomposable as

$$a = \sum_{i=1}^{n} a^i, \quad \operatorname{div} a^i = 0, \quad a^i_* a^i = 0, \quad i = 1, \cdots, n. \tag{9.1}$$

A special case is

$$a = (a_1, \cdots, a_m), \quad \frac{\partial a_k}{\partial x_k} = 0, \quad a^k = (0, \cdots, 0, a_k, 0, \cdots, 0), \quad k = 1, \cdots, m;$$

$$a = \sum_{k=1}^{m} a^k. \tag{9.2}$$

Example 1. Euler equation for free rigid body and Jacobian elliptic functions of modulus k,

$$\begin{aligned}\frac{dx_1}{dt} &= c_1 x_2 x_3 = a_1,\\ \frac{dx_2}{dt} &= c_2 x_1 x_3 = a_2,\\ \frac{dx_3}{dt} &= c_3 x_1 x_2 = a_3,\end{aligned} \qquad \begin{pmatrix} a_1 \\ a_2 \\ a_3 \end{pmatrix} = \begin{pmatrix} c_1 x_2 x_3 \\ 0 \\ 0 \end{pmatrix} + \begin{pmatrix} 0 \\ c_2 x_1 x_3 \\ 0 \end{pmatrix} + \begin{pmatrix} 0 \\ 0 \\ c_3 x_1 x_2 \end{pmatrix}$$

where (x_1, x_2, x_3) are angular momenta along principle axes of inertia with diagonal elements I_1, I_2, I_3, $c_1 = (I_2 - I_3)/I_2 I_3$, cyclic. Take $c_1 = 1$, $c_2 = -1$, $c_3 = -k^2$, it gives elliptic functions for $x_1(0) = 0$, $x_2(0) = x_3(0) = 1$, $x_1(t) = \operatorname{sn}(t, k)$, $x_2(t) = \operatorname{cn}(t, k)$, $x_3(t) = \operatorname{dn}(t, k)$.

Example 2. ABC flows [1]

$$\begin{aligned}\frac{dx_1}{dt} &= A \sin x_3 + C \cos x_2 = a_1,\\ \frac{dx_2}{dt} &= B \sin x_1 + A \cos x_3 = a_2,\end{aligned}$$

$$\frac{dx_3}{dt} = C\sin x_2 + B\cos x_1 = a_3,$$

$$\begin{aligned}\begin{pmatrix} a_1 \\ a_2 \\ a_3 \end{pmatrix} &= \begin{pmatrix} A\sin x_3 + C\cos x_2 \\ 0 \\ 0 \end{pmatrix} + \begin{pmatrix} 0 \\ B\sin x_1 + A\cos x_3 \\ 0 \end{pmatrix} \\ &\quad + \begin{pmatrix} 0 \\ 0 \\ C\sin x_2 + B\cos x_1 \end{pmatrix} \\ &= \begin{pmatrix} A\sin x_3 \\ A\cos x_3 \\ 0 \end{pmatrix} + \begin{pmatrix} 0 \\ B\sin x_1 \\ B\cos x_1 \end{pmatrix} + \begin{pmatrix} C\sin x_2 \\ 0 \\ C\cos x_2 \end{pmatrix}.\end{aligned}$$

All decompositions on the right satisfy (9.1), then (3.10) gives corresponding volume-preserving revertible algorithms.

In $\mathbf{R}^m$ all source-free fields have "vector potential" representations: under such representation, a field always decomposes into a sum of essentially 2-D source-free fields, for which essentially area-preserving algorithms can always be constructed, then one can apply the decomposition-composition method in §3 to obtain m-D volume-preserving algorithms [16, 39].

In $\mathbf{R}^2$, every source-free field (a_1, a_2) corresponds to a stream function or 2-D Hamiltonian ψ, unique up to a constant:

$$a_1 = -\frac{\partial \psi}{\partial x_2}, \qquad a_2 = \frac{\partial \psi}{\partial x_1}, \tag{9.3}$$

then area-preserving algorithms = 2-D symplectic algorithms.

In $\mathbf{R}^3$, every source-free field (a_1, a_2, a_3) corresponds to a vector potential (b_1, b_2, b_3), unique up to a gradient:

$$a = \operatorname{curl} b, \quad a_1 = \frac{\partial b_3}{\partial x_2} - \frac{\partial b_2}{\partial x_3}, \quad a_2 = \frac{\partial b_1}{\partial x_3} - \frac{\partial b_3}{\partial x_1}, \quad a_3 = \frac{\partial b_2}{\partial x_1} - \frac{\partial b_1}{\partial x_2}, \tag{9.4}$$

then one get source-free decomposition

$$\begin{aligned} a &= (a_1, a_2, a_3) \\ &= \Big(0, \frac{\partial b_1}{\partial x_3}, -\frac{\partial b_1}{\partial x_2}\Big) + \Big(-\frac{\partial b_2}{\partial x_3}, 0, \frac{\partial b_2}{\partial x_1}\Big) + \Big(\frac{\partial b_3}{\partial x_2}, -\frac{\partial b_3}{\partial x_1}, 0\Big) \\ &= a^1 + a^2 + a^3. \end{aligned} \tag{9.5}$$

Each field a^i is 2-D source-free and 0 in the 3rd dimension, apply area-preserving method P, say, with 3rd dimension fixed, then one get three volume-preserving maps

$$P_1^s \approx e^s(a^1) = e^s\Big(0, \frac{\partial b_1}{\partial x_3}, -\frac{\partial b_1}{\partial x_2}\Big), \qquad \text{ord 1, volume-preserving,}$$

$$
\begin{aligned}
&P_2^s \approx e^s(a^2) = e^s\Big(-\frac{\partial b_2}{\partial x_3}, 0, \frac{\partial b_2}{\partial x_1}\Big), \quad \text{ord 1, volume-preserving,}\\
&P_3^s \approx e^s(a^3) = e^s\Big(\frac{\partial b_3}{\partial x_2}, -\frac{\partial b_3}{\partial x_1}, 0\Big), \quad \text{ord 1, volume-preserving,}\\
&P_1^s = E^s\Big(0, \frac{\partial b_1}{\partial x_3}, 0\Big) I^s\Big(0, 0, -\frac{\partial b_1}{\partial x_2}\Big),\\
&\check{P}_1^s = E^s\Big(0, 0, -\frac{\partial b_1}{\partial x_2}\Big) I^s\Big(0, \frac{\partial b_1}{\partial x_3}, 0\Big),\\
&P_2^s = E^s\Big(0, 0, \frac{\partial b_2}{\partial x_1}\Big) I^s\Big(-\frac{\partial b_2}{\partial x_3}, 0, 0\Big),\\
&\check{P}_2^s = E^s\Big(-\frac{\partial b_2}{\partial x_3}, 0, 0\Big) I^s\Big(0, 0, \frac{\partial b_2}{\partial x_1}\Big),\\
&P_3^s = E^s\Big(\frac{\partial b_3}{\partial x_2}, 0, 0\Big) I^s\Big(0, -\frac{\partial b_3}{\partial x_1}, 0\Big),\\
&\check{P}_3^s = E^s\Big(0, -\frac{\partial b_3}{\partial x_1}, 0\Big) I^s\Big(\frac{\partial b_3}{\partial x_2}, 0, 0\Big),
\end{aligned}
\tag{9.6}
$$

then one get volume-preserving algorithms for field a

$$
\begin{aligned}
&f^s = P_1^s P_2^s P_3^s \approx e^s(a), \quad \text{ord 1, non-revertible,}\\
&g^s = P_1^{s/2} P_2^{s/2} P_3^{s/2} \circ \check{P}_3^{s/2} \check{P}_2^{s/2} \check{P}_1^{s/2} \approx e^s(a), \ \text{ord 2, revertible,}\\
&h^s = g^{\alpha s} g^{\beta s} g^{\alpha s} \approx e^s(a), \quad \text{ord 4, revertible.}
\end{aligned}
\tag{9.7}
$$

The basic elements of all these composites consist only in 6 explicit Euler and 6 implicit Euler in one variable appeared in (9.6).

On $\mathbf{R}^m$, to every source-free field (a_i), there exists, as a generalization of cases $m = 2, 3$, skew-symmetric tensor field of order 2, $b = (b_{ik})$, $b_{ik} = -b_{ki}$ so that

$$
a_i = \sum_{k=1}^{m} \frac{\partial b_{ik}}{\partial x_k}, \qquad i = 1, \cdots, m. \tag{9.8}
$$

By (9.8) we can decompose

$$
a = \sum_{i<k} a_{(ik)}, \ a_{(ik)} = \Big(0, \cdots, 0, \frac{\partial b_{ik}}{\partial x_k}, 0, \cdots, 0, -\frac{\partial b_{ik}}{\partial x_i}, 0, \cdots, 0\Big)^T, \ i < k. \tag{9.9}
$$

Every vector field $a_{(ik)}$ is 2-D source-free on the i-k plane and 0 in other dimensions so we can approximate the phase flow of $a_{(ik)}$ by area-preserving method which implies m-D volume-preserving, similar to the cases discussed above for $m = 3$. For more details see [16].

The tensor potential (b_{ik}) for a given (a_i) is by far not unique. For uniqueness one may impose normalizing conditions in many different ways. One way is to impose

$$
N_0: \quad b_{ik} = 0, \ |i - k| \geq 2, \tag{9.10}
$$

(this condition is ineffective for $m = 2$). The non-zero components are

$$b_{12} = -\,b_{21}, \quad b_{23} = -b_{32}, \quad \cdots \quad b_{m-1,m} = -b_{m,m-1}.$$

$$N_1: \quad b_{12}|_{x_1=x_2=0} = 0.$$

$$N_k,\ 2 \le k \le m-1: \quad b_{k,k+1}|_{x_k=0} = 0,$$

(this condition is ineffective for $m = 2$). Then all $b_{k,k+1}$ are uniquely determined by quadrature

$$b_{12} = \int_0^{x_2} a_1\,dx_2 - \int_0^{x_1} a_2\Big|_{x_2=0} dx_1,$$

$$b_{k,k+1} = -\int_0^{x_k} a_{k+1}\,dx_k, \quad 2 \le k \le m-1; \tag{9.11}$$

$$a = \sum_{k=1}^{m-1} a^k, \quad a^k = (0,\cdots,0,\frac{\partial b_{k,k+1}}{\partial x_{k+1}}, -\frac{\partial b_{k,k+1}}{\partial x_k}, 0,\cdots,0)^T. \tag{9.12}$$

When the vector field a is polynomial, then by (9.10) and (9.11) the tensor components b_{ik} are polynomials. Then the theorem on symplectic separability of all polynomial Hamiltonians (4.16) implies the source-free separability in the sense of (9.1) for all polynomial source-free systems. For numerical comparisons of 3D volume-preserving vs non-volume-preserving computations, see [10].

10. Formal dynamical systems and numerical algorithms

In order to facilitate the study of the problem $f^s \approx e^s_a$, the apparatus of formal power series is useful. We first introduce some notation. For $a = (a_1, \cdots, a_m)^T : \mathbf{R}^m \to \mathbf{R}^m$,

$$a^* := \sum a_i \frac{\partial}{\partial x_i} = \text{ linear differential operator of first order associated to } a.$$

The association $a \to a^*$ is linear, a^* operates on scalar functions $\phi : \mathbf{R}^m \to \mathbf{R}$ and on vector functions $b : \mathbf{R}^m \to \mathbf{R}^m$ as

$$a^*\phi = \sum a_i \frac{\partial \phi}{\partial x_i}$$

$$a^*b = a^*(b_1,\cdots,b_m)^T = (a^*b_1,\cdots,a^*b_m)^T = b_*a, \quad a^*1_m = a.$$

Multiple applications of linear differential operators are naturally defined such as a^*b^*, $(a^*b^*)c^*$, $a^*(b^*c^*)$, etc. The operations are multilinear, associative but non-commutative; thus powers can be defined

$$a^{*k} = a^*a^*\cdots a^*(k \text{ times}), \qquad a^k := a^{*k}1_m$$

The identity operator I operates on scalar and vector functions ϕ and b as $I\phi = \phi$, $Ib = b$. Lie bracket of vector functions a, b on $\mathbf{R}^m$ can be written as

$$[a,b] := a_*b - b_*a = b^*a - a^*b = (b^*a^* - a^*b^*)1_m.$$

Among the formal power series $\sum_0^\infty s^k a_k$, $a_k : \mathbf{R}^m \to \mathbf{R}^m$, we pick out two special classes. The first class consists of those with $a_0 = 0$, called near-0 formal vector fields; the second class consists of those with $a_0 = 1_m$, called near-1 formal maps (diffeomorphisms).

All near-0 formal vector fields $a^s = \sum\limits_1^\infty s^k a_k$ form a (∞-dim.) real Lie algebra $\mathbf{FV}_m$ under the Lie bracket

$$[a^s, b^s] = \Big[\sum_1^\infty s^k a_k, \sum_1^\infty s^k b_k\Big] := \sum_{k=2}^\infty s^k \sum_{i+j=k} [a_i, b_j].$$

The associated near-0 formal differential operators and their products are

$$(a^s)_* := \Big(\sum_1^\infty s^k a_k\Big)_* := \sum_1^\infty s^k a_{k*}$$

$$a^{s*} = \Big(\sum_1^\infty s^k a_k\Big)^* := \sum_1^\infty s^k a_k^*,$$

$$a^{s*} b^{s*} := \sum_2^\infty s^k \sum_{i+j=k} a_i^* b_j^*, \quad (a^{s*})^2 := a^{s*} a^{s*}, \text{ etc.}$$

For any vector function $a = (a_1, \cdots, a_m)^T : \mathbf{R}^m \to \mathbf{R}^m$ and any near-1 formal map $g^s = 1 + \sum s^k g_k$ we define the composition

$$(a \circ g^s)(x) = a(g^s(x)) = a(x) + \sum_1^\infty s^k (a \circ g)_k(x),$$

$$(a \circ g)_k = \sum_{n=1}^k \sum_{j_1+\cdots+j_n=k} \frac{1}{n!}(D^n a)(g_{j_1}, \cdots, g_{j_n}),$$

where

$$D^n a = (D^n a_1, \cdots, D^n a_m)^T,$$

$$D^n a_i(v_1, \cdots, v_n) = \sum_{j_1, \cdots, j_n = 1}^m \frac{\partial^n a_i}{\partial x_{j_1} \cdots \partial x_{j_n}} v_{1j_1} \cdots v_{nj_n}$$

is the usual n-th differential multi-linear form for n tangent vectors $v_j = (v_{j_1}, \cdots, v_{j_m})^T$, $j = 1, \cdots, n$ at point $x \in \mathbf{R}^m$ which is invariant under permutation of vectors. Using the identities

$$(D^1 a)(b) = b^* a,$$

$$(D^2 a)(b, c) = (c^* b^* - (c^* b)^*) a,$$

$$(D^3 a)(b, b, b) = (b^{*3} + 2b^{3*} - 3b^* b^{2*}) a,$$

we get in particular

$$
\begin{aligned}
(a \circ g)_1 &= g_1^* a, \\
(a \circ g)_2 &= g_2^* a + \frac{1}{2}(g_1^{*2} - g_1^{2*})a, \\
(a \circ g)_3 &= g_3^* a + ((g_2^* g_1^* - (g_2^* g_1)^*)a + \frac{1}{3!}(g_1^{*3} + 2g_1^{3*} - 3g_1^* g_1^{2*})a.
\end{aligned}
$$

For any two near-1 formal maps $f^s = 1_m + \sum s^k a_k$, $g^s = 1_m + \sum s^k g_k$, the composition $f^s \circ g^s$ is defined in a term by term way:

$$
\begin{aligned}
(f^s \circ g^s)(x) = f^s(g^s(x)) &= 1_m(g^s(x)) + \sum_{k=1}^{\infty} s^k f_k(g^s(x)) \\
&=: 1_m(x) + \sum_{k=1}^{\infty} s^k (f \circ g)_k(x),
\end{aligned}
$$

$$
\begin{aligned}
&(f \circ g)_1 = f_1 + g_1, \\
&(f \circ g)_k = f_k + g_k + \delta(f_1, \cdots, f_{k-1}; g_1, \cdots, g_{k-1}), \quad k \geq 2, \\
&\delta(f_1, \cdots, f_{k-1}; g_1, \cdots, g_{k-1}) = \sum_{n=1}^{k-1} \sum_{j=1}^{n} \sum_{i_1 + \cdots + i_j = n} \frac{1}{j!}(D^j f_{k-n})(g_{i_1}, \cdots, g_{i_j}),
\end{aligned}
$$

we get in particular,

$$
\begin{aligned}
(f \circ g)_2 =& f_2 + g_2 + g_1^* f_1, \\
(f \circ g)_3 =& f_3 + g_3 + g_1^* f_2 + g_2^* f_1 + \frac{1}{2}(g_1^{*2} - g_1^{2*})f_1, \\
(f \circ g)_4 =& f_4 + g_4 + g_1^* f_3 + g_2^* f_2 + g_3^* f_1 + \frac{1}{2}(g_1^{*2} - g_1^{2*})f_2 \\
&+ (g_2^* g_1^* - (g_2^* g_1)^*)f_1 + \frac{1}{3!}(g_1^{*3} + 2g_1^{3*} - 3g_1^* g_1^{2*})f_1.
\end{aligned}
$$

Under this composition rule, all near-1 formal maps $f^s = 1 + \sum_1^{\infty} s^k f_k$ form a (∞-dim) formal Lie group $\mathbf{FD}_m$. In group $\mathbf{FD}_m$ inverse elements, square roots, rational powers, etc., always exist, their coefficients can always be determined recursively by the defining composition relations. For example, the inverse $(f^s)^{-1} := 1 + \sum s^k h_k = h^s$ is defined by $(f^s \circ h^s) = 1_N$, hence

$$
f_1 + h_1 = 0, \quad f_k + h_k + \delta(f_1, \cdots, f_{k-1}; h_1, \cdots, h_{k-1}) = 0, \quad k \geq 2.
$$

In particular,

$$
\begin{aligned}
h_1 &= -f_1, \\
h_2 &= -f_2 + f_1^2, \\
h_3 &= -f_3 + f_1^* f_2 + (f_2^* - f_1^{2*})f_1 - \frac{1}{2}f_1^3 + \frac{1}{2}f_1^{2*} f_1.
\end{aligned}
$$

There is an obvious one-one correspondence between the Lie algebra $\mathbf{FV}_m$ and the Lie group $\mathbf{FD}_m$, established simply by $+1_m$ and by -1_m. However, the more significant one-one correspondence between them is given by exp and its inverse log.

$$\begin{aligned}
&\exp : \mathbf{FV}_m \to \mathbf{FD}_m \\
&a^s = \sum_1^\infty s^k a_k \to \exp a^s := 1_m + \sum_{k=1}^\infty \frac{1}{k!}(a^{s*})^k 1_m \\
&\qquad\qquad\qquad\qquad =: 1_m + \sum_1^\infty s^k f_k = f^s.
\end{aligned} \tag{10.1}$$

Note that

$$(a^{s*})^k = \left(\sum s^{i_1} a^*_{i_1}\right) \cdots \left(\sum s^{i_k} a^*_{i_k}\right) = \sum_{i_1, \cdots, i_k = 1}^\infty s^{i_1 + \cdots + i_k} a^*_{i_1} \cdots a^*_{i_k},$$

so we get easily

$$\begin{aligned}
f_k &= \sum_{n=1}^k \frac{1}{n!} \sum_{k_1 + \cdots + k_n = k} a^*_{k_1} \cdots a^*_{k_n} 1_m, \qquad k \geq 1, \quad f_1 = a_1, \\
f_k &= a_k + \sum_{n=2}^k \frac{1}{n!} \sum_{k_1 + \cdots + k_n = k} a^*_{k_1} \cdots a^*_{k_n} 1_m, \qquad k \geq 2, \quad f_2 = a_2 + \frac{1}{2} a_1^2.
\end{aligned} \tag{10.2}$$

Note that (10.2) provides a 2-way recursion formula from $a_1, \cdots, a_k$ to $f_1, \cdots, f_k$ and vice versa. Therefore exp maps $\mathbf{FV}_m$ one-one onto $\mathbf{FD}_m$ and its inverse, i.e., log is defined by the same (10.2):

$$\log = (\exp)^{-1} : \mathbf{FD}_m \to \mathbf{FV}_m, \quad \log \exp a^s = a^s, \quad \exp \log f^s = f^s.$$

In particular

$$\begin{aligned}
a_1 &= f_1, \quad a_2 = f_2 - \frac{1}{2} a_1^2, \quad a_3 = f_3 - \frac{1}{2}(a_1^* a_2 + a_2^* a_1) - \frac{1}{3!} a_1^3, \\
a_4 &= f_4 - \frac{1}{2}(a_1^* a_3 + a_2^2 + a_3^* a_1) \\
&\quad - \frac{1}{3!}(a_1^* a_1^* a_2 + a_1^* a_2^* a_1 + a_2^* a_1^* a_1) - \frac{1}{4!} a_1^4, \\
a_k &= f_k - \sum_{n=2}^{k-1} \frac{1}{n!} \sum_{k_1 + \cdots + k_n = k} a^*_{k_1} \cdots a^*_{k_n} 1_m - \frac{1}{k!} a_1^k, \quad k \geq 3.
\end{aligned} \tag{10.3}$$

An equivalent construction of $\log f^s = a^s$ is

$$\log f^s = \sum_{k=1}^\infty \frac{(-1)^{k-1}}{k} h_k^s, \tag{10.4}$$

where

$$h_1^s = f^s - 1_m, \quad h_k^s = h_{k-1}^s \circ f^s - h_{k-1}^s.$$

It is easy to compute

$$h_1^s = \sum_{k=1}^{\infty} s^k f_k = \sum_{k_1=1}^{\infty} s^{k_1}(1_m \circ f)_{k_1}$$

$$h_2^s = \sum_{k_1=k_2=1}^{\infty} s^{k_1+k_2}((1_m \circ f)_{k_1} \circ f)_{k_2},$$

$$h_3^s = \sum_{k_1,k_2,k_3=1}^{\infty} s^{k_1+k_2+k_3}(((1_m \circ f)_{k_1} \circ f)_{k_2} \circ f)_{k_3}, \text{ etc.}$$

Substituting in (10.4) and equating with $\sum_1^\infty s^k a_k$, we get

$$a_k = \sum_{n=1}^{k} \frac{(-1)^{n-1}}{n} \sum_{k_1+\cdots+k_n=k} (\cdots((1_m \circ f)_{k_1} \circ f)_{k_2} \cdots \circ f)_{k_n}. \tag{10.5}$$

It is easy to verify $\log \exp a^s = a^s$ for this log, so this is precisely the inverse of exp, thus agrees with the previous one.

We use the above construction (10.5) to establish the *formal* Campbell-Hausdorff formula:
For arbitrary near-1 formal maps f^s, g^s

$$\log(f^s \circ g^s) = \log f^s + \log g^s + \sum_{k=1}^{\infty} d_k(\log f^s, \log g^s), \tag{10.6}$$

where, for $\log f^s = a^s$, $\log g^s = b^s$,

$$d_k(a^s, b^s) = \frac{1}{k} \sum_{n=1}^{k} \frac{(-1)^{n-1}}{n} \sum_{\substack{p_1+q_1+\cdots+p_n+q_n=k \\ p_i+q_i \geq 1, p_i \geq 0, q_i \geq 0}} \frac{[(a^s)^{p_1}(b^s)^{q_1} \cdots (a^s)^{p_n}(b^s)^{q_n}]}{p_1!q_1! \cdots p_n!q_n!} \tag{10.7}$$

where

$$(x)^p = xx \cdots x \ (p\text{-times}), \quad [x_1 x_2 x_3 \cdots x_n] = [[\cdots[[x_1, x_2], x_3], \cdots], x_n].$$

In particular

$$d_1 = \frac{1}{2}[a^s, b^s], \quad d_2 = \frac{1}{12}([a^s b^s b^s] + [b^s a^s a^s]), \quad d_3 = -\frac{1}{24}[a^s b^s b^s a^s].$$

Let $\log(f^s \circ g^s) = c^s = \sum_1^\infty s^k c_k$. Then

$$c_1 = a_1 + b_1, \qquad c_2 = a_2 + b_2 + \frac{1}{2}[a_1 b_1],$$

$$c_3 = a_3 + b_3 + \frac{1}{2}([a_1 b_2] + [a_2 b_1]) + \frac{1}{12}([a_1 b_1 b_1] + [b_1 a_1 a_1]),$$

$$\begin{aligned}c_4 =& a_4 + b_4 + \frac{1}{12}([a_1b_3] + [a_2b_2] + [a_3b_1])\\&+ \frac{1}{12}([a_1b_1b_2] + [a_1b_2b_1] + [a_2b_1b_1] + [b_1a_1a_2]\\&+ [b_1a_2a_1] + [b_2a_1a_1]) - \frac{1}{24}[a_1b_1b_1a_1],\end{aligned}$$

etc.

Note that the classical CH formula is restricted to the composition of two one-parameter groups, where $\log f^s = sa_1$, $\log g^s = sb_1$.

The log transform reduces matters at the Lie group level to those at the easier level of Lie algebra. All properties of near-1 formal maps have their logarithmic interpretations. We list some of them, let $\log f^s = a^s = \sum_1^\infty s^k a_k$:

1. f^s is a phase flow, i.e., $f^{s+t} = f^s \circ f^t \Leftrightarrow \log f^s = sa_1$.
2. f^s is revertible, i.e., $f^s \circ f^{-s} = 1_m \Leftrightarrow \log f^s$ is odd in s.
3. f^s raised to real μth power $(f^s)^\mu \Leftrightarrow \log(f^s)^\mu = \mu \log f^s$. In particular $\log(f^s)^{-1} = -\log f^s$, $\log \sqrt{f^s} = \frac{1}{2}\log f^s$.
4. f^s scaled to $f^{\alpha s} \Leftrightarrow \log(f^{\alpha s}) = (\log f)^{\alpha s}$. In particular $\log(f^{-s}) = (\log f)^{-s}$.
5. $f^s - g^s = O(s^{p+1}) \Leftrightarrow \log f^s - \log g^s = O(s^{p+1})$.
6. $f^s \circ g^s = g^s \circ f^s \Leftrightarrow [\log f^s, \log g^s] = 0 \Leftrightarrow \log(f^s \circ g^s) = \log f^s + \log g^s$.
7. $(f^s \circ g^s) = h^s \Leftrightarrow \log h^s = \log(f^s \circ g^s) = \log f^s + \log g^s + \sum_1^\infty d_k(\log f^s, \log g^s)$.
8. f^s symplectic $\Leftrightarrow$ all a_k are Hamiltonian fields.
9. f^s contact $\Leftrightarrow$ all a_k are contact fields.
10. f^s volume-preserving $\Leftrightarrow$ all a_k are source-free fields.

The log transform has important bearing on dynamical systems with Lie algebra structure, the structure-preserving property of maps f^s at the Lie group $(\mathbf{G} \subset \mathbf{D}_m)$ level can be characterized through their logarithms at the associated Lie algebra $(\mathbf{L} \subset \mathbf{V}_m)$ level.

We return to the main problem of approximation to the phase flow for dynamical system $dx/dt = a(x)$.

$$f_a^s = f^s = 1_m + \sum s^k f_k \approx e_a^s = 1_m + \sum s^k e_k, \quad e_k = a^k/k!. \tag{10.8}$$

If $f_k = e_k$, $1 \le k \le p$, we say f_a^s is accurate to order $\ge p$, if moreover, $f_{p+1} \ne e_{p+1}$, we say it is accurate to order p.

Let $\log f^s = a^s = \sum s^k a_k$. Note that the first $p+1$ equations in (10.2) completely determine $a_1, a_2, \cdots, a_{p+1}$ and $f_1, f_2, \cdots, f_{p+1}$ each other. It is then easy to establish

$$\begin{aligned}&f_k = e_k, \quad 1 \le k \le p; \quad f_{p+1} \ne e_{p+1} \Longleftrightarrow\\&a = a_1 = e_1; \quad a_k = 0, \quad 1 < k \le p; \quad a_{p+1} = f_{p+1} - e_{p+1} \ne 0.\end{aligned} \tag{10.9}$$

So the orders of approximation for $f_a^s \approx e_a^s$ and for $\log f_a^s - sa$ are the same.

Moreover, note that we have a *formal field*

$$s^{-1} \log f^s = s^{-1} a^s = a + \sum_1^\infty s^k a_{k+1} = a + O(s^p) \tag{10.10}$$

which is equal to the original field a up to a near-0 perturbation and defines a *formal dynamical system*

$$\frac{dx}{dt} = (s^{-1} \log f^s)(x) = a(x) + \sum_1^\infty s^k a_{k+1}(x) \tag{10.11}$$

having a *formal phase flow* (in two parameters t and s with group property in t) $e^t_{s^{-1}a^s} = \exp ts^{-1}a^s$ whose *diagonal formal flow* $e^t_{s^{-1}a^s}|_{t=s}$ is exactly f^s. This means that any compatible algorithm f_a^s of order p gives perturbed solution of a right equation with field a; however, it gives the right solution of a perturbed equation with field $s^{-1} \log f_a^s = a + O(s^p)$. There could be many methods with the same formal order of accuracy but with quite different qualitative behavior. The problem is to choose among them those leading to allowable perturbations in the equation. For systems with geometric structure, the propositions 8, 9 and 10 provide guidelines for a proper choice. The structure-preservation requirement for the algorithms precludes all unallowable perturbations alien to the pertinent type of dynamics. Take, for example, Hamiltonian systems. A transition map f_a^s for Hamiltonian field a is symplectic if and only if all fields a_k are Hamiltonian, i.e., the induced perturbations in the equation are Hamiltonian [4, 15]. So symplectic algorithms are clean, inherently free from all kinds of perturbations alien to Hamiltonian dynamics (such as artificial dissipation inherent in the vast majority of conventional methods), this accounts for their superior performance. The situations are analogous for contact and volume-preserving algorithms. Propositions 8, 9 lead to the existence of formal energy for symplectic algorithms and that of formal contact energy for contact algorithms, see [8–9, 12–13, 37, and also 45]. For the incompatibility of symplectic structure preservation with energy conservation for algorithms, see [24, 26].

Appended Remarks. Our results on structure-preserving algorithms for dynamical systems and the constructive theory of generating functions for symplectic maps as surveyed above has been extended and further developed in various aspects as follows: For Poisson manifolds and Poisson maps, see [21–23, 25, 36]. For infinite-dimensional Hamiltonian systems, see [24, 27, 29, 35]. For quantum mechanical systems and unitary algorithms, see [30, 46]. For Hamiltonization of ordinary differential equations, see [39, 41, 42]. For comparative numerical studies for Hamiltonian vs non-Hamiltonian algorithms, see [4, 5, 14, 15, 34]. For the problem of symplecticity of multi-step methods, see [7, 17]. For KAM theorem for symplectic algorithms for Hamiltonian systems, see [32].

Acknowledgements. The authors would like to thank their collaborators Qin Meng-zhao, Li Wang-yao, Ge Zhong, Shang Zai-jiu, Wu Yu-hua, Li Chun-wang, Zhang Mei-qing, Tang Yi-fa, Zhu Wen-jie, Jiang Li-xin and Shu Hai-bing for fruitful cooperation.

References

1. V.I. Arnold, *Mathematical Methods of Classical Mechanics*, Springer, New York, 1978.
2. Feng Kang, *On difference schemes and symplectic geometry*, Proc. 1984 Beijing Symp. Diff. Geometry and Diff. Equations, Ed. Feng Kang, Science Press, Beijing, 1985, 42–58.
3. ———, *Difference schemes for Hamiltonian formalism and symplectic geometry*, J. Comp. Math., **4**:3(1986), 279–289.
4. ———, *The Hamiltonian way for computing Hamiltonian dynamics*, in Applied and Industrial Mathematics, 17–35, ed. R. Spigler, Kluwer, Netherlands, 1991.
5. ———, *How to compute properly Newton's equation of motion*, Proc. of 2nd Conf. on Numerical Methods for Partial Differential Equations, 1991, World Scientific, Singapore, 1992, 15–22.
6. ———, *Formal power series and numerical methods for differential equations*, Proc. of Inter. Conf. on Scientific Computation, Hangzhou, 1991, World Scientific, Singapore, 1992.
7. ———, *The step transition operator for multi-step methods of ODE's*, Preprint, A.S.C.C., 1991.
8. ———, *The calculus of generating functions and the formal energy for Hamiltonian algorithms*, Preprint, A.S.C.C., 1991.
9. ———, *The calculus of formal power series for diffeomorphisms and vector fields*, Preprint, A.S.C.C., 1991.
10. ———, *Symplectic, contact and volume preserving algorithms*, Proc. 1st China -Japan Conf. on Numer. Math, Beijing, 1992, World Scientific, Singapore, 1993, 1–28.
11. ———, *Formal dynamical systems and numerical algorithms*, Proc. Inter. Conf. on Computation of Differential Equations and Dynamical Systems, Beijing, 1992, World Scientific, Singapore, 1993, 1–10.
12. ———, *Contact algorithms for contact dynamical systems*, Preprint, A.S.C.C., 1992.
13. ———, *Theory of contact generating functions*, Preprint, A.S.C.C., 1992.
14. Feng Kang, Qin Meng-zhao, *The symplectic methods for computation of Hamiltonian equations*, Proc. Conf. on Numerical Methods for PDE's, Shanghai, 1986, ed. Zhu You-lan, Guo Ben-yu, Lect. Notes in Math 1297, Springer, 1987, 1–37.
15. ———, *Hamiltonian algorithms and a comparative numerical study*, Comput. Phys. Comm., **65**(1991), 173–187.
16. Feng Kang, Shang Zai-jiu, *Volume preserving algorithms for source-free dynamical systems*, Preprint, A.S.C.C., 1991.
17. Feng Kang, Tang Yi-fa, *Non-symplecticity of linear multi-step methods*, Preprint, A.S.C.C., 1991.
18. Feng Kang, Wang Dao-liu, *A note on conservation laws of symplectic difference schemes for Hamiltonian systems*, J. Comp. Math., **9**:3(1991), 229–237.
19. ———, *Variations on a theme by Euler*, Preprint, Acad. Sin. Comp. Ctr., 1991.
20. Feng Kang, Wu Hua-mo, Qin Meng-zhao, and Wang Dao-liu, *Construction of canonical difference schemes for Hamiltonian formalism via generating functions*, J. Comp. Math., **7**:1(1989), 71–96.
21. Ge Zhong, *Generating functions for the Poisson map*, A.S.C.C., Preprint, 1987.
22. ———, *Geometry of symplectic difference schemes and generating functions*, A.S.C.C., Preprint, 1988.
23. ———, *Generating functions, Hamilton-Jacobi equations and symplectic groupsoids on Poisson manifolds*, MSRI Preprint, Berkeley, 1988.
24. Ge Zhong, Feng Kang, *On the approximation of Hamiltonian systems*, *J. Comp. Math.*, **6**:1(1988), 88–97.

25. Ge Zhong, J. Marsden, *Lie-Poisson Hamilton-Jacobi theory and Lie-Poisson integrators*, Phys. Lett. A, **133**:3(1988), 137–139.
26. Li Chun-wang, *The incompatibility of symplectic structure preservation with energy conservation for numerical computation of Hamiltonian systems*, Preprint, A.S.C.C., 1987.
27. Li Chun-wang, Qin Meng-zhao, *Symplectic difference schemes for infinite dimensional Hamiltonian systems*, J. Comp. Math., **6**:2(1988), 164–174.
28. Qin Meng-zhao, Wang Dao-liu and Zhang Mei-qing, *Explicit symplectic difference schemes for separable Hamiltonian systems*, J. Comp. Math., **9:3**(1991), 211–221.
29. Qin Meng-zhao, Zhang Mei-qing, *Multi-stage symplectic schemes of two kind of Hamiltonian systems of wave equations*, Computer Math. Applic., **19**:10(1990), 51–62.
30. ———, *Explicit Runge-Kutta like unitary schemes to solve certain quantum operator equations of motion*, J. Stat. Phys., **60**(1990), 837–843.
31. Qin Meng-zhao, Zhu Wen-jie, *Construction of higher order symplectic schemes by composition*, Computing, **47**(1992), 309–321.
32. Shang Zai-jiu, *On KAM theorem for symplectic algorithms for Hamiltonian systems*, Ph. D. dissertation, A.S.C.C., 1991.
33. Shu Haibin, *A new approach to generating functions for contact systems*, Preprint, A.S.C.C., 1992.
34. Tang Yi-fa, *Hamiltonian systems and algorithms for geodesic flows on compact Riemannian manifolds*, Master Thesis, A.S.C.C., 1990.
35. Wang Dao-liu, *Semi-discrete Fourier spectral approximations of infinite dimensional Hamiltonian systems and conservation laws*, Computer Math. Applic., **21**:4 (1991), 63–75.
36. ———, *Poisson difference schemes for Hamiltonian systems on Poisson manifolds*, J. Comp. Math., **9**:2(1991), 115–124.
37. ———, *Some aspects of Hamiltonian systems and symplectic algorithms*, to appear in Physica D.
38. ———, *Symplectic difference schemes for perturbed Hamiltonian systems*, Preprint, ZIB, 1990.
39. ———, *Decomposition of vector fields and composition of algorithms*, Proc. Inter. Conf. on Computation of Differential Equations and Dynamical Systems, Beijing, 1992, World Scientific, Singapore, 1993, 179–184.
40. Wang Dao-liu, Wu Yu-hua, *Generating function methods for the construction of one-step schemes for ODE's*, A.S.C.C., Preprint, 1988.
41. Wu Yu-hua, *The generating function of the solution of ODE and its discrete methods*, Computer Math. Applic., **15**:12(1988), 1041–1050.
42. ———, *Symplectic transformations and symplectic difference schemes*, Chinese J. Numer. Math. Applic., **12**:1(1990), 23–31.
43. ———, *Discrete variational principle to the Euler-Lagrange equation*, Computer Math. Applic., **20**:8(1990), 61–75.
44. H. Yoshida, *Construction of higher order symplectic integrators*, Phys. Letters A, **150**(1990), 262–268.
45. ———, *Conserved quantities of symplectic integrators for Hamiltonian systems*, Preprint, 1990.
46. Zhang Mei-qing, *Explicit unitary schemes to solve quantum operator equations of motion*, J. Stat. Phys., **65**:3/4(1991),793–799.
47. Zhang Meiqing and Qin Mengzhao, *Explicit symplectic schemes to solve vortex systems*, Preprint, A.S.C.C., 1992.

Computing Center, Academia Sinica, P.O. Box 2719, Beijing 100080, China

Computing Center, Academia Sinica, P.O. Box 2719, Beijing 100080, China

Contemporary Mathematics
Volume **163**, 1994

Generalized Stability of Discretization and Its Applications to Numerical Solutions of Nonlinear Differential Equations

GUO BEN-YU (KUO PEN-YU)

1. Introduction

Richardson and Courant, Friedrichs, Lewy proposed the stability of finite difference schemes, which plays an important role in numerical solutions of partial differential equations (see [1, 33]). Lax and Richtmeyer developed the theory of stability and convergence of finite difference methods for initial value problems (see [31, 34]). On the other hand. Cantorovich considered the approximations of operator equations in abstract spaces (see [23]). But all of these results are applicable to linear problems usually. Stetter and Keller studied nonlinear stability and local stability respectively (see [22, 36]), which have been used successfully for nonlinear ordinary differential equations. In 1965, the author proposed generalized stability of difference schemes, which has been applied widely to numerical solutions of many nonlinear partial differential equations such as vorticity equations, Navier-Stokes equations, atmospheric equations and nonlinear wave equations (see [3, 4, 24–27]). Recently this theory was generalized to locally weak generalized stability, which is also suitable for nonlinear problems with several isolated solutions (see [5, 6]). In this paper, we present the main theoretical results and its applications. In Section 2, we introduce the basic idea of generalized stability. In Section 3, we focus on initial value problems. Then we explain how to use the above results for difference schemes of boundary value problems and initial value problems of nonlinear partial differential equations respectively in Section 4 and Section 5. Finally we show how to use the above theory to analyze the stability and the convergence of spectral method for nonlinear problems.

1991 *Mathematics Subject Classification.* 65M12, 65L20, 65M70, 65N12.

2. Generalized stability

Let B_q be Banach space with the norm $\|\cdot\|_q$, $q = 1, 2$.

$$S_q(u, R) = \{v \mid v \in B_q,\ \|u - v\|_q < R\}.$$

L is an operator mapping $u \in B_1$ to B_2. $f \in B_2$ is given. We consider the following operator equation

$$Lu = f. \tag{2.1}$$

Let h be a positive parameter. $B_{q,h}$ is a finite-dimensional Banach space with the norm $\|\cdot\|_{q,h}$, approximating B_q.

$$S_{q,h}(u_h, r) = \{v_h \mid v_h \in B_{q,h}, \|u_h - v_h\|_{q,h} < r\}.$$

Assume that $\mathrm{Dim}(B_{1,h}) = \mathrm{Dim}(B_{2,h})$. Let $\gamma_{q,h}$ be a continuous restriction operator from B_q to $B_{q,h}$, satisfying the following conditions

$$\lim_{h\to 0} \|\gamma_{1,h} u\|_{1,h} = \|u\|_1, \qquad \forall u \in B_1,$$
$$\lim_{h\to 0} \|\gamma_{2,h} g\|_{2,h} = \|g\|_2, \qquad \forall g \in B_2.$$

Let L_h and f_h be the approximations to L and f respectively. We consider the following approximate operator equation

$$L_h u_h = f_h. \tag{2.2}$$

The truncation error is defined as

$$R_h(u) = \gamma_{2,h} Lu - L_h \gamma_{1,h} u + f_h - \gamma_{2,h} f.$$

If u is a solution of (2.1), then $R_h(u) = f_h - L_h \gamma_{1,h} u$.

Let D be a subset of solutions of (2.1) whose corresponding data set F is dense in B_2. We say that (2.2) is consistent with (2.1), if and only if

$$\lim_{h\to 0} \|R_h(u)\|_{2,h} = 0, \quad \forall u \in D.$$

If for all u, the solutions of (2.1), there holds

$$\lim_{h\to 0} \|u_h - \gamma_{1,h} u\|_{1,h} = 0,$$

then we say that (2.2) is convergent.

The usual stability is as follows.

DEFINITION 1. *If there exist positive constants h_0 and C such that for all $h \le h_0$,*

$$\|u_h - v_h\|_{1,h} \le C\|L_h u_h - L_h v_h\|_{2,h}, \quad \forall u_h, v_h \in B_{1,h}, \tag{2.3}$$

then we say that (2.2) is stable.

According to the theory of Cantorovich (see [23]), we conclude that if (2.1) is well-posed, and (2.2) is consistent with (2.1), then the stability is equivalent to the convergence. But generally nonlinear problems do not satisfy (2.3). Indeed, such stability is very strong. Firstly (2.3) should hold for all $u_h \in B_{1,h}$. Secondly, C is an absolute constant. Finally, there is no restriction on $\|L_h u_h - L_h v_h\|_{2,h}$. Hence several authors generalized the stability in different ways. For instance, Strang served weak stability and Stetter developed nonlinear stability (see [36, 37]).

The author proposed another stability.

DEFINITION 2. *If there are non-negative constant $M(u_h, h)$ and positive constants h_0 and $N(u_h, h)$ such that for all $h \le h_0$ and $v_h \in B_{1,h}$, the inequality*

$$\|L_h u_h - L_h v_h\|_{2,h} \le N(u_h, h)$$

implies

$$\|u_h - v_h\|_{1,h} \le M(u_h, h)\|L_h u_h - L_h v_h\|_{2,h},$$

then we say that (2.2) is of weak generalized stability at u_h. Especially, if for all $v_h^{(q)} \in B_{1,h}$, the inequalities

$$\|L_h v_h^{(q)} - L_h u_h\|_{2,h} \le N(u_h, h), \quad q = 1, 2,$$

imply

$$\|v_h^{(1)} - v_h^{(2)}\|_{1,h} \le \bar{M}(u_h, h)\|L_h v_h^{(1)} - L_h v_h^{(2)}\|_{2,h},$$

then we say that (2.2) is of uniformly weak generalized stability at u_h.

Guo Ben-yu proved the following results (see [4]).

THEOREM 1. *Let u be the unique solution of (2.1), and assume that*
 (i) *L_h is defined and continuous in $S_{1,h}(\gamma_{1,h}u, R)$,*
 (ii) *(2.2) is of uniformly weak generalized stability at $\gamma_{1,h}u$,*
 (iii) $\|R_h(u)\|_{2,h} \le \min\{\bar{N}(\gamma_{1,h}u, h), \frac{R}{\bar{M}(\gamma_{1,h}u, h)}\}$.
Then (2.2) possesses a unique solution u_h.

THEOREM 2. *Let u and u_h be the unique solutions of (2.1) and (2.2) respectively, and assume that*
 (i) *L_h is defined and continuous in $S_{1,h}(\gamma_{1,h}u, R)$,*
 (ii) *(2.2) is of weak generalized stability at $\gamma_{1,h}u$,*
 (iii) $\|R_h(u)\|_{2,h} \le N(\gamma_{1,h}u, h)$.
Then

$$\|u_h - \gamma_{1,h}u\|_{1,h} \le M(\gamma_{1,h}u, h)\|R_h(u)\|_{2,h}.$$

The simplest case is so called generalized stability, or g-stability (see [2, 4]). It means that $M(u_h, h) = M(u_h)$, $N(u_h, h) = N(u_h)h^{\bar{s}}$. The infimum of such values $\bar{s}$ is called the index of generalized stability, denoted by s. But we can only get upper bound of s usually. Clearly if $\|R_h(u)\|_{2,h} = O(h^p)$, $s < p$ and $p > 0$, then (2.2) is convergent and $\|u_h - \gamma_{1,h}u\|_{1,h} = O(h^p)$. Furthermore, if $s \leq 0$, then (2.2) is convergent provided that (2.2) is consistent with (2.1). It can be seen that the smaller the index s, the stabler the scheme. In particular, if $s = -\infty$, then there is no restriction on the errors of data and so generalized stability almost turns to usual stability.

The above framework is useful only for problems with unique solution. Keller generalized the stability in another way. If there are constants C_R and R such that

$$\|v^{(1)} - v^{(2)}\|_1 \leq C_R\|Lv^{(1)} - Lv^{(2)}\|_2, \quad \forall v^{(q)} \in S_1(u, R),$$

then we say that L is stable in $S_1(u, R)$. If u is a solution of (2.1) and there exists positive constant R such that L is stable in $S_1(u, R)$, then we say that u is a stable solution. Obviously, such a solution is unique in the ball $S_1(u, R)$. Let $L'(u)$ be the Fréchet derivative of L at u. $L'(u)$ is nonsingular iff $L'(u)v = 0$ implies $v = 0$. If u is a solution of (2.1) and $L'(u)$ is nonsingular, then we say that u is an isolated solution of (2.1). It can be verified that

(i) If u is a stable solution of (2.1) and $L'(u)$ exists, then u is an isolated solution;

(ii) If u is an isolated solution of (2.1) and $L'(v)$ is lipschitz continuous in $S_1(u, R)$, then u is a stable solution of (2.1).

Keller proposed the following stability (see [22]).

DEFINITION 3. *If there are positive constants h_0, R and non-negative constant $M(u_h)$ such that for all $h \leq h_0$ and $v_h^{(q)} \in S_{1,h}(u_h, R)$,*

$$\|v_h^{(1)} - v_h^{(2)}\|_{1,h} \leq M(u_h)\|L_h v_h^{(1)} - L_h v_h^{(2)}\|_{2,h},$$

then we say that (2.2) is locally stable at u_h.

Hereafter L'_h denotes the Frèchet derivative of L_h. In [22], Keller showed that (2.2) has a unique solution u_h in $S_{1,h}(\gamma_{1,h}u, \rho_0)$, where ρ_0 is sufficiently small, provided that some conditions are fulfilled. He also pointed out that the local stability sometimes implies the convergence. Keller's theory has been successfully applied to approximate solutions of mildly nonlinear ordinary differential equations. But it is difficult to be used for quasi-linear differential equations and multi-dimensional problems. Thus we gave a new definition of stability.

DEFINITION 4. *If there are non-negative constant $M(u_h, h)$ and positive constants h_0, R and $N(u_h, h)$ such that for all $h \leq h_0$ and $v_h \in S_{1,h}(u_h, R)$, the inequality*

$$\|L_h u_h - L_h v_h\|_{2,h} \leq N(u_h, h)$$

implies

$$\|u_h - v_h\|_{1,h} \le M(u_h, h)\|L_h u_h - L_h v_h\|_{2,h},$$

then we say that (2.2) *is of locally weak generalized stability for* u_h *in* $S_{1,h}(u_h, R)$. *Especially, if for all* $h \le h_0$ *and* $v_h^{(q)} \in S_{1,h}(u_h, R)$, *the inequalities*

$$\|L_h v_h^{(q)} - L_h u_h\|_{2,h} \le \bar{N}(u_h, h), \quad q = 1, 2,$$

imply

$$\|v_h^{(1)} - v_h^{(2)}\|_{1,h} \le \bar{M}(u_h, h)\|L_h v_h^{(1)} - L_h v_h^{(2)}\|_{2,h},$$

then we say that (2.2) *is of uniformly local weak generalized stability.*

The following theorem gives a sufficient condition for such stability.

THEOREM 3. *If the following conditions are fulfilled*

(i) L_h *is defined and continuous in* $S_{1,h}(u_h, R)$,

(ii) *for all* $v_h \in S_{1,h}(u_h, R)$ *such that* $L_h v_h \in S_{2,h}(L_h u_h, \overline{\overline{N}}(u_h, h))$, *the inverse* $[L_h'(v_h)]^{-1}$ *exists and*

$$\|[L_h'(v_h)]^{-1}\| \le \overline{\overline{M}}(u_h, h), \tag{2.4}$$

then (2.2) *is of uniformly local weak generalized stability for* u_h *in* $S_{1,h}(u_h, R)$.

PROOF. Let

$$r_0(h) = \min(\overline{N}(u_h, h), \frac{R}{\overline{\overline{M}}(u_h, h)}).$$

By Implicit Function Theorem, L_h^{-1} exists in an open neighborhood of each $g_h \in S_{2,h}(L_h u_h, r_0(h))$. Moreover

$$(L_h^{-1})'(L_h v_h) = [L_h'(v_h)]^{-1}.$$

From (2.4), we have that for all $g_h \in S_{2,h}(L_h u_h, r_0(h))$,

$$\|(L_h^{-1})'(g_h)\| \le \overline{\overline{M}}(u_h, h). \tag{2.5}$$

Now, let

$$g_h^{(t)} = L_h v_h^{(2)} + t(L_h v_h^{(1)} - L_h v_h^{(2)}), \quad 0 \le t \le 1, \quad v_h^{(q)} \in S_{2,h}(L_h u_h, r_0(h)).$$

Because of the convexity of $S_{2,h}(L_h u_h, r_0(h))$, there exists $[L_h^{-1}(g_h^{(t)})]'$ satisfying (2.5). Therefore it follows from the mean value theorem that

$$\begin{aligned} \|v_h^{(1)} - v_h^{(2)}\|_{1,h} &= \|L_h^{-1}(L_h v_h^{(1)}) - L_h^{-1}(L_h v_h^{(2)})\|_{1,h} \\ &\le \overline{\overline{M}}(u_h, h)\|L_h v_h^{(1)} - L_h v_h^{(2)}\|_{2,h}. \end{aligned}$$

The conclusion follows then. □

For the proof of the existence of solutions of (2.2), we shall use the following lemma (see [36]).

LEMMA 1. *Suppose that the following conditions are satisfied*

(i) L_h *is defined and continuous in* $S_{1,h}(w_h, R)$,

(ii) *for all* $v_h^{(q)} \in S_{1,h}(w_h, R)$ *such that* $L_h v_h^{(q)} \in S_{2,h}(L_h w_h, r)$, *we have*

$$\|v_h^{(1)} - v_h^{(2)}\|_{1,h} \le M \|L_h v_h^{(1)} - L_h v_h^{(2)}\|_{2,h}.$$

Then for all $g_h \in S_{2,h}(L_h w_h, r_0)$, *we have a unique solution* v_h *in* $S_{1,h}(w_h, R)$ *such that*

$$L_h v_h = g_h, \quad r_0 = \min(r, \frac{R}{M}).$$

THEOREM 4. *If the following conditions are satisfied*

(i) L_h *is defined and continuous in* $S_{1,h}(\gamma_{1,h}u, R)$,

(ii) *problem* (2.2) *is of uniformly local weak generalized stability for* $\gamma_{1,h}u$ *in* $S_{1,h}(\gamma_{1,h}u, R)$,

(iii) $\|R_h(u)\|_{2,h} < r_0(h) = \min(\bar{N}(\gamma_{1,h}u, h), \dfrac{R}{\bar{M}(\gamma_{1,h}u, h)})$,

then (2.2) *possesses a unique solution* u_h *in* $S_{1,h}(\gamma_{1,h}u, R)$.

PROOF. Put $w_h = \gamma_{1,h}u$, $r_0 = r_0(h)$ in Lemma 1. Then L_h^{-1} exists in $S_{2,h}(L_h\gamma_{1,h}u, r_0(h))$. On the other hand, condition (iii) implies

$$\|f_h - L_h \gamma_{1,h} u\|_{2,h} < r_0(h)$$

and thus the conclusion follows. □

THEOREM 5. *Let* u *and* $u_h \in S_{1,h}(\gamma_{1,h}u, R)$ *be the solutions of* (2.1) *and* (2.2) *respectively, and assume that*

(i) (2.2) *is of locally weak generalized stability for* $\gamma_{1,h}u$ *in* $S_{1,h}(\gamma_{1,h}u, R)$,

(ii) $\|R_h(u)\|_{2,h} \le N(\gamma_{1,h}u, h)$.

Then

$$\|u_h - \gamma_{1,h}u\|_{1,h} \le M(\gamma_{1,h}u, h) \|R_h(u)\|_{2,h}.$$

PROOF. Let $\tilde{u}_h = u_h - \gamma_{1,h}u$. Then

$$\begin{cases} L_h(\gamma_{1,h}u) = f_h - R_h(u), \\ L_h(\gamma_{1,h}u + \tilde{u}_h) = f_h. \end{cases}$$

From condition (ii), we get

$$\|\tilde{u}_h\|_{1,h} \le M(\gamma_{1,h}u, h) \|R_h(u)\|_{2,h}.$$

□

We can evaluate u_h by the following Newton's procedure

$$u_h^{(p+1)} = u_h^{(p)} - [L_h'(u_h^{(p)})]^{-1}(L_h u_h^{(p)} - f_h), \quad p \ge 0, \tag{2.6}$$

or the simplified Newton's procedure

$$u_h^{(p+1)} = u_h^{(p)} - [L_h'(u_h^{(0)})]^{-1}(L_h u_h^{(p)} - f_h), \quad p \ge 0. \tag{2.7}$$

THEOREM 6. *Assume that the following conditions are satisfied*

(i) L'_h *exists in* $S_{1,h}(u_h^{(0)}, \rho)$ *and for all* $v_h^{(q)} \in S_{1,h}(u_h^{(0)}, \rho)$,

$$\|L'_h(v_h^{(1)}) - L'_h(v_h^{(2)})\| \leq K_h \|v_h^{(1)} - v_h^{(2)}\|_{1,h},$$

(ii) $g_h = [L'_h(u_h^{(0)})]^{-1}$ *exists, and*

$$\|g_h\| \leq b_h, \ \|g_h[L_h(u_h^{(0)}) - f_h]\| \leq \eta_h,$$

(iii) $a_h = b_h K_h \eta_h < \frac{1}{2}$, *and* $\rho_h = \dfrac{1 - \sqrt{1 - 2a_h}}{b_h K_h} \leq \rho$.

Then both (2.6) *and* (2.7) *are convergent, and the same limit* u_h *is a solution of* (2.2).

Remark 1. The theory of Stetter is a special case of generalized stability. The locally weak generalized stability is a generalization of Keller's theory.

3. Generalized stability of initial value problems

It is easy to use the theorems in the previous section to analyze stability and convergence of nonlinear boundary value problems. But it might be a little difficult for them to be used directly for nonlinear initial value problems. In this section, we present weak generalized stability of difference method for nonlinear initial value problems (see [7]). Such stability implies the convergence provided that the approximation error satisfies some reasonable conditions.

Let B_1 and B_2 be Banach spaces with the norms $\|\cdot\|_1$ and $\|\cdot\|_2$, $B_1 \subseteq B_2$. P and L are differential operators with respect to the variables in the spaces, mapping $u \in B_1$ to B_2. For simplicity, assume that P is linear, and that the inverse P^{-1} exists. Let B_3 be a Banach space with the norm $\|\cdot\|_3$, whose element f is a map from the interval $(0, T]$ to B_2. Let u_0 be a given element of B_1 describing the initial state. We consider the nonlinear operator equation

$$\begin{cases} \dfrac{d}{dt}(Pu(t)) = Lu(t) + f(t), & 0 < t \leq T, \\ u(0) = u_0. \end{cases} \tag{3.1}$$

We define a genuine solution of (3.1) as a one-parameter family $u(t)$ such that $u(t)$ is in the domains of P and L for each $t \in [0, T]$, and

$$\|\frac{1}{s}[Pu(t+s) - Pu(t)] - Lu(t) - f(t)\|_2 \to 0, \quad \text{as } s \to 0.$$

Let D_1 and D_3 be such subsets of B_1 and B_3 that for any $u_0 \in D_1$ and $f \in D_3$, (3.1) has a unique solution. We suppose that D_1 and D_3 are dense in B_1 and B_3.

Now, let h be a positive parameter. $B_{q,h}$ is a finite dimensional Banch space with the norm $\|\cdot\|_{q,h}$, approximating the space B_q. Assume that $\mathrm{Dim}(B_{1,h}) = \mathrm{Dim}(B_{2,h})$. Let $\gamma_{q,h}$ be a continuous restriction operator from B_q to $B_{q,h}$, and satisfies the following conditions

(i) for all $u \in B_1$, $\gamma_{1,h} u = \gamma_{2,h} u$,

(ii) for all $u \in B_q$, $\|\gamma_{q,h}u\|_{q,h} \to \|u\|_q$, as $h \to 0$, $q = 1, 2, 3$.

Next, let $d = d(h) > 0$ be the mesh spacing of the variable t, and

$$u_{h,t}(t) = \frac{1}{d}[u_h(t+d) - u_h(t)].$$

Let $P_h u_h(t)$ and $L_h(u_h(t), u_h(t+d))$ be the approximations to $Pu(t)$ and $Lu(t)$. Similarly, $f_h(t)$ and $u_{0,h}$ are the approximations to $f(t)$ and u_0 respectively. We consider the approximate equation

$$\begin{cases} (P_h u_h(t))_t = L_h(u_h(t), u_h(t+d)) + f_h(t), \\ u_h(0) = u_{0,h}. \end{cases} \tag{3.2}$$

Suppose that for each t, $u_h(t)$ is determined uniquely by (3.2). The simplest case is that P_h^{-1} exists and

$$L_h(u_h(t), u_h(t+d)) = \bar{L}_h(u_h(t)).$$

The approximation error for $t > 0$ is defined as

$$\begin{aligned} R_h(u(t)) = &- \gamma_{2,h}\frac{d}{dt}(Pu(t)) + (P_h\gamma_{1,h}u(t))_t + \gamma_{2,h}Lu(t) \\ &- L_h(\gamma_{1,h}u(t), \gamma_{1,h}u(t+d)) + \gamma_{2,h}f(t) - f_h(t). \end{aligned}$$

If $u(t)$ is the solution of (2.1), then

$$R_h(u(t)) = (P_h\gamma_{1,h}u(t))_t - L_h(\gamma_{1,h}u(t), \gamma_{1,h}u(t+d)) - f_h(t).$$

If the following conditions are fulfilled

(i) $\|\gamma_{1,h}u_0 - u_{0,h}\|_{1,h} \to 0$, as $h \to 0$,

(ii) for all $u(t) \in D_1$, $t \le T$ and $f \in D_3$,

$$\|R_h u(t)\|_{2,h} \to 0, \quad \text{as } h \to 0,$$

then we say that the approximation (3.2) is consistent with (3.1).

If for all $u(t)$, the solution of (3.1) and $u_h(t)$, the corresponding solution of (3.2),

$$\|\gamma_{1,h}u(t) - u_h(t)\|_{1,h} \to 0, \text{ as } h \to 0, \qquad 0 \le t \le T,$$

then we say that (3.2) is convergent.

We now turn to the stability. The usual stability is as follows.

DEFINITION 5. *If there are positive constants h_0 and $C(T)$ such that for all $h \le h_0$ and $t \le T$,*

$$\|u_h^{(1)}(t) - u_h^{(2)}(t)\|_{1,h} \le C(T)(\|u_{0,h}^{(1)} - u_{0,h}^{(2)}\|_{1,h} + \|f_h^{(1)} - f_h^{(2)}\|_{3,h}), \tag{3.3}$$

where $u_h^{(q)}$ is the unique solution of (3.2) with the data $u_{0,h}^{(q)}$ and $f_h^{(q)}$, then we say that (3.2) is stable.

According to Lax's theory (see [34]), we conclude that if (3.1) is well-posed with linear operators P and L, and (3.2) is consistent with (3.1), then the stability given by Definition 5 is equivalent to the convergence. Since such stability is not suitable for nonlinear problems, Guo proposed the following definition (see [7]).

DEFINITION 6. *If there are a non-negative constant $M(u_h, h, T)$ and positive constants h_0 and $N(u_h, h, T)$ such that for all $h \le h_0$, the inequality*

$$\|u_{0,h}^{(1)} - u_{0,h}^{(2)}\|_{1,h} + \|f_h^{(1)} - f_h^{(2)}\|_{3,h} \le N(u_h^{(1)}, h, T)$$

implies

$$\begin{aligned} \|u_h^{(1)}(t) - u_h^{(2)}(t)\|_{1,h} \le & M(u_h^{(1)}, h, T)(\|u_{0,h}^{(1)} - u_{0,h}^{(2)}\|_{1,h} \\ & + \|f_h^{(1)} - f_h^{(2)}\|_{3,h}), \qquad 0 \le t \le T, \end{aligned} \tag{3.4}$$

then we say that (3.2) is of weak generalized stability at $u_h^{(1)}$.

There is relation between the above stability and the convergence.

THEOREM 7. *Let $u(t)$ and $u_h(t)$ be the unique solutions of (3.1) and (3.2) respectively. If (3.2) is of weak generalized stability at $\gamma_{1,h}u(t)$, and*

$$\|\gamma_{1,h}u_0 - u_{0,h}\|_{1,h} + \|R_h(u)\|_{3,h} \le N(\gamma_{1,h}u, h, T),$$

then for all $t \le T$,

$$\|\gamma_{1,h}u(t) - u_h(t)\|_{1,h} \le M(\gamma_{1,h}u, h, T)(\|\gamma_{1,h}u_0 - u_{0,h}\|_{1,h} + \|R_h(u)\|_{3,h}).$$

PROOF. We have

$$\begin{cases} (P_h\gamma_{1,h}u(t))_t = L_h(\gamma_{1,h}u(t), \gamma_{1,h}u(t+d)) + f_h(t) + R_h(t), \\ \gamma_{1,h}u(0) = \gamma_{1,h}u_0. \end{cases} \tag{3.5}$$

By comparing (3.2) with (3.5), the conclusion follows from (3.4). □

The simplest case is also called generalized stability or g-stability (see [2]). It means that

$$M(u_h, h, T) = M(u_h, T)h^{\bar{s}}, \quad N(u_h, h, T) = N(u_h, T).$$

The infimum of such values $\bar{s}$ is also called the index of generalized stability, denoted by s. Clearly if $u_{0,h} = \gamma_{1,h}u_0$, $\|R_h(u)\|_{3,h} = O(h^p)$, $s < p$ and $p > 0$, then (3.2) is convergent with the accuracy $O(h^q)$.

Remark 2. It is not necessary that P is a linear operator.

Remark 3. It is not necessary that (2.1) is a differential equation of first order.

4. Applications to difference schemes of nonlinear boundary value problems

In this section, we show how to use the framework in Section 2 for difference schemes of nonlinear boundary value problems.

Let $I = \{x \mid 0 < x < 1\}$ and $\bar{I}$ be the closure of I. Define

$$B_1 = \{u \mid u \in C^2(I) \cap C^\circ(\bar{I})\}, \quad B_2 = \{g \mid g \in C^\circ(\bar{I})\},$$

equipped with the norms

$$\|u\|_1 = \max_{x\in\bar{I}} |u| + \sup_{x\in I} |\frac{du}{dx}| + \sup_{x\in I} |\frac{d^2u}{dx^2}|, \quad \|g\|_2 = \max_{x\in\bar{I}} |g|.$$

Let λ be a real number and $f(x,u) \in C^2(\bar{I} \times \mathcal{R})$. Suppose that $f' = \dfrac{\partial f}{\partial u} > 0$ for all x and u. Define

$$Lu(x) = \begin{cases} -\dfrac{d^2u}{dx^2} - \lambda f(x, u(x)), & x \in I, \\ u(x), & x = 0, 1, \end{cases}$$

and consider the equation

$$Lu(x) = 0, \quad x \in \bar{I}. \tag{4.1}$$

It is easy to see that

$$L'(v)w = \begin{cases} -\dfrac{d^2w}{dx^2} - \lambda f'(x, v)w, & x \in I, \\ w, & x = 0, 1. \end{cases} \tag{4.2}$$

The related eigenvalue problem is

$$\begin{cases} \dfrac{d^2w}{dx^2}(x) - \mu f'(x, v(x))w(x) = 0, & x \in I, \\ w(x) = 0, & x = 0, 1. \end{cases} \tag{4.3}$$

The eigenvalues of (4.3) are denoted by $\mu_j(v)$, $j = 1, 2, \cdots$. Suppose that $u \in C^4(I)$ is an isolated solution of (4.1). Then $\lambda \neq \mu_j(u)$, $j = 1, 2, \cdots$.

We take a uniform mesh of spacing h, $h = \dfrac{1}{N}$, N being an integer. Define

$$I_h = \{x = jh \mid 1 \le j \le N - 1\}.$$

The closure of I_h is denoted by $\bar{I}_h$. We shall use the following notations

$$u_{h,x}(x) = \frac{1}{h}(u_h(x+h) - u_h(x)), \quad u_{h,\bar{x}}(x) = u_{h,x}(x-h),$$

$$u_{h,x\bar{x}}(x) = \frac{1}{h^2}(u_h(x+h) - 2u_h(x) + u_h(x-h)).$$

Let $B_{1,h}$ and $B_{2,h}$ be discrete function spaces approximating B_1 and B_2 respectively. Define

$$\|u_h\|_{h,\infty} = \max_{x\in I_h} |u_h(x)|,$$
$$|u_h|_{h,\infty}^{(1)} = \max(\|u_{h,x}\|_{h,\infty}, \|u_{h,\bar{x}}\|_{h,\infty}),$$
$$|u_h|_{h,\infty}^{(2)} = \max(|u_{h,x}|_{h,\infty}^{(1)}, |u_{h,\bar{x}}|_{h,\infty}^{(1)}).$$

The norms of $B_{q,h}$ are

$$\|u_h\|_{1,h} = \|u_h\|_{h,\infty} + |u_h|_{h,\infty}^{(1)} + |u_h|_{h,\infty}^{(2)}, \quad \|g_h\|_{2,h} = \|g_h\|_{h,\infty}.$$

Besides $\gamma_{1,h}u = u$ for $x \in \bar{I}_h$.

We next define

$$L_h u_h(x) = \begin{cases} -u_{h,x\bar{x}}(x) - \lambda f(x, u_h(x)), & x \in I_h, \\ u_h(x), & x = 0, 1, \end{cases}$$

and consider the approximate problem

$$L_h u_h(x) = 0, \quad x \in \bar{I}_h. \tag{4.4}$$

We have

$$L_h'(v_h)w_h = \begin{cases} -w_{h,x\bar{x}} - \lambda f'(x, v_h)w_h, & x \in I_h, \\ w_h, & x = 0, 1. \end{cases} \tag{4.5}$$

The related eigenvalue problem is

$$\begin{cases} -w_{h,x\bar{x}}(x) - \mu_h f'(x, v_h(x))w_h(x) = 0, & x \in I_h, \\ w_h(x) = 0, & x = 0, 1. \end{cases} \tag{4.6}$$

The eigenvalues are denoted by $\mu_{h,j}(v_h)$, $j = 1, 2, \cdots$. Because of $f' > 0$,

$$\|w_h\|_{h,\infty} \le C_1(\min_j |\lambda - \mu_{h,j}(v_h)|)^{-1} \|L_h'(v_h)w_h\|_{2,h}, \tag{4.7}$$

where C_1 is a positive constant. On the other hand, $u(x)$ is an isolated solution of (4.1) and so $\lambda \neq \mu_j(u)$. By a minor extension of the arguments of Kuttler (see [30]), we have (also see [35]),

$$|\mu_{h,j}(u) - \mu_j(u)| = O(h^2).$$

Let R be a suitably small positive constant and $v_h \in S_{1,h}(\gamma_{1,h}u, R)$. Then

$$(\min_j |\lambda - \mu_{h,j}(v_h)|)^{-1} \le C_2(\gamma_{1,h}u)$$

and so from (4.7),

$$\|w_h\|_{h,\infty} \le C_1 C_2 \|L_h'(v_h)w_h\|_{2,h}.$$

Therefore

$$\|w_{h,x\bar{x}}\|_{2,h} \le C_3(\gamma_{1,h}u)\|L_h'(v_h)w_h\|_{2,h}.$$

By discrete Green function, we can prove that

$$\|w_{h,x}\|_{h,\infty} \le C_4 \|w_{h,x\bar{x}}\|_{h,\infty},$$

where C_4 is a positive constant. Hence

$$\|w_h\|_{1,h} \le C_5(\gamma_{1,h}u)\|L'_h(v_h)w_h\|_{2,h}.$$

It means that

$$\|[L'_h(v_h)]^{-1}\| \le C_5(\gamma_{1,h}u).$$

By Theorem 3, L_h is of uniformly local weak generalized stability for $\gamma_{1,h}u$ in $S_{1,h}(\gamma_{1,h}u, R)$ with

$$\bar{M}(\gamma_{1,h}u, h) = C_5(\gamma_{1,h}u) > 0.$$

Clearly $\|R_h(u)\|_{2,h} = O(h^2)$ and so for all sufficiently small h,

$$\|R_h(u)\|_{2,h} < r_0(h) = \frac{r}{C_5(\gamma_{1,h}u)}.$$

Finally by Theorem 4 and Theorem 5, the approximate problem (4.4) has a unique solution $u_h \in S_{1,h}(\gamma_{1,h}u, R)$ and $\|u_h - \gamma_{1,h}u\|_{1,h} = O(h^2)$.

5. Applications to difference schemes of initial-boundary value problems

In this section, we apply the idea of generalized stability to construction and error analysis of finite difference schemes for nonlinear initial-boundary value problems. We design schemes based on physical law and get better stability in the sense of Section 3. We also give the relation between the conservation and the generalized stability of nonlinear schemes.

We consider the following periodical problem

$$\begin{cases} \dfrac{\partial u}{\partial t} + u\dfrac{\partial u}{\partial x} - \nu\dfrac{\partial^2 u}{\partial x^2} = f, & x \in \mathcal{R}, \quad 0 < t \le T, \\ u(x,t) = u(x+1,t), & x \in \mathcal{R}, \quad 0 \le t \le T, \\ u(x,0) = u_0(x), & x \in \mathcal{R}, \end{cases} \tag{5.1}$$

where u_0 and f are given functions with the period 1.

Let h and τ be mesh spacings of the variables x and t respectively, and $\lambda = \dfrac{\tau}{h^2} < C_0$, C_0 being a positive constant. Let

$$I_h = \{x = jh \mid 0 \le j \le N-1\}, \quad S_\tau = \{t = k\tau \mid k = 0,1,2,\cdots\}.$$

The definitions of $u_{h,x}(x,t)$, $u_{h,\bar{x}}(x,t)$ and $u_{h,x\bar{x}}(x,t)$ are the same as before. Besides,

$$\begin{aligned} u_{h,\hat{x}}(x,t) &= \frac{1}{2}u_{h,x}(x,t) + \frac{1}{2}u_{h,\bar{x}}(x,t), \\ u_{h,t}(x,t) &= \frac{1}{\tau}(u_h(x,t+\tau) - u_h(x,t)). \end{aligned}$$

We introduce the discrete scalar product and norms as follows

$$(u_h(t), v_h(t)) = h \sum_{x \in I_h} u_h(x,t) v_h(x,t), \quad \|u_h(t)\|^2 = (u_h(t), u_h(t)),$$
$$|u_h(t)|_1^2 = \frac{1}{2}\|u_{h,x}(t)\|^2 + \frac{1}{2}\|u_{h,\bar{x}}(t)\|^2.$$

It is easy to verify that

$$2(u_h(t), u_{h,t}(t)) = \|u_h(t)\|_t^2 - \tau\|u_{h,t}(t)\|^2, \tag{5.2}$$

$$2(u_h(t), v_{h,x\bar{x}}(t)) + (u_{h,x}(t), v_{h,x}(t)) + (u_{h,\bar{x}}(t)), v_{h,\bar{x}}(t)) = 0, \tag{5.3}$$

$$2(u_{h,t}(t), u_{h,x\bar{x}}(t)) + (|u_h(t)|_1^2)_t - \tau|u_{h,t}(t)|_1^2 = 0 \tag{5.4}$$

and

$$\|u_h(t)v_h(t)\|^2 \leq \frac{1}{h}\|u_h(t)\|^2\|v_h(t)\|^2. \tag{5.5}$$

The key point of constructing schemes is to simulate physical laws. Indeed (5.1) possesses the conservation

$$\begin{aligned} &\int_0^1 u^2(x,t)dx + 2\nu\int_0^t\int_0^1 (\frac{\partial u}{\partial x}(x,t'))^2 dxdt' \\ &= \int_0^1 u_0^2(x)dx + 2\int_0^t\int_0^1 u(x,t')f(x,t')dxdt'. \end{aligned} \tag{5.6}$$

In order to simulate (5.6), we define the difference operator

$$J(u_h, v_h) = \frac{1}{3}v_h u_{h,\hat{x}} + \frac{1}{3}(v_h u_h)_{\hat{x}}.$$

It can be seen that (see [3])

$$(u_h(t), J(w_h(t), v_h(t))) + (w_h(t), J(u_h(t), v_h(t))) = 0. \tag{5.7}$$

A finite difference scheme for solving (5.1) is

$$\begin{cases} u_{h,t}(x,t) + J(u_h(x,t) + \delta\tau u_{h,t}(x,t), u_h(x,t)) & \\ \qquad - \nu u_{h,x\bar{x}}(x,t) - \nu\sigma\tau u_{h,x\bar{x}t}(x,t) = f(x,t), & x \in I_h,\ t \in S_\tau, \\ u_h(x+1,t) = u_h(x,t), & x \in I_h,\ t \in S_\tau, \\ u_h(x,0) = u_0(x), & x \in I_h \end{cases} \tag{5.8}$$

where δ and σ are parameters, and $0 \leq \delta,\ \sigma \leq 1$. We now check the conservation. For simplicity, let $\delta = \sigma = \frac{1}{2}$. By taking the scalar product of (5.8) with $u_h(x,t) + u_h(x,t+\tau)$, from (5.4) and (5.7) we have

$$\|u_h(t)\|_t^2 + \frac{\nu}{2}|u_h(t) + u_h(t+\tau)|_1^2 = (u_h(t) + u_h(t+\tau), f(t)).$$

Thus

$$\|u_h(t)\|^2 + \frac{\nu\tau}{2} \sum_{t'\in S_\tau, t'<t} |u_h(t') + u_h(t'+\tau)|_1^2$$
$$= \|u_0\|^2 + \tau \sum_{t'\in S_\tau, t'<t} (u_h(t') + u_h(t'+\tau), f(t')),$$

which is a reasonable analogy of (5.6).

We next analyze the stability. Let $\tilde{u}_h$ and $\tilde{f}$ be the errors of u_h and f respectively. Then

$$\begin{aligned}\tilde{u}_{h,t}(x,t) + & J(\tilde{u}_h(x,t) + \delta\tau\tilde{u}_{h,t}(x,t), u_h(x,t) + \tilde{u}_h(x,t)) + \\ & J(u_h(x,t) + \delta\tau u_{h,t}(x,t), \tilde{u}_h(x,t)) \\ & - \nu\tilde{u}_{h,x\bar{x}}(x,t) - \nu\sigma\tau\tilde{u}_{h,x\bar{x}t}(x,t) = \tilde{f}(x,t).\end{aligned} \tag{5.9}$$

Let m be an undetermined positive constant. By taking the scalar product of (5.9) with $2\tilde{u}_h(x,t) + m\tau\tilde{u}_{h,t}(x,t)$, from (5.2)–(5.4) we obtain that

$$\begin{aligned}\|\tilde{u}_h(t)\|_t^2 & + \tau(m-1-\varepsilon)\|\tilde{u}_{h,t}(t)\|^2 + 2\nu|\tilde{u}_h(t)|_1^2 + \nu\tau(\sigma + \frac{m}{2})(|\tilde{u}_h(t)|_1^2)_t \\ & + \nu\tau^2(m\sigma - \frac{m}{2} - \sigma)|\tilde{u}_{h,t}(t)|_1^2 \sum_{q=0}^{4} F_q(t) \\ & \le \|\tilde{u}_h(t)\|^2 + M_1(1 + \frac{m^2}{\varepsilon})\|\tilde{f}(t)\|^2\end{aligned} \tag{5.10}$$

where M_q denotes a positive constant depending only on u_h, $\varepsilon > 0$, and

$$\begin{aligned}F_0(t) =& 2(\tilde{u}_h(t), J(\tilde{u}_h(t), u_h(t) + \tilde{u}_h(t))) \\ & + m\delta\tau^2(\tilde{u}_{h,t}(t), J(\tilde{u}_{h,t}(t), u_h(t) + \tilde{u}_h(t))), \\ F_1(t) =& 2\delta\tau(\tilde{u}_h(t), J(\tilde{u}_{h,t}(t), u_h(t) + \tilde{u}_h(t))), \\ F_2(t) =& 2(\tilde{u}_h(t), J(u_h(t) + \delta\tau u_{h,t}(t), \tilde{u}_h(t))), \\ F_3(t) =& m\tau(\tilde{u}_{h,t}(t), J(\tilde{u}_h(t), u_h(t) + \tilde{u}_h(t))), \\ F_4(t) =& m\tau(\tilde{u}_{h,t}(t), J(u_h(t) + \delta\tau u_{h,t}(t), \tilde{u}_h(t))).\end{aligned}$$

By (5.7), the main nonlinear error terms vanish, i.e., $F_0(t) = 0$. Furthermore

$$F_1(t) + F_3(t) = \tau(m - 2\delta)(\tilde{u}_{h,t}(t), J(\tilde{u}_h(t), u_h(t) + \tilde{u}_h(t)))$$

and so from (5.5),

$$|F_1(t) + F_3(t)| \le \varepsilon\tau\|\tilde{u}_{h,t}(t)\|^2 + M_2(\|\tilde{u}_h(t)\|^2 + \nu h(m-2\delta)^2\|\tilde{u}_h(t)\|^2|\tilde{u}_h(t)|_1^2).$$

Moreover

$$|F_2(t) + F_4(t)| \le \varepsilon\tau\|\tilde{u}_{h,t}(t)\|^2 + \varepsilon\nu|\tilde{u}_h(t)|_1^2 + M_3\|\tilde{u}_h(t)\|^2.$$

Therefore (5.10) reads

$$\|\tilde{u}_h(t)\|_t^2 + \tau(m-1-3\varepsilon)\|\tilde{u}_{h,t}(t)\|^2 + \nu|\tilde{u}_h(t)|_1^2 + \nu\tau(\sigma+\frac{m}{2})(|\tilde{u}_h(t)|_1^2)_t$$
$$+\nu\tau^2(m\sigma-\frac{m}{2}-\sigma)|\tilde{u}_{h,t}(t)|_1^2 \le R_h(t) + M_4\|\tilde{f}(t)\|^2 \tag{5.11}$$

with

$$R_h(t) = M_5\|\tilde{u}_h(t)\|^2 + \nu(-1+\varepsilon+hM_2(m-2\delta)^2\|\tilde{u}_h(t)\|^2)|\tilde{u}_h(t)|_1^2.$$

Now let $p_0 > 0$ and choose the value of m in three different ways.

(i) If $\sigma > \frac{1}{2}$, then we take $m = m_1$,

$$m_1 = \max(1+3\epsilon+p_0, \frac{2\sigma}{2\sigma-1}).$$

Thus (5.11) leads to

$$\|\tilde{u}_h(t)\|_t^2 + p_0\tau\|\tilde{u}_{h,t}(t)\|^2 + \nu|\tilde{u}_h(t)|_1^2 + \nu\tau(\sigma+\frac{m}{2})(|\tilde{u}_h(t)|_1^2)_t$$
$$\le R_h(t) + M_4\|\tilde{f}(t)\|^2. \tag{5.12}$$

(ii) If $\sigma = \frac{1}{2}$, then we take $m = m_2$,

$$m_2 = 1+3\varepsilon+p_0+2\lambda\nu.$$

Since $\tau|\tilde{u}_{h,t}(t)|_1^2 \le 4\lambda\|\tilde{u}_{h,t}(t)\|^2$, We have

$$\tau(m-1-3\varepsilon)\|\tilde{u}_{h,t}(t)\|^2 + \nu\tau^2(m\sigma-\frac{m}{2}-\sigma)|\tilde{u}_{h,t}(t)|_1^2$$
$$\ge p_0\tau\|\tilde{u}_{h,t}(t)\|^2. \tag{5.13}$$

Thus (5.12) still holds.

(iii) If $\sigma < \frac{1}{2}$ and $\lambda < \frac{1}{2\nu(1-2\sigma)}$, then we take $m \ge m_3$,

$$m_3 = \frac{1+3\varepsilon+4\lambda\nu\sigma+p_0}{1+2\lambda\nu(2\sigma-1)}.$$

So (5.13) and (5.12) hold also.

In particular, if in addition

$$2\delta \ge \begin{cases} m_1, & \text{for } \sigma > \frac{1}{2}, \\ m_2, & \text{for } \sigma = \frac{1}{2}, \\ m_3, & \text{for } \sigma < \frac{1}{2}, \end{cases} \tag{5.14}$$

then we can take $m = 2\delta$ and thus $R_h(t) \le M_5\|\tilde{u}_h(t)\|^2$ for $\varepsilon \le 1$.

Let

$$E(t) = \|\tilde{u}_h(t)\|^2 + \tau \sum_{t'\in S_\tau,\ t'<t} (p_0\tau\|\tilde{u}_{h,t}(t')\|^2 + \nu|\tilde{u}_h(t')|_1^2),$$

$$\rho(t) = \|\tilde{u}_h(0)\|^2 + \tau \sum_{t'\in S_\tau,\ t'<t} \|\tilde{f}(t')\|^2.$$

By summing (5.12) for $t \in S_\tau$, We have

$$E(t) \le M_6\rho(t) + \tau \sum_{t' \in S_\tau,\ t' < t} R_h(t'). \tag{5.15}$$

LEMMA 2. *Suppose that the following conditions are fulfilled*

(i) *$Z(t)$ is a non-negative function on S_τ, and a, ρ and C_q are non-negative constants.*

(ii) *$F(Z)$ is such a function that if $Z \le C_0$, then $F(Z) \le 0$,*

(iii) *for all $t \in S_\tau$,*

$$Z(t) \le \rho + \tau \sum_{t' \in S_\tau,\ t' < t} (C_1 Z(t') + C_2 h^{-a} Z^2(t') + F(Z(t'))),$$

(iv) *$\rho e^{(C_1+C_2)T} \le \min(C_0, h^a)$ and $Z(0) \le \rho$.*

Then for all $t \in S_\tau$ and $t \le T$, we have

$$Z(t) \le \rho e^{(C_1+C_2)t}.$$

In particular, if $C_2 = 0$ and $F(Z) \le 0$ for all Z, then for all ρ and t, we have

$$Z(t) \le \rho e^{C_1 t}.$$

By (5.15) and lemma 2, we obtain the following conclusion.

THEOREM 8. *Let u_h be the solution of (5.8). If*

$$\sigma \ge \frac{1}{2}, \quad or \quad \lambda < \frac{1}{2\nu(1-2\sigma)},$$

then there exist positive constants M_7–M_9 depending only on u_h such that for all $t \le T$ and $\rho(T)e^{M_8 T} \le M_7 h^{-1}$, we have

$$E(t) \le M_9 \rho(t) e^{M_8 t}.$$

If the condition (5.14) also holds, then for all t and ρ, the above estimation is valid.

Remark 4. Scheme (5.8) possesses the index of generalized stability $s \le -0.5$. Thus it is convergent provided that the approximation (5.8) is consistent with (5.1). In particular, if (5.14) holds, then $s = -\infty$. It means that (5.8) possesses the stability which is very close to the stability given by Courant (see [1]).

Remark 5. The suitable value of parameter σ for linear term only weakens the restriction on the ratio λ. But the suitable value of parameter δ for nonlinear term can increase the generalized stability.

Remark 6. Since the nonlinear term $u\dfrac{\partial u}{\partial x}$ is approximated by suitable difference operator $J(u_h, u_h)$, the corresponding scheme possesses conservation. Also, the main nonlinear error terms vanish and so the index $s \le -0.5$. If we approximate the term $u\dfrac{\partial u}{\partial x}$ by

$$J'(u_h, u_h) = \alpha u_h u_{h,\hat{x}} + \frac{1-\alpha}{2}(u_h^2)_{\hat{x}}, \quad \alpha \ne \frac{1}{3},$$

then the conservation does not hold and the index $s \le 0.5$. This fact shows the relation between the generalized stability and the conservation of nonlinear finite difference schemes for fluid dynamics.

The technique in this paper has been widely used for finite difference methods in fluid dynamics, weather prediction, design of semiconductor devices and numerical study of nonlinear wave equations. The fruitful results can be found in [3–6, 24–27] and [8, 28, 29].

6. Applications to other numerical methods for nonlinear differential equations

The theory of generalized stability is also applicable to other numerical methods. For instance, it has been successfully used for spectral method and pseudospectral method (see [9-14, 32]). In this section, we take a pseudospectral scheme of K.D.V. equation as an example, to show this technique.

We consider the problem

$$\begin{cases} \dfrac{\partial u}{\partial t} + u\dfrac{\partial u}{\partial x} + \dfrac{\partial^3 u}{\partial x^3} = 0, & x \in \mathcal{R},\ 0 < t \le T, \\ u(x+1,t) = u(,xt), & x \in \mathcal{R},\ 0 \le t \le T, \\ u(x,0) = u_0(x), & x \in \mathcal{R}. \end{cases} \tag{6.1}$$

Let $(\cdot,\cdot)$ and $\|\cdot\|$ denote the inner product and norm of the space $L^2(I)$. For any positive integer n, the semi-norm and the norm of $H^n(I)$ are denoted by $|\cdot|_n$ and $\|\cdot\|_n$, respectively. Let $C_p^\infty(I)$ be the set of infinitely differentiable functions with period 1, defined on $\mathcal{R}$. $H_p^n(I)$ is the closure of $C_p^\infty(I)$ in $H^n(I)$. For any real $\sigma > 0$, define $H_p^\sigma(I)$ by complex interpolation between $H_p^{[\sigma]}(I)$ and $H_p^{[\sigma]+1}(I)$. Moreover, let A be a Banach space. $C(0,T;A)$ is a set of strongly continuous functions from $[0,T]$ to A, and $L^2(0,T;A)$ is a set of strongly measurable functions $u(t)$ from $(0,T)$ to A satisfying

$$\|u\|_{L^2(0,T;A)} = (\int_0^T \|u(t)\|_A^2 dt)^{1/2} < \infty.$$

Other similar notations have the usual meanings.

For any positive integer N, set

$$V_N = \text{span}\{e^{2\pi ikx} \mid |k| \le N\}$$

and let $\dot{V}_N$ be a subspace of V_N, containing all real-valued functions.

Let $h = 1/(2N+1)$ be the mesh size in variable x and $x_j = jh$ $(j = 0, 1, \cdots, 2N)$. The discrete inner product and norm are defined by

$$(u, v)_N = h \sum_{j=0}^{2N} u(x_j)\overline{v(x_j)}, \quad \|u\|_N = (u, u)_N^{1/2}.$$

Let $P_N : L^2(I) \to V_N$ be the orthogonal operator, i.e.,

$$(P_N u, v) = (u, v), \quad \forall v \in V_N, \tag{6.2}$$

and $P_C : C(\bar{I}) \to V_N$ be the interpolation operator such that

$$P_C u(x_j) = u(x_j), \quad 0 \le j \le 2N. \tag{6.3}$$

For any $u, v \in C(\bar{I})$, we have

$$(u, v)_N = (P_C u, P_C v)_N = (P_C u, P_C v). \tag{6.4}$$

As we know, pseudospectral method is less stable than spectral method due to the aliasing interaction. For remedying this deficiency, the author proposed a filtering technique (see [9]). Let $\alpha \ge 1$ and a_k be the Fourier coefficient of u. We define $R = R(\alpha)$ by

$$Ru = \sum_{|k| \le N} (1 - |\frac{k}{N}|^{\alpha}) a_k e^{2\pi i k x}.$$

In order to approximate the nonlinear term $u\frac{\partial u}{\partial x}$ reasonably, we define the operator $J_C : V_N \times V_N \to V_N$ as

$$J_C(u, v) = \frac{1}{3} P_C(\frac{\partial u}{\partial x} Rv) + \frac{1}{3}\frac{\partial}{\partial x}(P_C(uRv)).$$

If u, v, and $w \in \dot{V}_N$, then from (6.4) we have

$$(J_C(u, v), w) + (J_C(w, v), u) = 0. \tag{6.5}$$

The semi-discrete pseudospectral scheme for (6.1) is as follows

$$\begin{cases} \frac{\partial u_c}{\partial t} + J_C(u_c, u_c) + \frac{\partial^3 u_c}{\partial x^3} = 0, & x \in \mathcal{R},\ t > 0, \\ u_c(x+1, t) = u_c(x, t), & x \in \mathcal{R},\ 0 \le t \le T, \\ u_c(x, 0) = P_C u_0(x), & x \in \mathcal{R}. \end{cases} \tag{6.6}$$

By (6.5), scheme (6.6) possesses the conservation

$$\|u_c(t)\| = \|u_c(0)\|.$$

For analyzing errors, we list some lemmas (see [32]).

LEMMA 3. *If* $0 \le \mu \le \sigma$ *and* $u \in H_p^{\sigma}(I)$, *then*

$$\|P_N u - u\|_{\mu} \le C N^{\mu - \sigma} |u|_{\sigma}, \quad \|P_N u\|_{\sigma} \le C \|u\|_{\sigma}.$$

LEMMA 4. *If* $0 \le \mu \le \sigma$ *and* $u \in V_N$, *then*

$$\|u\|_\sigma \le CN^{\sigma-\mu}\|u\|_\mu.$$

LEMMA 5. *If* $0 \le \mu \le \sigma \le \gamma$, *then*

$$\|Ru - u\|_\mu \le CN^{\mu-\sigma}|u|_\sigma, \quad \forall u \in V_N$$

and

$$\|RP_N u - u\|_\mu \le CN^{\mu-\sigma}|u|_\sigma, \quad \forall u \in H_p^\sigma(I).$$

Now, let a_k and b_k be the Fourier coefficients of u and v respectively. Assume

$$a_{k+2N+1} = a_k, \quad b_{k+2N+1} = b_k,$$

and define the circle convolution

$$u * v = \sum_{|l|\le N}\sum_{|k|\le N} a_l b_{k-l} e^{2\pi i k x}.$$

It is easy to show that $u * v = v * u$.

LEMMA 6. *If* $u, v \in V_N$ *and* $w \in \dot{V}_N$, *then*

$$P_c(uv) = u * v, \quad (u * w, c) = (u, w * v).$$

LEMMA 7. *Let* $\varepsilon > 0$. *If* $u, v \in V_N$ *and* $w \in H_p^{3/2+\varepsilon}(I)$, *then*

$$|(\frac{\partial u}{\partial x} * Rv, w) + (u * R(\frac{\partial v}{\partial x}), w)| \le C_\varepsilon \gamma \|w\|_{3/2+\varepsilon}\|u\|\|v\|,$$

$$|(\frac{\partial u}{\partial x} * Ru, w) - (u * R(\frac{\partial u}{\partial x}), w)| \le C_\varepsilon \gamma \|w\|_{3/2+\varepsilon}\|u\|^2,$$

where C_ε *is a positive constant depending only on* ε.

The above lemma leads to the following conclusion.

LEMMA 8. *If* $\varepsilon > 0$, $u \in V_N$ *and* $w \in H_p^{\frac{3}{2}+\varepsilon}(I)$, *then*

$$|(\frac{\partial u}{\partial x} * Ru, w)| \le C_\varepsilon \gamma \|w\|_{\frac{3}{2}+\varepsilon}\|u\|^2.$$

We now turn to analyze the stability of (6.6), let $\tilde{u}_c$ and $\tilde{f}$ be the errors of u_c and the right side of (6.6) respectively. Then

$$\frac{\partial \tilde{u}_c}{\partial t} + J_C(\tilde{u}_c, u_c + \tilde{u}_c) + J_C(u_c, \tilde{u}_c) + \frac{\partial^3 \tilde{u}_c}{\partial x^3} = \tilde{f}. \qquad (6.7)$$

By Lemma 6,

$$J_C(u, v) = \frac{1}{3}\frac{\partial u}{\partial x} * Rv + \frac{1}{3}\frac{\partial}{\partial x}(u * Rv).$$

By taking the inner product of (6.7) with $2\tilde{u}_c$, we have from (6.5) that

$$\frac{\partial}{\partial t}\|\tilde{u}_c\|^2 + 2(J_C(u_c, \tilde{u}_c), \tilde{u}_c) = 2(\tilde{f}, \tilde{u}_c).$$

Let $\varepsilon > 0$. It follows Lemma 5 that

$$|(\frac{\partial u_c}{\partial x} * R\tilde{u}_c, \tilde{u}_c)| = |(\frac{\partial u_c}{\partial x} R\tilde{u}_c, \tilde{u}_c)_N|$$
$$\leq \|\frac{\partial u_c}{\partial x}\|_{L^\infty} \|R\tilde{u}_c\|_N \|\tilde{u}_c\|_N \leq C_\varepsilon \gamma \|u_c\|_{3/2+\varepsilon} \|\tilde{u}_c\|^2.$$

By Lemma 6 and Lemma 8,

$$|(\frac{\partial}{\partial x}(u_c * R\tilde{u}_c), \tilde{u}_c)| = |(u_c * R\tilde{u}_c, \frac{\partial \tilde{u}_c}{\partial x})|$$
$$= |(\frac{\partial \tilde{u}_c}{\partial x} * R\tilde{u}_c, u_c)| \leq C_\varepsilon \gamma \|u_c\|_{3/2+\varepsilon} \|\tilde{u}_c\|^2.$$

They lead to

$$|(J_C(u_c, \tilde{u}_c), \tilde{u}_c)| \leq C_\varepsilon \gamma \|u_c\|_{3/2+\varepsilon} \|\tilde{u}_c\|^2.$$

Therefore

$$\frac{\partial}{\partial t} \|\tilde{u}_c(t)\|^2 \leq C_\varepsilon \gamma \|u_c\|_{L^\infty(0,T;H^{3/2+\varepsilon})} \|\tilde{u}_c(t)\|^2 + \|\tilde{f}(t)\|^2,$$

and the conclusion follows.

THEOREM 9. *If $\varepsilon > 0$, then there exists a positive constant C depending on $\|u_c\|_{L^\infty(O,T;H^{3/2+\varepsilon})}$ such that for any $t \leq T$,*

$$\|\tilde{u}(t)\|^2 \leq e^{ct}\{\|\tilde{u}(0)\|^2 + \int_0^T \|(s)\|^2 ds\}.$$

Next consider the convergence of (6.6). Let $w = P_N u$ and $\tilde{e} = u_c - w$. By (6.1) and (6.6),

$$\frac{\partial \tilde{e}}{\partial t} + J_C(\tilde{e}, w + \tilde{e}) + J_C(w, \tilde{e}) + \frac{\partial^3 \tilde{e}}{\partial x^3} = f, \tag{6.8}$$

where

$$f = \frac{1}{3}[P_N(u\frac{\partial u}{\partial x} + \frac{\partial}{\partial x}(u^2)) - P_C(\frac{\partial w}{\partial x} Rw) - \frac{\partial}{\partial x}(P_C(wRw))].$$

From Lemma 3 and Lemma 5,

$$\|P_N(u\frac{\partial u}{\partial x}) - u\frac{\partial u}{\partial x}\| \leq CN^{1-\sigma}|u\frac{\partial u}{\partial x}|_{\sigma-1} \leq CN^{1-\sigma}\|u\|_\sigma^2,$$
$$\|u\frac{\partial u}{\partial x} - \frac{\partial w}{\partial x} Rw\| \leq \|u\frac{\partial u}{\partial x} - u\frac{\partial w}{\partial x}\| + \|\frac{\partial w}{\partial x} u - \frac{\partial w}{\partial x} Rw\|$$
$$\leq C(\|u\|_\sigma) N^{1-\sigma}$$

and

$$\|\frac{\partial w}{\partial x} Rw - P_C(\frac{\partial w}{\partial x} Rw)\| \leq CN^{1-\sigma}|\frac{\partial w}{\partial x} Rw|_{\sigma-1} \leq CN^{1-\sigma}\|u\|_\sigma^2.$$

Hence

$$\|P_N(u\frac{\partial u}{\partial x}) - P_C(\frac{\partial w}{\partial x} Rw)\| \leq C(\|u\|_\sigma) N^{1-\sigma}.$$

Similarly

$$\|P_N u^2 - u^2\| \le CN^{-\sigma}\|u\|_\sigma^2,$$
$$\|u^2 - wu\| \le CN^{-\sigma}\|u\|_{L^\infty}|u|_\sigma,$$
$$\|wu - wRw\| \le CN^{-\sigma}\|w\|_{L^\infty}|u|_\sigma,$$
$$\|wRw - P_C(wRw)\| \le CN^{-\sigma}\|u\|_\sigma^2.$$

Furthermore,

$$\begin{aligned}\|P_N\frac{\partial}{\partial x}(u^2) - \frac{\partial}{\partial x}(P_C(wRw))\| &= |P_N u^2 - P_C(wRw)|_1 \\ &\le CN\|P_N u^2 - P_C(wRw)\| \le C(\|u\|_\sigma)N^{1-\sigma}.\end{aligned}$$

So we get

$$\|f\| \le C(\|u\|_{L^\infty(0,T;H^\sigma)})N^{1-\sigma}.$$

Also

$$\|\tilde{e}(0)\| \le \|P_C u_0 - u_0\| + \|u_0 - P_N u_0\| \le CN^{-\sigma}|u_0|_\sigma.$$

Finally, by applying Theorem 9 to (6.8) and using Lemma 3 and the triangle inequality, we get the following result.

THEOREM 10. *If $3 \le \sigma \le \gamma$ and $u \in C(0,T;H^\sigma(I))$, then there exists a positive constant C depending only on $\|u\|_{L^\infty(0,T;H^\sigma)}$ such that for any $t \le T$,*

$$\|u_c(t) - u(t)\| \le CN^{1-\sigma}.$$

Recently we have used the theory of generalized stability to mixed Fourier-Chebyshev spectral (or pseudospectral) method, mixed finite element-spectral (or pseudospectral) method and finite difference-spectral (or pseudospectral) method (see [16–21]).

REFERENCES

1. Courant, R., Friedrichs. F. and Lewy. H., Math. Ann., **100**(1928), 32–74.
2. Griffiths, D., Bulletin of IMA, **18**(1982), 210–215.
3. Guo Ben-yu, Tech. Report of SUST (1965), also see Acta Math. Sinica, **17**(1974), 242–258.
4. Guo Ben-yu, Scientia Sinica, **25A**(1982), 702–715.
5. Guo Ben-yu, Scientia Sinica, **32A**(1989), 897–907.
6. Guo Ben-yu, J. Comput. Math. and Math. Phys., **32**(1992), 530–541.
7. Guo Ben-yu, HMS Bulletin, **33**(1992), 1–10.
8. Guo Ben-yu, Pascual, Pedro J., Rodriquez, Maria J. and Vazquez, L., Appl. Math. and Comput., **18**(1986), 1–14.
9. Guo Ben-yu, J. Comput. Math., **1**(1983), 353–362.
10. Guo Ben-yu, Acta Math. Sinica, **28**(1985), 1–15.
11. Guo Ben-yu and Manoranjan, V.S., IMA J. Numer. Anal., **5**(1985), 307–318.
12. Guo Ben-yu, Scientia Sinica, **28A**(1985), 1139–1153.
13. Guo Ben-yu and Ma He-ping, CALCOLO, **24**(1987), 263–282.
14. Guo Ben-yu and Cao Wei-ming, J. Comput. Phys., **74**(1988), 110–126.
15. Guo Ben-yu and Xiong Yue-shan, J. Comput. Phys., **84**(1989), 259–278.
16. Guo Ben-yu and Ma He-ping, SIAM J. Numer. Anal., **28**(1991), 113–132.
17. Guo Ben-yu and Cao Wei-ming, HMS Bulletin, **32**(1991), 83–108.

18. Guo Ben-yu and Zheng Jia-dong, Science in China, **35A**(1992), 1–13.
19. Guo Ben-yu and Cao Wei-ming, Math. Model. and Numer. Anal., **26**(1992), 469–491.
20. Guo Ben-yu, Ma He-ping, Cao Wei-ming and Huang Hui, J. Comp. Phys., **101**(1992), 207–217.
21. Guo Ben-yu and Cao Wei-ming, J. Comp. Phys., **101**(1992),375–385.
22. Keller, H.B., Math. Comp., **29**(1975), 464–476.
23. Krasnosel'skii, M.A., Vainikko, G.M., Zabreiko, P.P., Rutitskii, Ya.B. and Stetsenko, V.Ya., *Approximate solution of operator equations*, Wolters-Noordhoff Publishing Groningen, Netherlands, 1972.
24. Kuo Pen-yu, Scientia Sinica, **20**(1977), 287–304.
25. Kuo Pen-yu, Scientia Sinica, Math. Collection, **1**(1979), 39–52.
26. Kuo Pen-yu, C.R.Acad. Sc.Paris, Serie A, **291**(1980), 167–170.
27. Kuo Pen-yu, Wu Hua-mo, J. Math. Anal. and Applics., **81**(1981), 334–345.
28. Kuo Pen-yu and Sanz-Serna, J.M., IMA J. of Numer. Anal., **1**(1981), 215–221.
29. Kuo Pen-yu, COMPEL, **2**(1983), 57–75.
30. Kuttler, J.R., Finite difference approximations of eigenvalues of uniformly elliptic operators, SIAM J. Numerical Analysis, **7**(1970), 206–232.
31. Lax, P.D. and Richtmeyer, R.D., Comm. Pure and Appl. Math., **9**(1956), 267–293.
32. Ma He-ping and Guo Ben-yu, J. Comput. Phys., **65**(1986), 120–137.
33. Richardson, L.F., Phil. Trans. Roy. Soc. Lond., **210A**(1910), 307–357.
34. Richtmeyer, R.D. and Morton, K.W., *Difference Methods for Initial Value Problems*, Interscience, New York, 1967.
35. Simpson, R.B., *Existence and error estimates for solutions of a discrete analog of nonlinear eigenvalue problems*, Math. Comp., **26**(1972), 359–375.
36. Stetter, H.J., *Stability of nonlinear discretization algorithms*, in Numerical Solution of Differential Equations, 111–123, Ed. by Bramble, J.H., Academic Press, New York, 1966.
37. Strang, W.G., Duke Math. J., 27(1960), 221-230.

SHANGHAI UNIVERSITY OF SCIENCE, AND TECHNOLOGY, SHANGHAI, CHINA

Contemporary Mathematics
Volume **163**, 1994

The Boundary Element Method for Solving Variational Inequalities*

HAN HOU-DE

1. Introduction

The variational inequality type problems arise in many practical problems [5]. In literature, there are many authors who have studied these problems. For the existence, uniqueness and regularity results for variational inequalities we refer the reader to [3, 8]. The finite elements procedure for these problems are treated in [9–10]. We know that the solution of some variational inequalities satisfy the linear partial differential equation (or equations) in the domain even when the problems are nonlinear. For example, the fluid mechanics in media with semi-permeable boundaries, the elasticity with unilateral conditions, the elasticity with friction [5] and the electropaint process [1] are such variational inequalities. Hence it is natural and advantageous to apply the boundary element method to solve these variational inequalities.

In recent year we present the boundary element method for solving such problems. For instance, consider the Laplace equation with unilateral conditions [12, 14]

$$\begin{aligned}
&\triangle u = 0, \quad \text{in } \Omega,\\
&u = 0, \quad \text{on } \Gamma_0,\\
&u \geq 0, \quad \frac{\partial u}{\partial n} \geq g, \quad \text{on } \Gamma_1,\\
&u(\frac{\partial u}{\partial n} - g) = 0, \quad \text{on } \Gamma_1
\end{aligned} \tag{1.1}$$

where $\Omega \in \Re^n (n = 2 \text{ or } 3)$ is a bounded domain with boundary $\Gamma = \bar{\Gamma}_0 \cup \bar{\Gamma}_1$. In literature the problem (1.1) is usually called Signorini problem, which is

---

1991 *Mathematics Subject Classification.* 65N38, 49J40.

*This work was supported by the National Science Foundation of China.

characterized by the following variational inequality:

$$\text{Find } u \in K, \quad \text{such that}$$

$$a(u, v-u) \geq \int_{\Gamma_1} g(v-u)ds, \quad \forall v \in K \tag{1.2}$$

where $K = \{v | v \in H^1(\Omega),\ v = 0 \text{ on } \Gamma_0,\ v > 0 \text{ on } \Gamma_1\}$ is a nonempty convex set of Sobolev space $H^1(\Omega)$,

$$a(u, v) = \int_{\Omega} \nabla u \cdot \nabla v dx.$$

Firstly, we reduce problem (1.2) to a variational inequality on the boundary Γ based on the boundary reduction by use of the Calderon projector for the trace and normal derivative. The crucial point of this boundary reduction lies in the normal derivative of the double-layer as presented by the tangential derivative of the single layer [13]. The resulting boundary integro-differential operators, does not involve hypersingular boundary integral operator. The bilinear form arising in this boundary inequality is continuous and coercive on the suitable Sobolev spaces, this leads to existence and uniqueness for the solution of the boundary variational inequality. Furthermore we solve the boundary variational inequality with a uniform boundary element Galerkin method and get the optimal estimates for the Galerkin error. We will discuss the model problem (1.2) in detail in section 3 and section 4. For the special domain Ω such as a disk or a rectangle in $\Re^2$, we can use the Green function of the original equation with the Dirichlet boundary condition instead of the fundamental solution to reduce the problem (1.2) to a boundary variational inequality based on the canonical boundary reduction, which was proposed by Feng Kang in 1978 [6]. This framework was used to solve an electropaint process, the numerical results show the method is more effective (See section 6).

In the paper [14], an indirect boundary element method for solving Signorini problem has been presented. The boundary element approximation of a Signorini problem in dimension three has been considered in [11].

Han and Hsiao [16-17] concerned with the contact problem in linear elasticity. A boundary variational inequality is derived. Existence and uniqueness of the boundary variational inequality are established. Numerical approximations for the solution are considered and the error estimates are obtained.

We remark that the present approach also can be extended to another non-linear boundary value problems, such as the steady-state heat transfer problems with radiative boundary conditions. Details will be available in the forthcoming paper.

This method has three advantages:

(1) It reduces the dimensionality of the problem by one dimension and hence produces a much smaller system of inequalities.

(2) In general, the self-adjointness of the original problem is not preserved after the boundary reduction. From the computational point of view, the loss of the self-adjointness brings much trouble, such as more storage locations and more computer time. But the new boundary reduction mentioned above preserves the self-adjointness of the original problem.

(3) The result boundary variational inequalities do not involve hypersingular boundary integral operator, which is quite difficult to approximate.

We now return to the model problem (1.2).

2. The normal derivative of the double-layer potential

To begin with, we consider the case $\Omega \in \Re^2$ with a smooth boundary Γ. Let

$$u_1(x) = \int_\Gamma \mu(y) \ln |x-y| ds_y, \quad x \in \Omega, \tag{2.1}$$

$$u_2(x) = \int_\Gamma \varphi(y) \frac{\partial}{\partial n_y} \ln |x-y| ds_y, \quad x \in \Omega, \tag{2.2}$$

where $\mu(x)$, $\varphi(x)$ are given density functions on Γ, $n_y = (n_y^1, n_y^2)$ refers to the unit outward normal vector at point $y \in \Gamma$. As usual $H^m(\Omega)$ and $H^\alpha(\Gamma)$ denote the Sobolev spaces on Ω and Γ, m, α are two real numbers. Suppose that the density function $\mu(x)$, $\rho(x)$ are continuously differentiable on Γ. For the single-layer potential $u_1(x)$ and the double-layer potential $u_2(x)$ we know that [4]:

(i) The potential of a single layer and its tangential derivatives vary continuously as we pass through $x \in \Gamma$. However the normal derivative has a jump

$$\lim_{p \to x, p \in \Omega} \frac{\partial u_1(p)}{\partial n_x} \equiv \frac{\partial u_1(x)}{\partial n_x}^- = \int_\Gamma \mu(y) \frac{\partial}{\partial n_x} \ln |x-y| ds_y - \pi \mu(x), \quad \forall x \in \Gamma, \tag{2.3}$$

$$\lim_{p \to x, p \in \Omega^c} \frac{\partial u_1(p)}{\partial n_x} \equiv \frac{\partial u_1(x)}{\partial n_x}^+ = \int_\Gamma \mu(y) \frac{\partial}{\partial n_x} \ln |x-y| ds_y + \pi \mu(x), \quad \forall x \in \Gamma, \tag{2.4}$$

where $\Omega^c = \Re^2 \setminus (\Omega \cup \Gamma)$. The assumption on $\mu(x)$ can be weakened, for instance, suppose $\mu(x) \in H^{-\frac{1}{2}}(\Gamma)$ then the equalities (2.3) and (2.4) still hold in the weak sense.

(ii) Across the boundary Γ, the value of double layer potential at $x \in \Gamma$ has jump discontinuities given by

$$\lim_{p \to x, p \in \Omega} u_2(p) = \int_\Gamma \varphi(y) \frac{\partial}{\partial n_x} \ln |x-y| ds_y + \pi \varphi(x), \quad \forall x \in \Gamma, \tag{2.5}$$

$$\lim_{p \to x, p \in \Omega^c} u_2(p) = \int_\Gamma \varphi(y) \frac{\partial}{\partial n_x} \ln |x-y| ds_y - \pi \varphi(x), \quad \forall x \in \Gamma. \tag{2.6}$$

The assumption on $\varphi(x)$ can be weakened. Suppose $\varphi(x) \in H^{\frac{1}{2}}(\Gamma)$, then equalities (2.5) and (2.6) hold in the weak sense.

For the normal derivative of the double layer potential (2.2), we have the following lemma.

LEMMA 2.1. *The normal derivative of the double-layer $u_2(p)$ to the boundary Γ varies continuously when p penetrates the boundary along the normal at $x \in \Gamma$, and*

$$\frac{\partial u_2(x)}{\partial n_x} = \frac{d}{ds_x} \int_\Gamma \dot{\varphi}(y) \ln |x-y| ds_y, \quad \forall x \in \Gamma, \tag{2.7}$$

where $\dot{\varphi}(y) = d\varphi(y)/ds_y$ and ds_y is the arc-length along the boundary Γ.

PROOF. For any $p \in \Omega$ (or $x \in \Omega^c$) and an arbitrary unit vector $n_p = (n_p^1, n_p^2)$, we have

$$\frac{\partial u_2(p)}{\partial n_p} = \int_\Gamma \varphi(y) \frac{\partial^2}{\partial n_p \partial n_y} \ln |p-y| ds_y, \quad \forall p \in \Omega \text{ (or } p \in \Omega^c). \tag{2.8}$$

A computation shows that for every $p \neq y$, the following equality holds

$$\frac{\partial^2}{\partial n_p \partial n_y} \ln |p-y| = -\frac{\partial^2}{\partial \tau_p \partial \tau_y} \ln |p-y|, \tag{2.9}$$

where $\tau_p = (-n_p^2, n_p^1)$ is a unit vector perpendicular to n_p and $\tau_y = (-n_y^2, n_y^1)$ represents the unit tangent vector at point $y \in \Gamma$. Substituting (2.9) into (2.8) and integrating by parts, we obtain

$$\frac{\partial u_2(p)}{\partial n_p} = \frac{\partial}{\partial \tau_p} \int_\Gamma \dot{\varphi}(y) \ln |p-y| ds_y.$$

For any $x \in \Gamma$, we take $n_p = n_x$, n_x indicating the outward unit normal vector at $x \in \Gamma$ to the domain Ω and let p go to x, then we obtain

$$\frac{\partial u_2(x)}{\partial n_x} = \frac{d}{ds_x} \int_\Gamma \dot{\varphi}(y) \ln |x-y| ds_y, \quad \forall x \in \Gamma. \tag{2.10}$$

□

For the case $\Omega \in \Re^3$ with a smooth boundary Γ, we consider the single-layer and double-layer potentials

$$u_3(x) = \int_\Gamma \mu(y) \frac{1}{r(x,y)} ds_y, \tag{2.11}$$

$$u_4(x) = \int_\Gamma \varphi(y) \frac{\partial}{\partial n_y} \frac{1}{r(x,y)} ds_y, \tag{2.12}$$

where $r(x,y) = |x-y|$.

Analogous discontinuity relations for the potentials (2.11) and (2.12) can be found in [4]. Here we only study the normal derivatives of the double-layer potential (2.12) and have the following lemma.

LEMMA 2.2. *The normal derivative of the double layer potential $u_4(x)$ to the boundary Γ varies continuously as we pass through the boundary Γ, and*

$$\frac{\partial u_4(x)}{\partial n_x} = \frac{\partial}{\partial \tau_x}\int_\Gamma \frac{\partial \varphi(y)}{\partial \tau_y}\frac{1}{r(x,y)}ds_y + \frac{\partial}{\partial \sigma_x}\int_\Gamma \frac{\partial \varphi(y)}{\partial \sigma_y}\frac{1}{r(x,y)}ds_y + \frac{\partial}{\partial \rho_x}\int_\Gamma \frac{\partial \varphi(y)}{\partial \rho_y}\frac{1}{r(x,y)}ds_y, \quad \forall x \in \Gamma. \tag{2.13}$$

Here $n_x = (n_x^1, n_x^2, n_x^3)$ refers to the unit outward normal vector at point $x \in \Gamma$ to the domain Ω and

$$\tau_x = (0, n_x^3, -n_x^2), \quad \tau_y = (0, n_y^3, -n_y^2),$$
$$\sigma_x = (-n_x^3, 0, n_x^1), \quad \sigma_y = (-n_y^3, 0, n_y^1),$$
$$\rho_x = (n_x^2, -n_x^1, 0), \quad \rho_y = (n_y^2, -n_y^1, 0).$$

PROOF. For any $p \in \Omega$ (or $p \in \Omega^c$) and an arbitrary unit vector $n_p = (n_p^1, n_p^2, n_p^3)$, we have

$$\frac{\partial u_4(p)}{\partial n_p} = \int_\Gamma \varphi(y)\frac{\partial^2}{\partial n_p \partial n_y}\frac{1}{r(p,y)}ds_y. \tag{2.14}$$

A computation shows

$$\frac{\partial^2}{\partial n_p \partial n_y}\frac{1}{r(p,y)} = -\frac{\partial^2}{\partial \tau_p \partial \tau_y}\frac{1}{r(p,y)} - \frac{\partial^2}{\partial \sigma_p \partial \sigma_y}\frac{1}{r(p,y)} - \frac{\partial^2}{\partial \rho_p \partial \rho_y}\frac{1}{r(p,y)}, \tag{2.15}$$

where

$$\tau_p = (0, n_p^3, -n_p^2),$$
$$\sigma_p = (-n_p^3, 0, n_p^1),$$
$$\rho_p = (n_p^2, -n_p^1, 0).$$

The vectors τ_p, σ_p, ρ_p are perpendicular to n_p. In general, they are not unit vectors and

$$\frac{\partial W(p)}{\partial \tau_p} = \nabla W(p) \cdot \tau_p,$$
$$\frac{\partial W(p)}{\partial \sigma_p} = \nabla W(p) \cdot \sigma_p,$$
$$\frac{\partial W(p)}{\partial \rho_p} = \nabla W(p) \cdot \rho_p,$$

for given function $W(p)$. Substituting (2.15) into (2.14), we obtain

$$\frac{\partial u_4(p)}{\partial n_p} = -\frac{\partial}{\partial \tau_p}\int_\Gamma \varphi(y)\frac{\partial}{\partial \tau_y}\frac{1}{r(p,y)}ds_y - \frac{\partial}{\partial \sigma_p}\int_\Gamma \varphi(y)\frac{\partial}{\partial \sigma_y}\frac{1}{r(p,y)}ds_y - \frac{\partial}{\partial \rho_p}\int_\Gamma \varphi(y)\frac{\partial}{\partial \rho_y}\frac{1}{r(p,y)}ds_y. \tag{2.16}$$

Then we know that for every $W(y) \in H^{\frac{1}{2}}(\Gamma)$, then

$$\int_\Gamma \frac{\partial W(y)}{\partial \tau_y} ds_y = 0,$$
$$\int_\Gamma \frac{\partial W(y)}{\partial \sigma_y} ds_y = 0,$$
$$\int_\Gamma \frac{\partial W(y)}{\partial \rho_y} ds_y = 0.$$

Hence the equality (2.16) is reduced to

$$\begin{aligned} \frac{\partial u_4(p)}{\partial n_p} =& \frac{\partial}{\partial \tau_p} \int_\Gamma \frac{\partial \varphi}{\partial \tau_y} \frac{1}{r(p,y)} ds_y + \frac{\partial}{\partial \sigma_p} \int_\Gamma \frac{\partial \varphi}{\partial \sigma_y} \frac{1}{r(p,y)} \\ &+ \frac{\partial}{\partial \rho_p} \int_\Gamma \frac{\partial \varphi}{\partial \rho_y} \frac{1}{r(p,y)} ds_y. \end{aligned} \tag{2.17}$$

For any $x \in \Gamma$, we take $n_p = n_x$, n_x indicating the outward unit normal vector at $x \in \Gamma$ to the domain Ω and let p go to x, then from (2.17) we obtain

$$\begin{aligned} \frac{\partial u_4(x)}{\partial n_x} =& \frac{\partial}{\partial \tau_x} \int_G m \frac{\partial \varphi}{\partial \tau_y} \frac{1}{r(x,y)} ds_y + \frac{\partial}{\partial \sigma_x} \int_\Gamma \frac{\partial \varphi}{\partial \sigma_y} \frac{1}{r(x,y)} \\ &+ \frac{\partial}{\partial \rho_x} \int_\Gamma \frac{\partial \varphi}{\partial \rho_y} \frac{1}{r(x,y)} ds_y, \quad \forall x \in \Gamma. \end{aligned} \tag{2.18}$$

We can see that the normal derivative of the double-layer potential can be represented as a sum of the tangential derivatives of the three single-layer potentials, hence the normal derivative of the double-layer potential is continuous as x passes through the boundary Γ. □

3. The equivalent boundary variational inequality for problem (1.2)

Let Ω be a bounded domain in $\Re^n (n = 2$, or $n = 3)$ with a smooth boundary Γ. Suppose $\Gamma = \bar{\Gamma}_0 \cup \bar{\Gamma}_1$ with $\Gamma_0 \neq \emptyset$. We define

$$\begin{aligned} &\tilde{H}^1(\Omega) = \{v \in H^1(\Omega), \quad v|_{\Gamma_0} = 0\}, \\ &\tilde{H}^{\frac{1}{2}}(\Gamma) = \{v \in H^{\frac{1}{2}}(\Gamma), \quad v|\Gamma_0 = 0\}, \\ &\tilde{K} = \{v \in H^{\frac{1}{2}}(\Gamma), \quad v \geq 0 \text{ a.e. on } \Gamma_1\}, \\ &\tilde{H}^{-\frac{1}{2}}(\Gamma) = \{\mu \in H^{-\frac{1}{2}}(\Gamma), \quad \int_\Gamma \mu ds = 0\}, \\ &V = \tilde{H}^{\frac{1}{2}}(\Gamma) \times \tilde{H}^{-\frac{1}{2}}(\Gamma), \quad \text{with norm} \\ &\qquad \|(v,\mu)\|_V^2 = \|v\|^2_{\frac{1}{2},\Gamma} + \|\mu\|^2_{-\frac{1}{2},\Gamma}. \end{aligned}$$

In the first place, we consider the case $n = 2$. Suppose that u is the solution of the problem (1.2), then in the domain Ω, $\triangle u = 0$. Let $\lambda = \frac{\partial u}{\partial n}|_\Gamma \in H^{-\frac{1}{2}}(\Gamma)$.

From the Green's formula we have

$$u(x) = \int_\Gamma \frac{\partial E(x,y)}{\partial n_y} u(y) ds_y - \int_\Gamma E(x,y)\lambda(y) ds_y, \quad \forall x \in \Omega, \tag{3.1}$$

where $E(x,y) = \frac{1}{2\pi} \ln |x-y|$. By the properties of the single-layer and double-layer potentials, we obtain the first relationship between λ and $u|_\Gamma$:

$$\frac{1}{2} u(x) = \int_\Gamma \frac{\partial E(x,y)}{\partial n_y} u(y) ds_y - \int_\Gamma E(x,y)\lambda(y) ds_y, \quad \forall x \in \Gamma. \tag{3.2}$$

Furthermore, using the behavior of the normal derivatives of the single-layer and double-layer potentials, we get

$$\frac{1}{2}\lambda(x) = \int_\Gamma \frac{\partial^2 E(x,y)}{\partial n_x \partial n_y} u(y) ds_y - \int_\Gamma \frac{\partial E(x,y)}{\partial n_x} \lambda(y) ds_y, \quad \forall x \in \Gamma,$$

where

$$\int_\Gamma \frac{\partial^2 E(x,y)}{\partial n_x \partial n_y} u(y) ds_y = \frac{d}{ds_x} \int_\Gamma E(x,y) \dot{u}(y) ds_y, \quad \forall x \in \Gamma. \tag{3.3}$$

Multiplying (3.2) by a function $\mu \in \tilde{H}^{-\frac{1}{2}}(\Gamma)$, we have

$$\begin{aligned} -\frac{1}{2}\int_\Gamma u(x)\mu(x) ds_x + \int_\Gamma\int_\Gamma \frac{\partial E(x,y)}{\partial n_y} u(y)\mu(x) ds_y ds_x \\ - \int_\Gamma\int_\Gamma E(x,y)\lambda(y)\mu(x) ds_y ds_x = 0. \end{aligned} \tag{3.4}$$

Let

$$\begin{aligned} a_0(\mu,\lambda) &= -\int_\Gamma\int_\Gamma E(x,y)\lambda(y)\mu(x) ds_y ds_x, \\ b(\mu,u) &= \frac{1}{2}\int_\Gamma u(x)\mu(x) ds_x - \int_\Gamma\int_\Gamma \frac{\partial E(x,y)}{\partial n_y} u(y)\mu(x) ds_y ds_x. \end{aligned}$$

Then equation (3.4) is reduced to

$$-b(\mu,u) + a_0(\mu,\lambda) = 0, \quad \forall \mu \in \tilde{H}^{-\frac{1}{2}}(\Gamma). \tag{3.5}$$

For any $v \in H^1(\Omega)$ and u satisfying $\triangle u = 0$, we obtain

$$a(u,v) = \int_\Omega \nabla u \cdot \nabla v dx = \int_\Gamma \lambda v ds. \tag{3.6}$$

Inserting (3.3) into (3.6) and integrating by parts, we have

$$\begin{aligned} a(u,v) = &-\int_\Gamma\int_\Gamma E(x,y)\dot{u}(y)\dot{v}(x) ds_y ds_x \\ &- \int_\Gamma\int_\Gamma \frac{\partial}{\partial n_x} E(x,y)\lambda(y) v(x) ds_y ds_x + \frac{1}{2}\int_\Gamma \lambda v ds \\ =& a_0(\dot{u},\dot{v}) + b(\lambda, v). \end{aligned}$$

Hence the variational inequality (1.2) is reduced to the following boundary variational inequality:

$$\begin{aligned} &\text{Find } (u,\lambda)\in \tilde{K}\times \tilde{H}^{-\frac{1}{2}}(\Gamma) \text{ such that} \\ &a_0(\dot{u},\dot{v}-\dot{u})+b(\lambda,v-u)\geq \int_{\Gamma_1} g(v-u)ds, \quad \forall v\in \tilde{K}, \\ &-b(\mu,u)+a_0(\mu,\lambda)=0, \quad \forall \mu\in \tilde{H}^{-\frac{1}{2}}(\Gamma); \end{aligned} \tag{3.7}$$

or

$$\begin{aligned} &\text{Find } (u,\lambda)\in \tilde{K}\times \tilde{H}^{-\frac{1}{2}}(\Gamma) \text{ such that} \\ &A(u,\lambda;v-u,\mu)\geq \int_{\Gamma_1} g(v-u)ds, \quad \forall v\in \tilde{K},\ \forall \mu\in \tilde{H}^{-\frac{1}{2}}(\Gamma), \end{aligned} \tag{3.8}$$

where

$$A(u,\lambda;v,\mu)=a_0(\dot{u},\dot{v})+b(\lambda,v)-b(\mu,u)+a_0(\mu,\lambda)$$

is a bilinear form on $V\times V$.

We now consider the case $n=3$, in which $\Omega\subset\Re^3$ and the smooth boundary Γ is a closed surface. Similarly the variational inequality (1.2) in three dimension can be reduced to the following variational inequality on the closed surface Γ [11]:

$$\begin{aligned} &\text{Find } (u,\lambda)\in \tilde{K}\times \tilde{H}^{-\frac{1}{2}}(\Gamma) \text{ such that} \\ &a_2(u,v-u)+b_1(\lambda,v-u)\geq \int_{\Gamma_1} g(v-u)ds, \quad \forall v\in \tilde{K}, \\ &-b_1(\mu,u)+a_1(\mu,\lambda)=0, \quad \forall \mu\in \tilde{H}^{-\frac{1}{2}}(\Gamma); \end{aligned} \tag{3.9}$$

where

$$\begin{aligned} a_1(\mu,\lambda) &= \int_\Gamma\int_\Gamma \frac{1}{4\pi r(x,y)}\mu(x)\lambda(y)ds_x ds_y, \\ a_2(u,v) &= \int_\Gamma\int_\Gamma [\frac{\partial u(y)}{\partial \tau_y}\frac{\partial v(x)}{\partial \tau_x}+\frac{\partial u(y)}{\partial \sigma_y}\frac{\partial v(x)}{\partial \sigma_x}+\frac{\partial u(y)}{\partial \rho_y}\frac{\partial v(x)}{\partial \rho_x}]\frac{1}{4\pi r(x,y)}ds_x ds_y, \\ b_1(\mu,v) &= \int_\Gamma\int_\Gamma \frac{\partial}{\partial n_x}(\frac{1}{4\pi r(x,y)})\mu(y)v(x)ds_x ds_y+\frac{1}{2}\int_\Gamma \mu(x)v(x)ds_x. \end{aligned}$$

For the boundary variational inequality (3.8) we have the following lemmas.

LEMMA 3.1. *There is a constant* $\alpha_0>0$ *such that*

$$\|\dot{v}\|_{-\frac{1}{2},\Gamma}\geq \alpha_0\|v\|_{\frac{1}{2},\Gamma}, \quad \forall v\in \tilde{H}^{\frac{1}{2}}(\Gamma). \tag{3.10}$$

LEMMA 3.2. $A(u,\lambda;v,\mu)$ *is a bounded bilinear form on* $V\times V$*; that is there exists a constant* $M>0$ *such that*

$$|A(u,\lambda;v,\mu)|\leq M\|(u,\lambda)\|_V\|(v,\mu)\|_V, \quad \forall (u,\lambda),\ (v,\mu)\in V. \tag{3.11}$$

Furthermore, there is a constant $\beta > 0$ *such that*

$$A(v,\mu;v,\mu) \geq \beta \|(v,\mu)\|_V^2, \quad \forall (v,\mu) \in V. \tag{3.12}$$

LEMMA 3.3. *Suppose* $u \in \tilde{H}^{\frac{1}{2}}(\Gamma) \cap H^{\frac{1}{2}+\alpha}(\Gamma)$ *and* $\lambda \in \tilde{H}^{-\frac{1}{2}}(\Gamma) \cap H^{-\frac{1}{2}+\alpha}(\Gamma)$ $(0 \leq \alpha \leq 1)$, *then there exists a constant* $M_\alpha > 0$ *such that*

$$\begin{aligned}|A(u,\lambda;v,\mu)| \leq & M_\alpha(\|u\|^2_{\frac{1}{2}+\alpha,\Gamma} + \|\lambda\|^2_{-\frac{1}{2}+\alpha,\Gamma})^{\frac{1}{2}} \cdot \\ & (\|v\|^2_{\frac{1}{2}-\alpha,\Gamma} + \|u\|^2_{-\frac{1}{2}-\alpha,\Gamma})^{\frac{1}{2}}, \quad \forall (v,\mu) \in V.\end{aligned}$$

The proofs of Lemma 3.1-3.3 can be found in [12].

The inequality (3.10) shows that $\|\dot{v}\|_{-\frac{1}{2},\Gamma}$ is an equivalent norm in the space $\tilde{H}^{-\frac{1}{2}}(\Gamma)$. By Lemma 3.2, an application of Theorem 2.1 in [10, Chapter I] yields the following result.

THEOREM 3.1. *Suppose that* $g \in H^{-\frac{1}{2}}(\Gamma_1)$, *then the variational inequality* (3.8) (*or* (3.7)) *has a unique solution* $(u,\lambda) \in \tilde{K} \times \tilde{H}^{-\frac{1}{2}}(\Gamma)$.

An analogous result for variational inequality (3.9) can be obtained [11]:

THEOREM 3.2. *Suppose that* $g \in H^{-\frac{1}{2}}(\Gamma_1)$, *then variational inequality* (3.9) *on the closed surface* Γ *has a unique solution* $(u,\lambda) \in \ddot{K} \times \ddot{H}^{-\frac{1}{2}}(\Gamma)$.

4. The boundary element approximation of the boundary variational inequality (3.8)

Assume that the boundary Γ of Ω is represented as

$$x_1 = x_1(s), \quad x_2 = x_2(s), \quad 0 < s < L, \tag{4.1}$$

and $x_j(0) = x_j(L)$, $j = 1,2$. Furthermore, Γ is divided into segments $\{T\}$ by the points $x^i = (x_1(s_i), x_2(s_i))$, $i = 1, 2, \ldots, N$, that include the two end points of Γ_0, with $s_1 = 0$, $s_{N+1} = L$; we define

$$h = \max_{1 \leq l \leq N} |s_{l+1} - s_l|.$$

This partition of Γ is denoted by J_h. Let

$$S_h = \{v_h \in C^{(0)}(\Gamma),\ v_h|_T \text{ is a linear function}, \quad \forall T \in J_h \text{ and } v_h|_{\Gamma_0} = 0\}, \tag{4.2}$$

$$\tilde{K}_h = \{v_h \in S_h,\ v_h \geq 0 \text{ on } \Gamma_1\}, \tag{4.3}$$

$$M_h = \{\mu_h|_T \text{ is a constant}, \forall T \in J_h \text{ and } \int_\Gamma \mu_h ds = 0\}. \tag{4.4}$$

Obviously, S_h and M_h are two finite dimensional subspaces of $\tilde{H}^{\frac{1}{2}}(\Gamma)$ and $\tilde{H}^{-\frac{1}{2}}(\Gamma)$, respectively; $\tilde{K}_h$ is a closed convex subset of S_h and it is nonempty.

The subspaces S_h and M_h are two regular finite element spaces in the sense of Babuska and Aziz [2] that satisfy the following approximation property:

$$\inf_{v_h\in\tilde{K}_h}\{\|u-v_h\|^2_{\frac{1}{2},\Gamma}+\|u-v_h\|_{\frac{1}{2}-\alpha,\Gamma}\}$$
$$\leq C^1_\alpha h^{2\alpha}\{\|u\|^2_{\frac{1}{2}+\alpha,\Gamma}+\|u\|_{\frac{1}{2}+\alpha,\Gamma}\}, \tag{4.5}$$
$$\inf_{\mu_h\in M_h}\{\|\lambda-\mu_h\|^2_{-\frac{1}{2},\Gamma}+\|\lambda-\mu_h\|_{-\frac{1}{2}-\alpha,\Gamma}\}$$
$$\leq C^2_\alpha h^{2\alpha}\{\|\lambda\|^2_{-\frac{1}{2}+\alpha,\Gamma}+\|\lambda\|_{-\frac{1}{2}+\alpha,\Gamma}\}. \tag{4.6}$$

We now consider the discrete problem:

$$\text{Find}\quad (u_h,\lambda_h)\in\tilde{K}_h\times M_h\quad\text{such that}$$
$$A(u_h,\lambda_h;v_h-u_h,\mu_h)\geq\int_{\Gamma_1}g(v_h-u_h)ds,\quad\forall v_h\in\tilde{K}_h,\ \mu_h\in M_h. \tag{4.7}$$

It is straightforward to prove:

THEOREM 4.1. *The problem* (4.7) *has a unique solution* $(u_h,\lambda_h)\in\tilde{K}_h\times M_h$.

Furthermore, we obtain the error estimate stated in the following theorem.

THEOREM 4.2. *Suppose that the solution of the problem* (3.8), (u,λ) *satisfies* $u\in\tilde{K}\cup H^{\frac{1}{2}+\alpha}(\Gamma)$, $\lambda\in H^{-\frac{1}{2}+\alpha}(\Gamma)$, *and* $g\in H^{-\frac{1}{2}+\alpha}(\Gamma_1)$ *with* $0<\alpha\leq 1$; *then the following error estimate holds:*

$$\|(u-u_h,\lambda-\lambda_h)\|^2_V\leq C_\alpha h^{2\alpha}\{\|u\|^2_{\frac{1}{2}+\alpha,\Gamma}+\|u\|_{\frac{1}{2}+\alpha,\Gamma}$$
$$+\|\lambda\|^2_{-\frac{1}{2}+\alpha,\Gamma}+\|\lambda\|_{-\frac{1}{2}+\alpha,\Gamma}\}, \tag{4.8}$$

where C_α *is a constant independent of* h.

PROOF. Taking $v=u_h$ and $\mu=\lambda_h-\lambda$ in (3.8), we have

$$-A(u,\lambda;u_h-u,\lambda_h-\lambda)\leq\int_{\Gamma_1}g(u_h-u)ds, \tag{4.9}$$

and from (4.7) we get

$$-A(u_h,\lambda_h;v_h-u_h,\mu_h-\lambda_h)\leq-\int_{\Gamma_1}g(v_h-u_h)ds,$$
$$\forall v_h\in\tilde{K}_h,\ \mu_h\in M_h. \tag{4.10}$$

On the other hand

$$\|(u-u_h,\lambda-\lambda_h)\|^2_V\leq\frac{1}{\beta}A(u-u_h,\lambda-\lambda_h;u-u_h,\lambda-\lambda_h)$$
$$=\frac{1}{\beta}\{A(u-u_h,\lambda-\lambda_h;u-v_h,\lambda-\mu_h)$$
$$+A(u,\lambda;v_h-u,\mu_h-\lambda)$$
$$-A(u,\lambda_h;v_h-u_h,\mu_h-\lambda_h)$$

$$
\begin{aligned}
&- A(u, \lambda; u_h - u, \lambda_h - \lambda)\} \\
\leq &\frac{1}{\beta}\{A(u - u_h, \lambda - \lambda_h; u - v_h, \lambda - \mu_h) \\
&+ A(u, \lambda; v_h - u, \mu_h - \lambda) - \int_{\Gamma_1} g(v_h - u)ds\}, \\
&\forall v_h \in \tilde{K}_h, \ \mu_h \in M_h.
\end{aligned} \tag{4.11}
$$

Combining (4.11), (3.11) and (3.12) we obtain

$$
\begin{aligned}
\|(u - u_h, \lambda - \lambda_h)\|_V^2 \leq &\frac{1}{\beta}\{M\|(u - u_h, \lambda - \lambda_h)\|_V \|(u - v_h, \lambda - \mu_h)\|_V \\
&+ M_\alpha(\|u\|^2_{\frac{1}{2}+\alpha,\Gamma} + \|\lambda\|^2_{-\frac{1}{2}+\alpha,\Gamma})^{\frac{1}{2}} \\
&\cdot (\|u - v_h\|^2_{\frac{1}{2}-\alpha,\Gamma} + \|\lambda - \mu_h\|^2_{-\frac{1}{2}-\alpha,\Gamma})^{\frac{1}{2}} \\
&+ \|g\|_{-\frac{1}{2}+\alpha,\Gamma_1} \|u - v_h\|_{-\frac{1}{2}-\alpha,\Gamma}\} \\
&\forall v_h \in \tilde{K}_h, \quad \mu_h \in M_h.
\end{aligned}
$$

Finally from approximation properties (4.5) and (4.6), the error estimate (4.8) follows immediately. □

5. The canonical boundary element method

For the regular domain Ω such as a disk or a rectangle in $\Re^2$, we can use the Green function of the original equation with the Dirichlet boundary condition instead of the fundamental solution to reduce the problem (1.2) to a boundary variational inequality based on the canonical boundary reduction. In this section we only consider the case $n = 2$. Let $G(x, y)$ denote the Green function of Laplace equation $\Delta u = 0$ with the boundary condition $u|_\Gamma = 0$, then for any $u \in H^1(\Omega)$ satisfying $\Delta u = 0$ in Ω, we have [7]

$$
\left.\frac{\partial u(x)}{\partial n_x}\right|_\Gamma = -\int_\Gamma \frac{\partial^2 G(x, y)}{\partial n_x \partial n_y} u(y) ds_y . \tag{5.1}
$$

Furthermore for $u, v \in H^1(\Omega)$ and u is a weak solution of $\Delta u = 0$, we get

$$
\begin{aligned}
a(u, v) &= \int_\Gamma \frac{\partial u(x)}{\partial n_x} v(x) ds_x \\
&= -\int_\Gamma \int_\Gamma \frac{\partial^2 G(x, y)}{\partial n_x \partial n_y} u(y) v(x) ds_y ds_x = B(u, v).
\end{aligned} \tag{5.2}
$$

Hence the problem (1.2) is reduced to the following boundary variational inequality:

$$
\text{Find } \ u \in \tilde{K} \ \text{ such that}
$$

$$
B(u, v - u) \geq \int_{\Gamma_1} g(v - u) ds, \quad \forall v \in \tilde{K}. \tag{5.3}
$$

The boundary variational inequality (5.3) is equivalent to the problem (3.7), but the problem (5.3) does not include the unknown function $\lambda = \frac{\partial u}{\partial n}|_\Gamma$. We now consider the discrete problem:

$$\text{Find } u_h \in \tilde{K}_h \text{ such that}$$
$$B(u_h, v_h - u_h) \geq \int_{\Gamma_1} g(v_h - u_h)ds, \quad \forall v_h \in \tilde{K}_h. \tag{5.4}$$

We have [14]

LEMMA 5.1. *$B(u,v)$ is a bounded bilinear form on $H^{\frac{1}{2}}(\Gamma) \times H^{\frac{1}{2}}(\Gamma)$, that is, there exists a constant $M > 0$ such that*

$$|B(u,v)| \leq M\|u\|_{\frac{1}{2},\Gamma}\|v\|_{\frac{1}{2},\Gamma}, \quad \forall u, \ v \in H^{\frac{1}{2}}(\Gamma).$$

Furthermore, there is a constant $\beta > 0$ such that

$$B(v,v) \geq \beta\|v\|^2_{\frac{1}{2},\Gamma}, \quad \forall v \in H^{\frac{1}{2}}(\Gamma).$$

THEOREM 5.1. *Suppose that $g \in H^{-\frac{1}{2}}(\Gamma_1)$, then the variational inequality (5.3) has a unique solution $u \in \tilde{K}$.*

THEOREM 5.2. *The problem (5.4) has a unique solution $u_h \in \tilde{K}_h$ and the following error estimate holds*

$$\|u - u_h\|^2_{\frac{1}{2},\Gamma} \leq C(u,g)h$$

where $C(u,g)$ is a constant independent of h and $u \in H^1(\Gamma) \cap \tilde{K}$ is the solution of problem (5.3).

6. Numerical example

A simple time-independent model was proposed for the process of painting a metal surface by electrodeposition by J. M. Aitchison etc. [1]. Consider a rectangle with height b and width $2a$. Three sides on the interior of this rectangle make up the workpiece, the remaining side (the base which is of length $2a$) is the cathode and the bath occupied by electrolyte is the interior of the rectangle. Since the symmetry of the problem we need only consider the right half of the rectangle. The problem to be solved can be written as follows:

$$\begin{gathered}
\Delta\varphi = 0 \ \text{ in } \Omega, \\
\frac{\partial\varphi}{\partial n} = 0 \ \text{ in } \Gamma_3, \\
\varphi = 1 \ \text{ on } \Gamma_2, \\
\varphi \geq 0, \ \frac{\partial\varphi}{\partial n} + \varepsilon \geq 0 \text{ and } \varphi \cdot (\frac{\partial\varphi}{\partial n} + \varepsilon) = 0 \text{ on } \Gamma_1,
\end{gathered} \tag{6.1}$$

where $\varepsilon > 0$ is the dimensionless disolution curicent and

$$\begin{aligned}
\Omega &= \{(x_1, x_2) \mid \ 0 < x_1 < a, \ 0 < x_2 < b\}, \\
\Gamma_3 &= \{(0, x_2) \mid \ 0 \le x_2 \le b\}, \\
\Gamma_2 &= \{(x_1, 0) \mid \ 0 \le x_1 \le a\}, \\
\Gamma_1 &= \{(x_1, b) \mid \ 0 \le x_1 \le a\} \cup \{(a, x_2) \mid \ 0 \le x_2 \le b\}.
\end{aligned}$$

Let $u = \varphi - 1$ and

$$K_1 = \{v \mid \ v \in H^1(\Omega), \ v = 0 \text{ on } \Gamma_2 \text{ and } v \ge -1 \ on \ \Gamma_1\},$$

then the problem (6.1) is reduced to the following variational inequality:

$$\text{Find} \ \ u \in K_1 \ \ \text{such that}$$

$$a(u, v - u) \ge -\varepsilon \int_{\Gamma_1} g(v - u) ds, \quad \forall v \in K_1. \tag{6.2}$$

We have gotten the numerical solution for this problem using the method described in §5 [15]. Four meshes were used in this computation for $a = \frac{1}{2}$, $b = 1$, $\varepsilon = 0.7,\ 0.55,\ 0.5,\ 0.4$. Table 1 shows the values of φ on the nodes $x^i (i = 1, 2, \ldots, 15)$ on Γ_1 for various values of ε and various meshes. Here $x^i = (0.5, 0.1i)$ for $i = 1, 2, \ldots, 10$ and $x^i = (1.5 - 0.1i, 1)$ for $i = 11, 12, \ldots, 15$. From the numerical results in Table 1 we see that when h goes to zero the boundary element approximation solution converges to the solution of the problem (6.1) quite fast. When $\varphi > 0$ on Γ_1 a paint layer has been deposited a thickness of the paint layer is given by

$$\delta = \varphi / \varepsilon.$$

Figure 1 in [15] shows the painted surface for $b = 1$, $a = \frac{1}{2}$ and various ε. It is seen that the paint film thickness decreases as ε increase. Moreover for $\varepsilon \le 0.5$ the surface is completely painted and for $\varepsilon \approx 0.7$ we see that the top is totally free of paint.

References

1. Aitchison, J. M., Laccy, A. A. and Schillor, M., *A model for an electropaint process*, IMA J. Appl. Math., **33**(1984), 17–31.
2. Babuska, I. and Aziz, A. K., *Survey lectures on the mathematical foundations of the finite element method*, The Mathematical Foundations of the Finite Element Method with Applications to Partial Differential Equations (A. K. Aziz ed.), Academic Press, New York, 1972, 5–359.
3. Brezis, H., *Problems unlaterux*, J. de Math. Pures et Appliquess, **51**(1972), 1–168.
4. Courant, R. and Hilbert, D., *Methods of Mathematical Physics*, Volume **II**, Interscience Publishers, 1962.
5. Duvant, G., Lions, J. L., *Les inequitions en mecanaque et en physique*, Dunod, Paris, 1972.
6. Feng K., *Canonical boundary reduction and finite element methods*, Proceedings of Symposium on Finite Element Method, Scinence Press, Beijing, China, 1982.
7. Feng, K. and Yu, D., *Canonical integral equations of elliptic boundary value problems and their numerical solutions*, Proceedings of the China-France Symposium on Finite Element Methods, Science Press, Beijing, 1983, 211–252.

8. Friedman, A., *Variational Principles and Free Boundary Problems*, John Wiley and Sons, New York, 1982.
9. Glowinski, R., *Numerical Methods for Nonlinear Variational Problems*, Springer-Verlag, New York, 1982.
10. Glowinski, R., Lions, J. L. and Tremolieses, R., *Numerical Analysis of Variational Inequalities*, North-Holland, Amsterdam, 1981.
11. Han, H., *A boundary element method for Signorini problems in three dimensions*, Numer. Math, **60**(1991), 63–75.
12. Han, H., *A direct boundary element methods for Signorini problems*, Math. Comp., **55**(1990), 115–128.
13. Han, H., *Boundary integro-differential equations of elliptic boundary value problems and their numerical solutions*, Scientia Sinica, **10**(1988), 1153–1165.
14. Han, H., *The boundary finite element methods for Signorini problems*, in Lecture Notes in Mathematics, No. 1297, Springer-Verlag, Berlin, Heidelberg, 1987, 38–49.
15. Han, H., Guan, Z. and Yu, C., *The canonical boundary element analysis of a model for an electropaint process-The canonical boundary element approximation of a Signorini problem*, Appl. Math. A Journal of Chinese Universities, **3**(1988), 101–111.
16. Han, H., and Hsiao, G. C., *The boundary element method for a contact problem*, in Theory and Applications of Boundary Element Methods, Tsinghua Press, 1988, 33–38.
17. Han, H. and Hsiao, G. C., *The boundary element method for contact problems in linear elasticity*, to appear.

Table 1.

(1) $\varepsilon = 0.7$

φ	$h = \frac{1}{10}$	$h = \frac{1}{20}$	$h = \frac{1}{30}$	$h = \frac{1}{40}$
x^1	0.697	0.740	0.715	0.717
x^2	0.571	0.572	0.567	0.567
x^3	0.429	0.431	0.430	0.430
x^4	0.310	0.308	0.310	0.310
x^5	0.206	0.204	0.206	0.206
x^6	0.118	0.116	0.119	0.119
x^7	0.049	0.048	0.051	0.051
x^8	0.000	0.002	0.006	0.006
x^9	0.000	0.000	0.000	0.000
x^{10}	0.000	0.000	0.000	0.000
x^{11}	0.000	0.000	0.000	0.000
x^{12}	0.000	0.000	0.000	0.000
x^{13}	0.000	0.000	0.000	0.000
x^{14}	0.000	0.000	0.000	0.000
x^{15}	0.000	0.000	0.000	0.000

(2) $\varepsilon = 0.55$

φ	$h = \frac{1}{10}$	$h = \frac{1}{20}$	$h = \frac{1}{30}$	$h = \frac{1}{40}$
x^1	0.728	0.769	0.744	0.746
x^2	0.616	0.616	0.611	0.611
x^3	0.484	0.485	0.483	0.483
x^4	0.370	0.368	0.368	0.368
x^5	0.268	0.264	0.266	0.266
x^6	0.178	0.175	0.176	0.176
x^7	0.102	0.098	0.100	0.100
x^8	0.040	0.038	0.039	0.039
x^9	0.000	0.000	0.000	0.000
x^{10}	0.000	0.000	0.000	0.000
x^{11}	0.000	0.000	0.000	0.000
x^{12}	0.003	0.003	0.003	0.003
x^{13}	0.017	0.014	0.013	0.013
x^{14}	0.025	0.022	0.019	0.020
x^{15}	0.027	0.022	0.018	0.016

(3) $\varepsilon = 0.5$

φ	$h = \frac{1}{10}$	$h = \frac{1}{20}$	$h = \frac{1}{30}$	$h = \frac{1}{40}$
x^1	0.740	0.780	0.755	0.757
x^2	0.634	0.634	0.629	0.628
x^3	0.507	0.507	0.505	0.505
x^4	0.396	0.393	0.393	0.392
x^5	0.296	0.292	0.292	0.292
x^6	0.206	0.203	0.203	0.202
x^7	0.128	0.125	0.125	0.125
x^8	0.062	0.060	0.060	0.060
x^9	0.006	0.011	0.011	0.011
x^{10}	0.000	0.000	0.000	0.000
x^{11}	0.000	0.003	0.002	0.002
x^{12}	0.025	0.023	0.021	0.020
x^{13}	0.043	0.040	0.036	0.036
x^{14}	0.054	0.049	0.044	0.043
x^{15}	0.057	0.046	0.038	0.033

(4) $\varepsilon = 0.4$

φ	$h = \frac{1}{10}$	$h = \frac{1}{20}$	$h = \frac{1}{30}$	$h = \frac{1}{40}$
x^1	0.771	0.807	0.781	0.783
x^2	0.686	0.679	0.672	0.671
x^3	0.577	0.567	0.562	0.561
x^4	0.481	0.464	0.461	0.459
x^5	0.392	0.372	0.368	0.366
x^6	0.312	0.288	0.284	0.282
x^7	0.240	0.214	0.209	0.207
x^8	0.177	0.148	0.142	0.139
x^9	0.118	0.148	0.142	0.139
x^{10}	0.071	0.029	0.016	0.012
x^{11}	0.115	0.081	0.073	0.071
x^{12}	0.144	0.112	0.103	0.100
x^{13}	0.165	0.132	0.121	0.118
x^{14}	0.172	0.138	0.125	0.121
x^{15}	0.178	0.126	0.103	0.088

Department of Applied Mathematics, Tsinghua University, Beijing 100084, China

Contemporary Mathematics
Volume **163**, 1994

The Study of Numerical Analysis of Matrix Eigenvalue Problem in China

JIANG ER-XIONG

I. A series studies for QL algorithm of symmetric tridiagonal matrices

Let T_1 be an unreducible tridiagonal matrix. In order to find its eigenvalues, we choose a sequence $\{\sigma_k\}$, where σ_k is a shift.

From

$$T_k - \sigma_k I = Q_k L_k,$$
$$T_{k+1} = L_k Q_k + \sigma_k I,$$

we form symmetric tridiagonal matrix sequence $\{T_k\}$, where Q_k are orthogonal matrices, L_k are lower triangular matrices.

If the diagonal elements of T_k are $\alpha_i^{(k)}$, $i = 1, 2, \cdots, n$, and $\beta_i^{(k)}$, $i = 1, 2, \cdots, n-1$, are the subdiagonal elements, then we have following results.

1. When $\beta_1^{(k)} \to 0$, $\alpha_1^{(k)}$ can be taken as an approximate eigenvalue of T_1, but $\alpha_1^{(k)}$ is not assured to converge.

From the perturbation estimation we know that there is an eigenvalue $\lambda(k)$ of T_1 which gives

$$|\alpha_1^{(k)} - \lambda(k)| \le |\beta_1^{(k)}|.$$

But there is a problem, i.e., whether the eigenvalue $\lambda(k)$ depends on k, when k is sufficiently large. We can prove that if the shift σ_k satisfies

$$\alpha_1^{(k)} - \sigma_k \to 0,$$

then, when

$$\beta_1^{(k)} \to 0,$$

1991 *Mathematics Subject Classification.* 65F15.

0271-4132/94 $1.00 + $.25 per page

we have

$$\alpha_1^{(k)} \to \lambda,$$

where λ is an eigenvalue of T_1.

More generally, if

$$\alpha_1^{(k)} - \sigma_k \to 0,$$

then, when

$$\beta_j^{(k)} \to 0, \quad j = 1, 2, \cdots, p,$$

we have

$$\alpha_j^{(k)} \to \lambda_{i_j}, \quad j = 1, 2, \cdots, p,$$

where λ_{i_j} are p eigenvalues of T_1.

The condition

$$\alpha_1^{(k)} - \sigma_k \to 0$$

is valid for both Wilkinson-shift and Rayleight Quotient shift, see [4].

2. We proposed a correct relation between $\beta_1^{(k+1)}$ and $\beta_1^{(k)}$ from which it is easy to deduce the rate of asymptotic convergence of various shifts [4].

3. $\sigma = \alpha_1 - \operatorname{sign}\delta \cdot \dfrac{(\beta_1)^2}{|\delta| + (\delta^2 + \beta_1^2)^{1/2}}$ is called Wilkinson-shift, where $\delta = (\alpha_2 - \alpha_1)/2$.

J. Wilkinson proved that if one uses Wilkinson-shift for any unreducible symmetric tridiagonal matrix T_1, then $\beta_1^{(k)} \to 0$, and the asymptotic convergence rate of $\beta_1^{(k)}$ is quadratic. Wilkinson also conjectured that the asymptotic convergence rate of $\beta_1^{(k)}$ is cubic. But so far this guess has not been proved.

Yu Chonghua [11] and Zhang Guodong [13] gave the following result in their theses.

If

$$\alpha_1^{(k)} \to \lambda_{i_1},$$

and

$$\lambda_{i_1} - \lambda_{i_1 - 1} \neq \lambda_{i_1 + 1} - \lambda_{i_1},$$

then

$$\beta_2^{(k)} \to 0,$$

therefore we conclude that the convergence rate of $\beta_1^{(k)}$ is cubic.

4. Jiang Erxiong and Zhang Zhenyue put forth a new shifting strategy called RW-shift [9]. When σ_k is taken such value, then $\beta_1^{(k)} \to 0$ and $\beta_1^{(k)}$ has cubic asymptotic convergence rate for any T_1. In [5] Jiang also proved that in this case we have $|\beta_1^{(k+1)}| < |\beta_1^{(k)}|$.

5. The Rayleigh Quotient shift is $\sigma_k = \alpha_1^{(k)}$, while Wilkinson-shift σ_k is one eigenvalue of 2×2 matrix $\begin{bmatrix} \alpha_1^{(k)} & \beta_1^{(k)} \\ \beta_1^{(k)} & \alpha_2^{(k)} \end{bmatrix}$.

If we take one eigenvalue of 3×3 matrix

$$T_{1,3}^{(k)} = \begin{bmatrix} \alpha_1^{(k)} & \beta_1^{(k)} & 0 \\ \beta_1^{(k)} & \alpha_2^{(k)} & \beta_2^{(k)} \\ 0 & \beta_2^{(k)} & \alpha_3^{(k)} \end{bmatrix}$$

as shift σ_k, what result we can get?

In [13] Zhang Guodong proved that if we take shift σ_k, one of the three roots of $\det(T_{1,3}^{(k)} - \lambda I) = 0$ which satisfies

$$|\alpha_1^{(k)} - \sigma_k| \leq |\alpha_3^{(k)} - \sigma_k|,$$

such σ_k always exists, then for arbitrary $\epsilon > 0$ we can find K, when $k > K$

$$|\beta_1^{(k)} \cdot \beta_2^{(k)}| < \epsilon \quad \text{or} \quad |\beta_1^{(k+1)} \cdot \beta_2^{(k+1)}| < \epsilon$$

at least one holds, and we have asymptotic estimate

$$\beta_1^{(k+1)} \cdot \beta_2^{(k+1)} = O(\beta_1^{(k)^3} \beta_2^{(k)^3} \beta_3^{(k)^2}).$$

6. There are two schemes for QL algorithm implementation, one is explicit and other is implicit, both have its advantages respectively. Yu Chonghua proposed an implicit-explicit scheme [11], which has the above two advantages and gives good numerical results.

7. Let $T_{1,s}^{(k)}$ is the $s \times s$ leading principal submatrix of T_k, and let its s eigenvalues be denoted by $\sigma_1^{(k)}, \sigma_2^{(k)}, \cdots, \sigma_s^{(k)}$, the following procedure of producing $\{T_k\}$:

$$\begin{aligned} T_k - \sigma_1^{(k)} I &= Q_k L_k, \\ T_{k+1/s} &= L_k Q_k + \sigma_1^{(k)} I, \\ T_{k+1/s} - \sigma_2^{(k)} I &= Q_{k+1/s} L_{k+1/s}, \\ T_{k+2/s} &= L_{k+1/s} Q_{k+1/s} + \sigma_2^{(k)} I, \\ &\cdots\cdots \\ T_{k+(s-1)/s} - \sigma_s^{(k)} I &= Q_{k+(s-1)/s} L_{k+(s-1)/s}, \\ T_{k+1} &= L_{k+(s-1)/s} Q_{k+(s-1)/s} + \sigma_s^{(k)} I, \end{aligned}$$

is called s-shift QL algorithm.

In [7] the following result was given:

If $n \geq 2s + 1$, and $\{T_k\}$ is produced by s-shift QL algorithm, then

(i) $\{|\beta_1 \beta_2 \cdots \beta_s|\}$ is a monotone nonincreasing sequence.

(ii) $|\beta_1 \beta_2 \cdots \beta_s| \to 0$ otherwise
$|\beta_{s+1} \beta_{s+2} \cdots \beta_{2s}| \to 0$ and $|q_{s+1,1}| \to 1$, $|q_{1,s+1}| \to 1$.

The superscript k is omitted for the sake of simplicity, where $q_{i,j}$ is the i-th component of q_j, and q_j is the j-th column of $\tilde{Q}_k = Q_k Q_{k+1/s} \cdots Q_{k+(s-1)/s}$.

II. Calculation of generalized eigenvalue problems

Consider the generalized eigenvalue problem

$$Ax = \lambda Bx,$$

where A, B are $n \times n$ symmetric matrices.

1. If B is positive, there is a well known algorithm to find eigenpairs called Rayleigh Quotient Iteration (RQI). But RQI does not converge for some special initial vector, and it converges very slowly when a bad vector is taken as initial vector, and there are a lot of bad vectors for any eigenvalue problem. In [5] and [6] Jiang Erxiong proposed a modified Rayleigh Quotient Iteration (MRQI), which overcomes this drawback, and not only always converges but also is faster than RQI.

2. Chao Zhihao [2] pointed out the shape of Kronecker canonical form of matrix $A - \lambda B$ when B is semipositive.

3. In [2] Chao Zhihao also extended the Fix-Heiberger method to the case where matrices A, B are only symmetric.

III. Two-parameter eigenvalue problem

For equations

$$\begin{cases} (\lambda A_1 + \mu B_1 + C_1)x = 0, \\ (\lambda A_2 + \mu B_2 + C_2)x = 0, \end{cases}$$

it is asked to find certain special λ and μ such that the equations have nonzero solution x. This is a two-parameter eigenvalue problem.

Ji Xingzhi gave an effective method for finding λ and μ when A_i and B_i are diagonal matrices, C_i are Jacobi matrices in his Ph.D. dissertation [3]. That method is similar to the sectioning method, and could be called planar sectioning method.

IV.

For an eigenvalue problem of $n \times n$ matrix it is interesting to know how to split it into two eigenvalue problems of $p \times p$ and $q \times q$ matrices, where $p + q = n$. This challenge is put forward to numerical analysis of eigenvalue problem by MIMD computer.

In his Ph.D. dissertation Zhang Zhenyue realized this splitting which can be done continuously when matrix is symmetric and tridiagonal and the problem is to find extreme eigenvalues [14].

V.

Bai Zhaojun studied the generalized singular values problem. In his Ph.D. dissertation he considered GSVD theory, GSVD direct method and its parallel implementation [1].

Let the common singular values of A be $\sigma_1 \geq \sigma_2 \geq \sigma_3 \geq \cdots \geq \sigma_n \geq 0$. From Eckart and Young theorem we have

$$\sigma_k = \min\{\|E\|_2 \mid \text{rank}\,(A+E) \leq k-1\}.$$

Zha Hongyuan proved that the generalized singular values proposed by Van Loan, C.C. Paige and M.A. Sounders satisfy

$$r_k = \min\{\|E\|_2 \mid \text{rank}\,(A+BE) \leq k-1\}$$

or

$$r_k' = \min\{\|E\|_2 \mid \text{rank}\,(A+EB) \leq k-1\}.$$

Furthermore, he proposed a theory of the restricted singular value decomposition (RSVD) [12], and the restricted singular values

$$\omega_k = \min\{\|E\|_2 \mid \text{rank}\,(A+BEC) \leq k-1\}.$$

VI. Linear optimal control problem

Solutions of algebraic Riccati equations are very important, but there are few effective numerical methods. It is known that the solution of the continuous-time algebraic Riccati equation can be gotten from an associated invariant subspace of a Hamiltonian matrix M. According to this properties, Xu Hongguo gives a new way to find the invariant subspace in his Ph.D. thesis [10]. It shows that the matrix F satisfying $F^2 = M^2$ and $\text{Re}\lambda(F) < 0$ exists and is unique and skew-Hamiltonian. Further, the first n columns of $M+F$ form the associated invariant subspace. With this idea a more effective structure-preserving algorithm can be designed. There are similar properties and algorithm for the discrete-time algebraic Riccati equations.

VII. Eigenvalue perturbation of matrix

In [8], there is an eigenvalue perturbation result as follows. Let C, B and E be $n \times n$ matrices and $C = B - E$. Let $J = FBF^{-1}$ be the Jordan form of B,

$$J = \begin{bmatrix} J_1(m_1, \lambda_1) & & & \\ & J_2(m_2, \lambda_2) & & \\ & & \ddots & \\ & & & J_k(m_k, \lambda_k) \end{bmatrix},$$

where

$$J_i(m_i, \lambda_i) = \begin{bmatrix} \lambda_i & 1 & & 0 & \\ & \lambda_i & 1 & & \\ & & \ddots & \ddots & \\ & 0 & & \ddots & 1 \\ & & & & \lambda_i \end{bmatrix}$$

are $m_i \times m_i$ Jordan blocks, and denote $m = \max_i\{m_i\}$. If

$$\|FEF^{-1}\| \leq \frac{1}{m+1},$$

then for any eigenvalue μ of C there exists an eigenvalue λ of B, such that

$$|\lambda - \mu| \leq \frac{m}{m+1}(m+1)^{1/m}\|FEF^{-1}\|^{1/m}.$$

Let m_0 be the order of the largest Jordan block to which λ belongs, then we also have

$$|\lambda - \mu| \leq \frac{m_0}{m_0+1}(m_0+1)^{1/m_0}\|FEF^{-1}\|^{1/m_0}.$$

REFERENCES

1. Bai Zhaojun, *The Direct GSVD Algorithm and Its Parallel Implementation*, Ph.D. Dissertation, Department of Mathematics, Fudan University, 1987.
2. Cao Zhihao, *On a deflation method for the symmetric generalized eigenvalue problem*, LAA, **92**(1987), 187–196.
3. Ji Xingzhi, *Multi-Parameter Eigenvalue Problem and Jointed Eigenvalue Problem*, Ph.D. Dissertation, Department of Mathematics, Fudan University, 1987.
4. Jiang Erxiong, *On the convergence of diagonal elements and asymptotic convergence rates for the shift tridiagonal QL algorithm*, J. Comput. Math., **3**:3(1985), 252–261. (M.R. 87k:65046;p. .M. 81142, 1986).
5. Jiang Erxiong, *The modified Rayleigh Quotient Iteration*, J. Comput. Math., **6**:1(1988), 80–87. (M.R. 90a:65085)
6. Jiang Erxiong, *An algorithm for finding generalized eigenpaires of a symmetric definite matrix pencil*, LAA, **132**(1990), 65–91. (M.R. 91f:65074).
7. Jiang Erxiong, *A note on the Double-Shift QL Algorithm*, LAA, **171**(1992), 121-132.
8. Jiang Erxiong, *Bounds for the smallest singular value of a Jordan block with an application to eigenvalue perturbation*, to appear in LAA.
9. Jiang Erxiong and Zhang Zhenyue, *A new shift of the QL algorithm for irreducible symmetric tridiagonal matrices*, LAA, **65**(1985), 261–272. (M.R. 86g:65082,1986).
10. Xu Hongguo, *Solving Algebraic Riccati Equations via Skew-Hamiltonian Matrices*, Ph.D. Dissertation, Department of Mathematics, Fudan University, 1991.
11. Yu Chonghua, *On QL algorithm for symmetric matrices*, Ph.D. Dissertation, Department of Mathematics, Fudan University, 1987.
12. Zha Hongyuan, *The Restricted Singular Value Decomposition of Matrix Triplets*, SIAM J. Matrix Anal. Appl., **12**:1(1991), 172–194.
13. Zhang Guodong, *Two researches on the QL algorithm for the symmetric tridiagonal matrices*, M.S. Dissertation, Department of Mathematics, Fudan University, 1986.
14. Zhang Zhenyue, *Tridiagonal Matrices, Eigenvalue Problem, Inverse Eigenvalue Problem and the Parallel Computation of Extreme Eigenpairs*, Ph.D. Dissertation, Department of Mathematics, Fudan University, 1989.

DEPARTMENT OF MATHEMATICS, FUDAN UNIVERSITY, SHANGHAI 200433, CHINA

Contemporary Mathematics
Volume **163**, 1994

Numerical Analysis and Scientific Computation in Hong Kong

KWOK YUE-KUEN, HUANG HONG-CI AND CHAN RAYMOND H.

1. Introduction

This paper is a survey of the research works in numerical analysis and scientific computation by Hong Kong mathematicians and engineers. It is written in the hope that researchers in similar fields in other parts of the world can obtain an easy accessible information on the research activities in Hong Kong academic community. The paper cannot be expected to be comprehensive but every effort is taken to make it as representive as possible.

The research areas are quite diversified. We manage to group the research activities into five subfields, namely,

(i) Numerical Linear Algebra
(ii) Numerical Methods for Differential Equations
(iii) Finite Element Methods
(iv) Parallel Computation
(v) Computational Fluid Dynamics

2. Numerical linear algebra

R.H. Chan and his collaborators have made contributions in the areas of Iterative Toeplitz Solvers and Iterative Methods for Queuing Models. D. Ho et al published several works on Pole Assignments. H.C. Huang threw some new insight about the concept of Condition Number. Y.K. Kwok et al proposed a numerical algorithm for Inversion of Laplace Transform. The details of their contributions are presented below:

1991 *Mathematics Subject Classification.* 65F, 65M, 65W, 65C, 73K, 76F, 76N, 76R, 76T.
Key words and phrases. Numerical Analysis, Scientific Computation, Hong Kong.

2.1. Iterative Toeplitz solvers. Toeplitz systems appear in many fields of applied mathematics. Examples are solutions of convolution type integral equations, signal reconstruction and signal restoration in signal processing, time series analysis for auto-regressive time series. Traditionally, people use direct Toeplitz solvers for solving these systems and the complexity is of $O(n \log^2 n)$ where n is the size of the Toeplitz matrix.

Chan and Strang [3] showed that if the so-called Strang's circulant preconditioner is used in conjunction with the preconditioned conjugate gradient method, then the system can be solved in $O(n \log n)$ operations, provided that the generating function f of the Toeplitz matrix is an even, positive function in the Wiener class.

Since then, many preconditioners have been proposed and used. In [5], a new circulant preconditioner was introduced which performs at the same rates as Strang's preconditioner. Chan [8] invented the band-Toeplitz preconditioner which can handle the case where f is nonnegative. Also, Chan and Ng [17] designed band-Toeplitz preconditioner for solving Toeplitz-plus-band systems and Chan and Tang [19] developed fast band-Toeplitz preconditioner which performs as good as circulant ones.

Analysis of preconditioner proposed by others has also been carried out. In [5] and [12], Chan et al analyzed the convergence rate of Strang's circulant preconditioner more closely. In [5], Chan gave an explicit bound on the convergence rate and showed that Strang's preconditioner is optimal in the l_1 norm. Using Jackson Theorem in approximation theory, Chan and Yeung [12] showed that the method converges for the class of Lipschitz continuous functions.

Chan [4] analyzed T. Chan's circulant preconditioner and showed that it performs at the same rate as Strang's preconditioner for f that are in the Wiener class. Later, Chan and Yeung [11, 16] enlarged the class to 2π-periodic continuous functions. The convergence rate of the method for Tyrtyshinov's circulant preconditioner was given in [7]. The T. Chan's and Tyrtyshinov's circulant preconditioners were studied from the operator point of view in [6]. Chan and Yeung [14] unified the analysis of all these circulant preconditioners from the view-point of convolution products and they were able to construct general circulant preconditioners and analyzed preconditioners proposed by Ku and Kuo and Huckles easily.

Applications of these methods to different fields of mathematics have been carried out. Chan [10] applied the methods to solve integral equation derived from the inverse heat equation. Also, Chan and Jin [9] applied the methods to solve skew-Hermitian type Toeplitz systems arising frequently in solving first order hyperbolic systems. The method was used to solve elliptic problem of second order [13] and the results were then generalized to second order hyperbolic and parabolic equations [21].

In signal processing application, Chan and Jin [15] have analyzed the convergence rate of a block circulant preconditioner which has been used successfully in noise reduction model. One-dimensional signal restoration problems usually give rise to Toeplitz least squares problems. Chan et al [18] have analyzed the possible use of circulant preconditioners for such problems. The results are being generalized to two-dimension [20].

2.2. Iterative methods for queuing models. Chan [1, 2] considered the solution of steady-state probability distribution of a Markovian queuing network with some overflow capacity built-in. The problem is equivalent to finding a null vector for a singular matrix. First, the matrix was transformed to a non-singular one by preconditioning it with a singular preconditioner. The resulting system was then solved by the conjugate gradient method. The preconditioner was chosen by considering the problem from the view point of numerical partial differential equations. Convergence analysis was given for the case of two-queue, one-way overflow network. Domain decomposition techniques were used in designing preconditioners for queuing networks with irregular state-spaces [22].

2.3. Pole assignments. Ho et al [23] considered the bounds on output feedback gain matrix in the multivariable linear control design. A variety of sensitivity analysis for the problem was presented and some precautions in choosing eigenvectors were given.

Some geometric properties of controllability subspaces were used by Ho [25] to design new algorithm for selecting a mixture of left and right eigenvectors to satisfy certain constraints. The technique, based on orthogonal properties of controllability and observability subspaces, can assign all the desired poles exactly or change to new pole locations without lengthily iteration process. The problem of arbitrary pole assignment was also discussed in [24]. Necessary and sufficient conditions were given together with a computational procedure based on these results.

2.4. Condition number. For an arbitrary n-by-n complex matrix A, its condition number $Cond(A)$ for various possible norms was considered by Huang [26]. The main results are

(i) If A is not a multiple of identity, then there is no finite upper bound for $Cond(A)$ on U, the set of all norms defined on C^n.

(ii) If A is nonsingular, then

$$\inf_{\|\cdot\| \in U} cond(A) = \frac{\max_i |\lambda_i|}{\min_i |\lambda_i|}$$

where λ_i are eigenvalues of A.

(iii) If A is nonsingular, then

$$\|A\| \, \|A^{-1}\| = \frac{\max_i |\lambda_i|}{\min_i |\lambda_i|}$$

for some norm if and only if A has no Jordan block corresponding to the largest and smallest eigenvalues in magnitude.

In another work, Huang and Gui [27] proved that the condition number of mixed finite element approximation for a biharmonic equation is of order $O(h^{-2})$. This is more favorable than the direct discretization method which gives a coefficient matrix with condition number $O(h^{-4})$.

2.5. Numerical inversion of Laplace transform. Kwok et al [28] proposed a new algorithm for the numerical inversion of Laplace transform. It involves the approximation of the kernel in the Bromwich integral by a two-parameter function. The method is believed to work for a wider range of functions since it is not based on any assumption on representing $f(t)$ by certain class of basis functions.

3. Numerical methods for differential equations

J. Caldwell et al developed several techniques for solving Burgers' equation. The techniques include method of lines, moving node finite element method, variational-iterative scheme and Fourier Transform. T.M. Shih, C.B. Liem et al constructed fourth order finite difference methods for solving boundary and eigenvalue problems of elliptic equations, and the methods of operator combination for solving second order elliptic equations. Y.K. Kwok et al developed a systematic procedure to derive stability conditions for three-level difference schemes for initial-boundary problems for multi-dimensional convective-diffusion equations. K.M. Liu et al have made contributions on the Tau method for numerical solution of eigenvalue problems for differential equations. The details are presented below:

3.1. Solution of Burgers' equation. The one-dimensional Burgers' equation

$$\frac{\partial u}{\partial t} + u\frac{\partial u}{\partial x} = \frac{1}{R}\frac{\partial^2 u}{\partial^2 x}$$

where R is the Reynolds number, often arises in the mathematical modeling in fluid dynamics involving turbulence. A numerical approach adopted by Caldwell et al [29] for solving Burgers' equation used the method of lines involving finite Fourier series. In this method, a set of ordinary differential equations is solved for the amplitudes of the sine terms. It is observed that numerical difficulties arise in the solution for large Reynolds number. To obtain higher accuracy, Caldwell et al [30] used a moving node finite element method to obtain a solution of Burgers' equation under certain prescribed conditions.

Furthermore, Caldwell et al [31] demonstrated how a variational-iterative scheme based on complementary variational principles can be applied to solve non-linear partial differential equations. The aim is to isolate a linear self-adjoint operator(e.g. the Laplacian) from the rest of the equation. A variationally equivalent formulation is then constructed which can be solved iteratively. The test example chosen is the steady-state Burgers' equation.

Recently, Caldwell [32] developed a numerical procedure where the space derivatives are computed with very high accuracy by means of Fourier Transform methods. Using this approach a forward marching problem involves discrete time steps but space derivatives are accurate within the limit to which a distribution can be defined on a finite set of meshpoints. The numerical computation can be carried out efficiently by using the Fast Fourier Transform algorithm. This method of computing the space derivatives gives results that are substantially more accurate than those obtained from finite difference expressions.

Shih, Liem and Lu [33] constructed difference schemes for solving eigenproblem of the Sturm-Liouville equation. Using certain correction procedure, they showed that the errors for each eigenvalue and its corresponding eigenfunction are $O(h^4)$. The method was generalized to boundary and eigenvalue problems of divergence type linear second order elliptic partial differential equations [34]. Also, Shih and Liem [35] developed a class of operator combination formulas of order four suitable for solving second order elliptic equations. Numerical experiments indicated that their formulas perform better than other algorithms based on correction methods.

By reducing a quadratic simple von Neumann polynomial to a linear polynomial, Kwok et al [36] developed a systematic procedure to derive stability conditions for three-level difference schemes for multi-dimensional convective-diffusion equation. Discrepancies on time step restriction between von Neumann and matrix analyses for boundary value problems were discussed.

Liu and Ortiz's research works cover two areas [37-41]. Initiated in 1982, the first area is concerned with the analysis and application of recent formulations of the Tau Method to the numerical solution of systems of ordinary and partial differential equations and related eigenvalue problems. The second area, closely related to the former, is the development of numerical techniques for the approximate solution of two- and three-dimensional crack problems in fracture mechanics.

Liu has also produced a FORTRAN Tau software called TAUSYS2 for mixed-order systems of linear ordinary differential equations with supplementary (initial, boundary and mixed) conditions.

4. Finite element methods

On the theoretical side, in 1966, H.C. Huang gave an error estimate of finite element method for solving plane elasticity problem [42]. The paper represents one of the earliest works in mathematical finite elements. Later, Huang et al continued to develop some theory related to multigrid techniques, a posteriori error estimate and numerical stability of discretized systems derived from boundary value problems. S.S. Chow et al extended a postprocessing technique for boundary flux calculations to higher dimensional boundary value problems solved by finite element methods. Chow also studied the finite element solutions to a class of second order elliptic problem and the corresponding error estimates.

On the applied side, Y.K. Cheung and his collaborators developed the Semi-Analytical Finite Strip Method and the Higher Order Finite Element Method for the analysis of tall buildings. The group also developed the Finite Strip Method, Incremental Harmonic Balance Method, and the U-Transformation Method for various structural, mechanical and dynamical applications in engineering. Y.C. Hon studied a storm surge model using finite element method. W.M. Xue et al developed an automated quasi $3D$ finite element grid generation method and applied it to a biomedical problem. The details of their works are discussed below:

4.1. Extrapolation, multigrid technique and a posteriori error estimate. Huang et al [43] introduced extrapolation in finite element method with non-uniform meshes, which was showed to be as efficient as in finite difference method with uniform meshes. Two types of extrapolation were suggested for solving elliptic boundary value problems by successively refining meshes: one for gaining a higher order approximation to the true solution of the partial differential equation, and the other for gaining a good initial guess in iteration from level to level. A posteriori error estimate and other multigrid technique were also considered. In [44], [45], Huang et al presented some algorithms combining the multigrid method with extrapolation to speed up convergence.

In [46], Huang et al proved an asymptotic error expansion which exists for problem of Poisson equation $\Delta u = f$, assuming $f \in C^2(\Omega)$ and homogeneous boundary value on a rectangular domain Ω. It is known that the solution u of the problem generally belongs to $H^{3-\epsilon}(\Omega)$, independent of the smoothness of the right hand function f. According to available theories, including very strong smooth assumptions on the true solution u, no asymptotic error expansion can be obtained in this situation. In [47-49], Huang et al presented some developments on a posteriori error analysis for h version or h-p version of finite element methods in one or two dimensional boundary value problems.

4.2. New measure on numerical stability. In the solution of boundary value problems by finite element and finite difference methods, it is commonly thought that if triangles with small angles or narrow rectangles appear in the grid, the numerical stability for the discretized system will be poor. Such a point of view is due to the measure of numerical stability by the condition number of the coefficient matrix. However, it has been found by some researchers that even quite flat triangles appear in a triangulation, the practical computation of the finite element method does not encounter any numerical difficulty.

Using theoretical analysis, Huang et al [50] revealed that it is misleading to measure numerical stability for a discretized system of a boundary value problem by the condition number of the coefficient matrix. They also gave a new reasonable measure which is dependent only on the partial differential equation but independent of the grid size and the grid shape. In practical computations, a satisfactory estimated value of the measure defined in [50] can be obtained not too costly.

4.3. Boundary-flux calculations. Chow et al [51] described a postprocessing technique for determining derivatives (fluxes or stresses) from finite element solutions to boundary value problems. The approach is a generalization to higher dimension of a procedure known to give highly accurate (in fact superconvergent) results in one dimension. Formally, the extension of the procedure to higher dimension is direct. However, the resulting flux relation now involves a contour integral rather than point values and the implementation is more complex. Moreover, new difficulties arise from the presence of corners in polygonal domains.

In [51], a formulation and implementation of the boundary flux calculation technique for a class of scalar field problems was presented. The accuracy of the results and rates of convergence of the method in numerical experiments on representative model problems were also shown. The formulation can be applied to nonlinear problems with little extra effort since the postprocessing projection is a linear technique.

4.4. Non-linear elliptic equations of monotone type. Chow [52] considered the application of the finite element method to a class of second order elliptic boundary value problems of divergence form and with gradient nonlinearity in the principal coefficient, and the derivation of error estimates for the finite element approximations. Such problems arise in many practical situations - for example, in shock-free airfoil design, seepage through coarse grained porous media, and in some glaciological problems.

By exploiting the algebraic properties of the nonlinear coefficient, the abstract operators accompanying such problems may be shown to satisfy certain continuity and monotonicity inequalities. With the aid of these inequalities and some standard results from approximation theory, Chow showed how one may derive error estimates for the finite element approximations in the energy norm.

4.5. Analysis of tall buildings. Cheung et al [53, 54] developed the Semi-Analytical Finite Strip Method and the Higher Order Finite Element Method for the analysis of tall buildings. Both methods have been shown to be superior to the standard finite element method because of the drastic reduction in data input and computational effort.

4.6. Spline finite strip method. The Finite Strip Method had been extended by Cheung et al using spline function in place of the continuously differentiable functions such as beam eigenfunctions, stability functions, etc. [55-57]. As a result of this extension, the Finite Strip Method can be conveniently applied to problems with any boundary condition, concentrated loading, multi-spans and arbitrary shapes.

4.7. Incremental harmonic balance method. The Incremental Harmonic Balance Method has been developed by a group under Cheung's supervision to deal with strongly nonlinear vibration problems since the traditional perturbation methods are only valid for weakly nonlinear phenomena [58-60]. This method has been applied to a large number of beams, plates and shell problems and has been found to be one of the best tools in nonlinear dynamic analysis by researchers both local and abroad.

4.8. U-transformation method. The U-Transformation Method was first derived by Cheung et al from the mode method for rotational periodic structures [61-63]. The static and dynamic equation for cyclic periodic structures can be uncoupled in the domain of a single substructure by U-transformation. It is then extended to the double U-transformation method for structures with cyclic periodicity in two directions. The method has been applied to continuous beams, grillage, double layer grids, plates, etc.

4.9. Storm surge model. Hon [64] investigated an operational model to study the storm surge in Tolo harbour of Hong Kong. Finite element method with triangular and quadrilateral elements under various computational schemes were used to solve the two-dimensional vertically integrated hydrodynamical equations. Parallel processing technique was also used to increase the efficiency of computations.

4.10. Grid generation. Xue et al [65] developed an automated quasi $3D$ finite element grid generation method for particular $3D$ complex connected domain, across which some are simply-connected $2D$ regions and some are multi-connected $2D$ regions. A sub-division algorithm based on the variational principle has been developed to ascertain the smoothness and orthogonality of the generated grid in any $2D$ cross sections. Smooth transition between the simply and multi-connected $2D$ regions is maintained. The method was applied to generate a finite element grid for human knee stump.

5. Parallel computation

The papers in this field concentrate on solving elliptic boundary value problems by the domain decomposition method, C.B. Liem et al proposed and proved the convergence of several parallel algorithms of domain decomposition. S. Zhang et al developed a domain decomposition method based on the preconditioned conjugate gradient method and a parallel iterative domain decomposition method with many subdomains and cross points. J. Zou et al developed a class of algebraic subproblem decomposition methods and parallel algorithms for solving nonsymmetric and weakly nonlinear elliptic boundary value problems. The details of their works are presented below:

5.1. Linear problems. Liem et al [66] proposed two synchronous parallel algorithms of domain decomposition: one based on the minimum modulus principle and the other based on the discrete maximum principle. The convergence of both algorithms were proved. Using Lions' framework, Liem et al [67] relaxed some conditions imposed on the boundaries of the subdomains and proved the convergence of the theorems in [66] again. The convergence rate is shown to be $O(1/\log \frac{H}{h})$, where H and h are the diameters of a typical subdomain and an element respectively.

Zhang et al [68] developed a domain decomposition method based on the preconditioned conjugate gradient method. If N is the number of subdomains, the condition number of the preconditioned system does not exceed $O((1+\log N)^3)$. It is completely independent of the mesh size. The number of iterations required to decrease the energy norm of the error by a fixed factor is proportional to $O((1+\log N)^{3/2})$.

Zhang et al [69] also proposed a parallel iterative domain decomposition method with many subdomains and cross points. A subdomain is selected to cover each cross point, which is called covering domain. Covering domains do not intersect each other. The proposed method includes solving Dirichlet problems or mixed problems on subdomains and solving Dirichlet problems on covering domains. It is highly parallelizable and has local communication property.

5.2. Non-linear problems. So far, in the field of domain decomposition method, little work has been done on the non-symmetric elliptic boundary value problems and non-linear elliptic problems.

Zou et al [70] developed a class of algebraic subproblem decomposition methods and parallel algorithms to solve the discretized systems which arise from symmetric or non-symmetric elliptic boundary value problems, weakly nonlinear elliptic problems, linear or nonlinear network problems, Stefan problem, etc. For properly chosen initial iterative vectors with a small cost, the methods can produce monotone decreasing or increasing convergent sequences componentwise. Such property is much useful for error estimates in practical computations.

Liem et al [71] presented two parallel algorithms based on domain decomposition for solving a variational inequality over a closed convex cone and proved their convergence.

6. Computational fluid mechanics

W.H. Hui et al developed a generalized Lagrangian method of description for steady flows. The method was shown to be superior to the Eulerian description in many aspects. With the help of symbolic computation, they applied group theory to obtain similarity solutions for unsteady boundary layer equations and other classical equations. In addition, they studied the behavior of Stokes' series for large amplitude waves using symbolic and numerical computation. J.H. W. Lee et al have made contributions on environmental hydraulics and water quality modeling. The applications include multiple shallow water jet-induced flow, mixing of a round buoyant jet in a current and modeling of tidal circulation and marine water quality. Y.K. Kwok et al analyzed the nature and location of singularities of a perturbation series extended by a computer. The use of the technique to physical problems include subcritical flows over a 2-D airfoil and the transonic controversy. Also, Kwok developed finite difference algorithms for solving hydrodynamic model for the simulation of gas-particle flows in fluidized bed reactors. J.C.H. Fung et al performed calculations on kinematic simulation of flow fields. The details of their works are presented below:

6.1. Generalized lagrangian methods. Steady supersonic flow are traditionally studied using the Eulerian method of describing fluid motion which tends to resolve discontinuities poorly, especially sliplines. Hui et al [72-75] developed a generalized Lagrangian method of description for steady flow. Systematic comparisons between the two methods showed that the generalized Lagrangian method is superior to the Eulerian one in that:

(i) it resolves sliplines crisply,
(ii) its shock resolution improves greatly with increasing Mach number,
(iii) it requires no grid generation, and
(iv) its inherent parallelism of the flow field makes it better suited than the Eulerian one for adopting to parallel computation.

Extension to three-dimensional flow and to subsonic flow are being carried out.

6.2. Similarity solutions via symbolic computation. Similarity solutions to linear or nonlinear partial differential equations can be obtained systematically using group theory. The main idea is to find a Lie group of transformation leaving a given system of PDE invariant. This requires solving a much larger system of linear PDE which is most easily and comprehensively done via symbolic computation, e.g. the Maple. Hui et al [76] have performed the calculations which include Burgers' equation and KdV equation. One significant result is the generation of the whole family of similarity solutions for unsteady boundary layer equations.

6.3. Stokes water waves of large amplitude. In his classic study of finite amplitude periodic gravity water waves, Stokes represented the coordinates x and y as double series in terms of the velocity potential ϕ and the stream function ψ. His method has since been followed by most researchers for over a century. Hui et al [77] studied the behavior of Stokes' series for large amplitude waves, in which a combination of symbolic computation and accurate numerical computation performed on a Cray X-MP was used. Clear evidence was presented showing that the Stokes double series appears to diverge for large amplitude waves. In contrast, it was shown that the new formulation introduced recently by Hui et al [78], in which $y = y(x, \psi)$, produces a series solution that is so well behaved that accurate results of all the wave properties can be obtained with relatively low order partial sums.

6.4. Multiple shallow water jet-induced flow. The predominantly two-dimensional inviscid flow generated by a series of water jets in shallow water, acting as a line momentum source, is computed by a semi-analytical vortex model. The unknown vortex strength along unknown slip streamlines is governed by two nonlinear integral equations. Lee et al [79, 80] obtained the solution by an iterative collocation method and the results showed that the momentum-induced flow separates near, but not exactly at, the two ends of the line source. The computations provided a basic understanding of the mixing induced by such jet groups in shallow coastal waters – often used as an efficient means of discharging cooling water from condenser cooling systems of large thermal power plants. For situations where the flow is misaligned with the jet momentum, the flow has also been successfully computed by a fractional step method with the use of higher order finite elements [81, 82].

6.5. Mixing of a round buoyant jet in a current. A general Lagrangian jet model, JETLAG, has been developed by Lee et al [83, 84] to predict the mixing of a buoyant jet discharge into a stratified tidal current. The unknown three-dimensional jet trajectory, width, velocity, and tracer concentration are simulated by applying discrete mass and momentum balances to a plume element. The essence of the model lies in the accurate determination of the forced entrainment from the ambient flow into the vortex pair flow in the current-dominated regime – the far field.

6.6. Modeling of tidal circulation and marine water quality. Lee et al [85, 86] developed a semi-implicit and a characteristic Galerkin finite element method to reliably compute advection-dominated tidal transport problems. They also developed a general model for simulation of dissolved oxygen and phytoplankton dynamics. The model has been verified for a coastal bay in Hong Kong and successfully applied elsewhere [87, 88]. Finite element models have been applied to predict the diurnal change in the vertical oxygen structure – the computation can often be gainfully used to interpret on-line telemetric data, and hence greatly save the field work required [89]. An expedient means of coupling hydrodynamic and water quality has also been proposed [90].

6.7. Computer extension of perturbation series. Nowadays, we relegate the tedious computation of extending a perturbation series to a computer. When more terms of the perturbation series are available, it becomes plausible to analyze the series to reveal the location and structure of the singularities of the series. Very often, the nature of singularities have strong resemblances to the physics of the problem.

Since the 1940's there has been a controversy over the commencement of shocks in a transonic flow whenever the critical Mach number is exceeded. It is usually assumed that the breakdown of convergence of a perturbation series for some flow quantity is taken to imply the non-existence of smooth solution. In the solution of the complete potential equation for the velocity potential, Kwok et al [91] showed that the phenomenon of shockless transonic flow is dependent on the imposed boundary condition. Kwok [92] also performed the computer extension of the Janzen-Rayleigh series to solve for the steady, inviscid, irrotational subcritical flow over an arbitrary 2-D airfoil up to $O(M^6)$, where M is the Mach number. In addition, Kwok [93] calculated the incompressible flow over a perturbed body and revealed that the location and structure of the nearest singularity of a perturbation series do change with the field point of evaluation and the choice of the physical quantities.

6.8. Simulation of gas-particle flows. The hydrodynamic model for two-phase gas-particle flow in fluidized bed reactor is developed from the basic principles of conservation of mass and momentum. In the model, the two phases are under a common pressure field with non-differential exchange terms to describe the exchanges of mass and momentum between phases. Kwok [94] developed a new finite difference algorithm using the primitive variables formulation to solve for the non-linear system of partial differential equations governing fluidization phenomena.

6.9. Kinematic simulation of flow fields. The eddy structure in time and space and the motion of particles cannot be derived directly from the two-point spectra and correlation functions. They can only be investigated using the velocity field either computed directly (if the Reynolds number is low) or by using an approximate high Reynolds number simulation of velocity field. Fung et al [95] have constructed an isotropic field, given the spectrum $E(k)$. The resulting velocity field can be simulated fairly cheaply and can be tested against experiments and direct numerical simulation.

Acknowledgements We would like to thank our colleagues who have contributed to the contents of this paper. Also, without the encouragement of Professor Chung-Chun Yang, this paper may not come into existence. The first author thanks Miss Odissa Wong for her careful typing of the manuscript.

References

1. R.H. Chan, *Iterative methods for overflow queuing models I*, Numer. Math., **51**(1987), 143–180.
2. R.H. Chan, *Iterative methods for overflow queuing models II*, Numer. Math., **54**(1988), 57–78.
3. R.H. Chan and G. Strang, *Toeplitz equations by conjugate gradient with circulant preconditioner*, SIAM J. Sci. Stat. Comput., **10**(1989), 104–119.
4. R.H. Chan, *The spectrum of a family of circulant preconditioned Toeplitz systems*, SIAM J. Numer. Anal., **26**(1989), 503–506.
5. R.H. Chan, *Circulant preconditioners for Hermitian Toeplitz systems*, SIAM J. Matrix Anal. Appl., **10**(1989), 542–550.
6. R.H. Chan, X.Q. Jin and M.C. Yeung, *The circulant operator in the Banach algebra of matrices*, Linear Algebra and Its Appl., **149**(1991), 41–53.
7. R.H. Chan, X.Q. Jin and M.C. Yeung, *The spectra of super-optimal circulant preconditioned Toeplitz systems*, SIAM J. Numer. Anal., **28**(1991), 871–879.

8. R.H. Chan, *Toeplitz preconditioners for Toeplitz systems with nonnegative generating functions*, IMA J. Numer. Anal., **11**(1991), 333–345.
9. R.H. Chan and X.Q. Jin, *Circulant and skew-circulant preconditioners for skew-Hermitian type Toeplitz systems*, BIT, **31**(1991), 632–646.
10. R.H. Chan, *Numerical solutions for the inverse heat problems in* R^N., to appear in SEA Bull. Math., 1992.
11. R.H. Chan and M.C. Yeung, *Circulant preconditioners for Toeplitz matrices with positive continuous generating functions*, Math. Comput., **58**(1992), 233–240.
12. R.H. Chan and M.C. Yeung, *Jackson's theorem and circulant preconditioned Toeplitz systems*, to appear in J. Approx. Theory, 1992.
13. R.H. Chan and T.F. Chan, *Circulant preconditioners for elliptic problems*, to appear in J. Numer. Linear. Alg. Appl., 1992.
14. R.H. Chan and M.C. Yeung, *Circulant preconditioners constructed from kernels*, to appear in SIAM J. Numer. Anal., 1992.
15. R.H. Chan and X.Q. Jin, *A family of block preconditioners for block systems*, to appear in SIAM J. Sci. Stat. Comput., 1992.
16. R.H. Chan and M.C. Yeung, *Circulant preconditioners for complex Toeplitz matrices*, submitted.
17. R.H. Chan and K.P. Ng, *Fast iterative solvers for Toeplitz-plus-band systems*, submitted.
18. R.H. Chan, J. Nagy and R. Plemmons, *Circulant preconditioned Toeplitz least squares iterations*, submitted.
19. R.H. Chan and P.P. Tang, *Fast band-Toeplitz preconditioners for Toeplitz systems*, submitted.
20. R.H. Chan, J. Nagy and R. Plemmons, *Block circulant preconditioners for 2-dimensional deconvolution problems*, in preparation.
21. X.Q. Jin and R.H. Chan, *Circulant preconditioners for second order hyperbolic and parabolic problems*, to appear in BIT.
22. R.H. Chan, *Domain decomposition methods for queuing networks with irregular statespaces*, to appear in Proc. IMA Workshop on Linear Algebra, Markov Chains and Queuing Models, Minneapolis, Minnesota, 1992.
23. D. Ho and L.R. Fletcher, *Perturbation theory of output feedback pole assignment*, Int. J. Control., **48**(1988), 1075–1088.
24. D. Ho and L.R. Fletcher, *Extending partial pole assignment by output feedback*, in preparation.
25. D. Ho, *A geometric approach and approximation methods in output feedback pole assignment*, submitted.
26. H.C. Huang, *Bounds on condition number of a matrix*, J. Comput. Math., **2**(1984), 356–360.
27. H.C. Huang and W.Z. Gui, *The condition number of mixed finite element approximation for biharmonic equation*, Mathematica Numerica Sinica, **6**(1984), 444–448.
28. Y.K. Kwok and D. Barthez, *An algorithm for the numerical inversion of Laplace transforms*, Inverse Problem, **5**(1989), 1089–1095.
29. J. Caldwell and P. Wanless, *A Fourier series approach to Burgers' equation*, J. Phys. A, Math. Gen., **14**(1981), 1029–1037.
30. J. Caldwell, *Solution of Burgers' equation for large Reynolds number using finite elements with moving nodes*, Appl. Math. Modeling, **11**(1987), 211–214.
31. R. Saunders, J. Caldwell and P. Wanless, *A variational-iterative scheme applied to Burgers' equation*, IMA J. Numer. Anal., **4**(1984), 349–362.
32. J. Caldwell, *Solution of Burgers' equation by Fourier transform methods*, Proc. IMA Conf. on Wavelets, Fractals and Fourier Transforms, Oxford University Press, 1991.
33. T.M. Shih, C.B. Liem and T. Lu, *A fourth order method for the Sturm-Liouville eigenproblem*, SEA Bull. Math., **13**(1989), 115–121.
34. T. Lu, C.B. Liem and T.M. Shih, *A fourth order finite difference method for the boundary value and eigenvalue problems of divergence type elliptic equations*, to appear in SEA Bull. Math.

35. T.M. Shih and C.B. Liem, *The method of operator combination for solving second order elliptic equations*, J. Numer. Meth. Comput. Appl., **7**(1986), 125–129.
36. Y.K. Kwok and K.K. Tam, *Stability analysis of three-level difference schemes for initial-boundary problems for multi-dimensional convective-diffusion equations*, to appear in Comm. in Appl. Numer. Methods.
37. K.M. Liu and E.L. Ortiz, *Numerical solution of eigenvalue problems for partial differential equations with the Tau-lines method*, Comput. Math. Appl., **12**(1986), 1153–1168.
38. K.M. Liu and E.L. Ortiz, *Tau method approximate solution of high order differential eigenvalue problems defined in the complex plane, with an application to Orr-Sommerfeld stability equation*, Comm. in Appl. Numer. Methods, **3**(1987), 187–194.
39. K.M. Liu and E.L. Ortiz, *Tau method approximation of differential eigenvalue problems where the spectral parameter enters nonlinearly*, J. Comput. Phys, **72**(1987), 299–310.
40. K.M. Liu, *Numerical solution of differential eigenvalue problems with variable coefficients with the Tau collocation method*, Math. Comput. Modeling, **11**(1988), 672–675.
41. K.M. Liu and E.L. Ortiz, *Numerical solution of ordinary and partial functional differential eigenvalue problem with the Tau method*, Computing, **41**(1989), 205–217.
42. H.C. Huang, J.X. Wang, J.Z. Cui, Z.K. Lin and J.F. Zhao, *Variational difference method for solving plane elasticity problem*, Appl. Comput. Math. **3**:2(1966).
43. H.C. Huang and G.Y. Liu, *Solving elliptic boundary value problems by successive mesh subdivisions*, Mathematica Numerica Sinica, (1978), No.2, 41–52; No.3, 28–35.
44. H.C. Huang, W.N. E and M. Mu, *Extrapolation combined with MG method for solving finite element equation*, J. Comput. Math., **4**(1986), 362–367.
45. M. Mu and H.C. Huang, *MG method with extrapolation for calculating stress intensity factors on re-entrant domains*, Chinese J. Numer. Math. Appl., **12**(1990), 34–41.
46. H.C. Huang, M. Mu and W.M. Han, *Asymptotic expansion for numerical solution of less regular problems in a rectangle*, Mathematica Numerica Sinica, **8**(1986), 217–224.
47. H.C. Huang and W.N. E, *A posteriori error analysis of the F.E.M. for 1D boundary value problems*, Chinese Quarterly J. Math., **2**(1987), 43–47.
48. W.N. E and H.C. Huang, *A posteriori error analysis of the F.E.M. for 2D boundary value problems*, Chinese Quarterly J. Math., **3**(1988), 97–107.
49. J. Zou and H.C. Huang, *A posteriori error analysis of the h-p version of F.E.M. in 1D*, Chinese J. Numer. Math. Appl., **12**(1990), 75–90.
50. M. Mu and H.C. Huang, *A new measure of numerical stability for the discretized system of elliptic boundary value problems*, Mathematica Numerica Sinica, **11**(1989), 298–302.
51. S.S. Chow, G.F. Carey and M.K. Seager, *Approximate boundary-flux calculations*, Computer Methods in Appl. Mech. Eng., **45**(1985), 107–120.
52. S.S. Chow, *Finite element error estimate for nonlinear elliptic equations of monotone type*, Numer. Math., **54**(1989), 373–383.
53. Y.K. Cheung and S. Swaddiwudhipong, *Analysis of frame shear wall structures using finite strip elements*, Proc. Institution of Civil Engineers, Part 2, **65**(1978), 517–535.
54. H.C. Chan and Y.K. Cheung, *Lateral and torisonal analysis of spatial wall systems using higher order elements*, The Structural Engineer, **58B**:3(1990), 67–71.
55. Y.K. Cheung and S.C. Fan, *Static analysis of right box girder bridges by spline finite strip method*, Proc. Institution of Civil Engineers, Part 2, Research and Theory, Technical Note 363, 75(1983), 311–323.
56. W.Y. Li, Y.K. Cheung and L.G. Tham, *Spline finite strip analysis of general plates*, J. Engineering Mechanics, ASCE, **112**(1986), 43–54.
57. D.S. Zhu and Y.K. Cheung, *Postbuckling analysis of shells by spline finite strip method*, Computers and Structures, **31**(1989), 357–364.
58. S.L. Lau and Y.K. Cheung, *Amplitude incremental variational principle for nonlinear vibration of elastic systems*, Trans. Amer. Soc. Mech. Engineers, J. Appl. Mech., **48**(1981), 959–964.
59. S.L. Lau, Y.K. Cheung and S.Y. Wu, *Incremental harmonic balance method with multiple time scales for aperiodic vibration of nonlinear systems*, Trans. Amer. Soc. Mech. Engineers, J. Appl. Mech., **50**(1983), 871–876.

60. S.L. Lau and Y.K. Cheung, *Incremental Hamilton's principle with multiple time scales for nonlinear aperiodic vibrations of shells*, Trans. Amer. Soc. Mech. Engineers, J. Appl. Mech., **53**(1986), 456–466.
61. C.W. Cai, Y.K. Cheung and H.C. Chan, *Dynamic response of infinite continuous beams subjected to a moving force — an exact method*, J. Sound and Vibration, **123**(1988), 461–472.
62. Y.K. Cheung, H.C. Chan and C.W. Cai, *Exact method for static analysis of periodic structures*, J. Engineering Mechanics, ASCE, **115**(1989), 415–434.
63. C.W. Cai, Y.K. Cheung and H.C. Chan, *Uncoupling of dynamic equations for periodic structures*, J. Sound and Vibration, **139**(1990), 253–263.
64. Y.C. Hon, *Typhoon surge in Tolo Harbour of Hong Kong*, Chinese J. Numer. Math. Appl., **13**(1991), 173–179.
65. W.M. Xue, Y.C. Hon, H.C. Huang and D.T. Huang, *An algorithm of finite element grid generation for three-dimensional complex connected domain*, to appear in Acta Mechanica Solida Sinica, **5**:4(1992).
66. C.B. Liem, T.M. Shih and T. Lu, *Parallel algorithms for solving partial differential equations*, SIAM Proc. Second Inter. Symposium on Domain Decomposition Methods, LA. Calif., 1989.
67. C.B. Liem, T.M. Shih and T. Lu, *A synchronous domain decomposition method with an estimate of convergence rate*, submitted.
68. S. Zhang and H.C. Huang, *Multigrid multi-level domain decomposition*, J. Comput. Math., **9**(1991), 17–27.
69. S. Zhang and H.C. Huang, *Parallel iterative domain decomposition method for elliptic equations*, Scientia Sinica, (1992), 1233–1242.
70. J. Zou and H.C. Huang, *Algebraic subproblem decomposition methods and parallel algorithms with monotone convergence*, to appear in J. Comput. Math.
71. C.B. Liem, T.M. Shih and T. Lu, *Parallel algorithms for variational problems based on domain decomposition*, submitted.
72. C.Y. Loh and W.H. Hui, *A new Lagrangian method for steady supersonic flow computation, Part I: Godunov scheme*, J. Comput. Physics, **89**(1990), 207–240.
73. W.H. Hui and C.Y. Loh, *A new Lagrangian method for steady supersonic flow computation, Part II: slip-line resolution*, to appear in J. Comput. Physics, 1992.
74. W.H. Hui and C.Y. Loh, *A new Lagrangian method for steady supersonic flow computation, Part III: strong shocks*, to appear in J. Comput. Physics 1992.
75. W.H. Hui and Y. Zhao, *A generalized Lagrangian method for gas dynamics*, Proc. 2nd North American-Soviet Union Workshop on Comput. Aerodynamics, Montreal, Canada, 1991.
76. P.K.H. Ma and W.H. Hui, *Similarity solutions of the two-dimensional unsteady boundary layer equations*, J. Fluid Mechanics, **216**(1990), 537–559.
77. W.H. Hui and G. Tenti, *A new approach to steady flows with free surfaces*, ZAMP, **33**(1982), 569–589.
78. W.M. Drennan, W.H. Hui and G. Tenti, *Accurate calculations of Stokes water waves of large amplitude*, ZAMP, **43**(1992).
79. J.H.W. Lee and M. Greenberg, *Line momentum source in shallow inviscid fluid*, J. Fluid Mechanics, **145**(1984), 287–304.
80. J.H.W. Lee, *Boundary effects on a submerged jet group*, J. Hydraulic Research, **22**(1984), 199–216.
81. Y.K. Cheung, J.H.W. Lee and C.W. Li, *Flow induced by a fanned out jet group*, Proc. 2nd Int. Conf. on Advances in Numerical Methods, Swansea, **2**(1987), T42-1/T42-8.
82. C.W. Li and J.H.W. Lee, *Line momentum source in crossflow*, Int. J. Engineering Science, **29**(1991), 1409–1418.
83. J.H.W. Lee and V. Cheung, *Generalized Lagrangian model for buoyant jets in a current*, J. Environmental Engineering, ASCE, **116**(1990), 1085–1106.
84. W.T. Lee and J.H.W. Lee, *Computation of an axisymmetric vertical buoyant jet in lateral confinement*, Proc. Asian Conf. Comput. Mech., Hong Kong, **2**(1991), 1713–1719.

85. C.W. Li, J.H.W. Lee and Y.K. Cheung, *Mathematical model study of tidal circulation in Tolo Harbour, Hong Kong: development and verification of a semi-implicit finite element scheme*, Proc. Institution of Civil Engineers, Part 2, **81**(1986), 569–592.
86. J.H.W. Lee , J. Peraire and O.C. Zienkiewicz, *The characteristic Galerkin method for advection-dominated problem - an assessment*, Computer Meths. Appl. Mech. Eng., **61**(1987), 359–369.
87. J.H.W. Lee , R. Wu and Y.K. Cheung, *Forecasting of dissolved oxygen in a marine fish culture zone*, J. Environmental Engineering, ASCE, **117**(1991), 816–833.
88. J.H.W. Lee, A. Hirayama and H.W. Lee, *Short term dissolved oxygen dynamics in eutrophic semi-enclosed bay*, J. Coastal Engineering, Japan Society of Civil Engineers, **38**(1991), 861–865.
89. J.H.W.Lee, Y.K.Cheung and H.S. Lee, *Simulation of vertical structure of dissolved oxygen in a coastal bay*, Proc. Int. Symposium on Environmental Hydraulics, Hong Kong, **2**(1991), 1057–1062.
90. K.W. Chan, J.H.W. Lee and O.C. Zienkiewicz, *Tidal transport calculations with the characteristic Galerkin method*, Proc. Int. Symposium on Environmental Hydraulics, Hong Kong, **2**(1991), 1029–1034.
91. Y.K. Kwok and L. Sirovich, *On some aspects of the transonic controversy*, SIAM J. Appl. Math., **47**(1987), 279–295.
92. Y.K. Kwok, *A regular perturbation method for subcritical flow over a two-dimensional airfoil*, IMA J. Appl. Math., **43**(1989), 71–80.
93. Y.K. Kwok, *Structures of the nearest singularities of a regular perturbation series*, Comm. in Appl. Numer. Math., **7**(1991), 19–28.
94. Y.K. Kwok, *Theoretical considerations for finite difference algorithms for simulation of gas-particle flows*, Proc. Asian Pacific Conf. on Comput. Mech. **2**(1991), 1583–1588.
95. J.C.H. Fung, J.C.R. Hunt, N.A. Malik and R.J. Perkins, *Kinematics simulation of homogeneous turbulent flows generated by random Fourier modes*, to appear in J. Fluid Mechanics.

Department of Mathematics, Hong Kong University of Science and Technology, Clear Water Bay Road, Hong Kong

Department of Mathematics, Hong Kong Baptist College, Waterloo Road, Hong Kong

Department of Mathematics, Hong Kong University, Pokfulam Road, Hong Kong

Contemporary Mathematics
Volume **163**, 1994

Interpolated Finite Elements and Global Error Recovery

LIN QUN

ABSTRACT. This report surveys some theoretical and applicable results on the interpolated finite elements in China. The optimal meshes and a higher interpolation of lower finite element data are suggested to recover the global convergent orders of the finite element solution and its gradient.

1. Introduction

The finite element method (FEM) is nothing but the partial differential equation of a weak form restricted on a finite element space. See Feng Kang [13] in 1965 for a systematic study of conforming FEM and see Shi Zhongci in this volume for the new development of nonconforming FEM. This report is concerned with the conforming FEM.

There are many reports concerning the analysis of conforming FEM (c.f. Strang and Fix [65] in 1973, Shaidurov [64] in 1989, Babuska and Szabo [1] and Ciarlet [12] in 1991). But, there still remain some fundamental problems such as

(i) What is the optimal shape of mesh structure and whether it is better to use parallelograms or triangles?

(ii) How to interpolate the finite element data and whether it is better to interpolate a higher polynomial from lower finite element data?

Let us consider the first problem. Actually, we have shown in [31, 28] that different mesh shapes give the different convergent orders of the finite element solution. The irregular mesh leads only a rudimentary convergent order, even divergence. In order to recover the global convergent order of the finite element solution, we defined the optimal meshes [28, 31], which are discussed below.

1991 *Mathematics Subject Classification.* **65N30, 65N50, 65N15, 65B05.**

1.1. Superconvergence for interpolation error functionals. The convergent order of FEM is related to certain interpolation error functionals, e.g.,

$$\int_\Omega (u - i_h u)_x \varphi_x, \ \int_\Omega (u - i_h u)_x \varphi, \ \int_\Omega (u - i_h u)_x \varphi_y, \tag{1}$$

where Ω is a domain, u a smooth function, $i_h u$ the finite element interpolation and φ the finite element test function. Their convergent orders depend on the mesh shape. There are two extremes. For the irregular mesh, functionals (1) do not even converge: e.g., in the bilinear case

$$\int_\Omega (u - i_h u)_x \varphi_x = O(h)||\varphi||_1 = O(1)||\varphi||_0 \,.$$

For the mesh of rectangular shape, i.e.,

0) *Rectangular mesh*: functionals (1) have order two superconvergence, e.g.,

$$\int_\Omega (u - i_h u)_x \varphi_x = O(h^2)||\varphi||_0.$$

For the meshes of different shapes, functionals (1) have different orders:

1) *Deformed rectangular mesh*, i.e., mapped from a rectangular mesh: functionals (1) have order one superconvergence;

2) *Piecewise deformed rectangular mesh*, i.e., composed of several deformed rectangular meshes: functionals (1) have nearly order one superconvergence;

3) *Mostly (deformed) rectangular mesh*, i.e., the (deformed) rectangular mesh inserting a few layers of ir-rectangular elements: functionals (1) have a half order superconvergence.

We call them optimal meshes since they guarantee the superconvergence of functionals (1), which are observed in [24, 29–31] by constructing certain integral identities.

We list below, in the bilinear case, how the different mesh shapes give the different orders of functionals (1), bounded in L_2- and in H_1-norms, respectively.

Error Mesh	$\int (i_h u - u)_x \varphi_x$	$\int (i_h u - u)_x \varphi$	$\int (i_h u - u)_x \varphi_y$
0)	$O(h^2)\|\|\varphi\|\|_0$ if $\frac{\partial u}{\partial n} = 0$ (or $\varphi = 0$) on $\partial\Omega$	$O(h^2)\|\|\varphi\|\|_0$ if $\varphi = 0$ on $\partial\Omega$ and almost uniform	$O(h^2)\|\|\varphi\|\|_0$ if $\varphi = 0$ on $\partial\Omega$ and almost uniform
1) a.u.	$O(h)\|\|\varphi\|\|_0$	$O(h^2)\|\|\varphi\|\|_0$ if $\varphi = 0$ on $\partial\Omega$	$O(h)\|\|\varphi\|\|_0$
2) a.u.	$O(h)\|\|\varphi\|\|_0$	$O(h^{1.5})\|\|\varphi\|\|_0$	$O(h)\|\|\varphi\|\|_0$
3)	$O(h^{0.5})\|\|\varphi\|\|_0$	$O(h^{1.5})\|\|\varphi\|\|_0$	$O(h^{0.5})\|\|\varphi\|\|_0$
Arbitrary	$O(1)\|\|\varphi\|\|_0$	$O(h)\|\|\varphi\|\|_0$	$O(1)\|\|\varphi\|\|_0$

Error / Mesh	$\int (i_h u - u)_x \varphi_x$	$\int (i_h u - u)_x \varphi$	$\int (i_h u - u)_x \varphi_y$
0)	$O(h^2)\|\|\varphi\|\|_1$	$O(h^2)\|\|\varphi\|\|_1$	$O(h^2)\|\|\varphi\|\|_1$ if $\varphi = 0$ on $\partial\Omega$
1)	$O(h^2)\|\|\varphi\|\|_1$ if $\varphi = 0$ on $\partial\Omega$	$O(h^2)\|\|\varphi\|\|_1$	$O(h^2)\|\|\varphi\|\|_1$ if $\varphi = 0$ on $\partial\Omega$
2)	$O(h^{1.5})\|\|\varphi\|\|_1$, $O(h^2 \lvert \ln h \rvert^{1/2})\|\|\varphi\|\|_1$ if a.u.	$O(h^2)\|\|\varphi\|\|_1$	$O(h^{1.5})\|\|\varphi\|\|_1$, $O(h^2 \lvert \ln h \rvert^{1/2})\|\|\varphi\|\|_1$ if a.u.
3)	$O(h^{1.5})\|\|\varphi\|\|_1$	$O(h^{1.5})\|\|\varphi\|\|_1$	$O(h^{1.5})\|\|\varphi\|\|_1$
Arbitrary	$O(h)\|\|\varphi\|\|_1$	$O(h)\|\|\varphi\|\|_1$	$O(h)\|\|\varphi\|\|_1$

The proofs will be given in Sections 2 and 3.

1.2. Global recovery of finite element solution. Above superconvergence for functionals (1) recovers the global L_2-convergent orders of various finite element solutions u_h, v_h and P_h: in the bilinear case,

Mesh	Biharmonic	Hyperbolic	Stokes
	$v_h - \nabla^2 u$	$u_h - u$	$P_h - P$
Almost uniform rectangular	$O(h^2)$	$O(h^2)$	$O(h^2)$
A.U. deformed rectangular	$O(h)$	$O(h^2)$	$O(h^{1.5})$
Piecewise a.u. deformed rectangular	$O(h)$	$O(h^{1.5})$	$O(h^{1.5})$
Mostly rectangular	$O(h^{1/2})$	$O(h^{1.5})$	$O(h^{1.5})$
Arbitrary	$O(1)$	$O(h)$	$O(h)$

The proofs will be given in Section 4.2.

The above table answers the first problem.

Now we consider the second problem.

1.3. Global recovery of finite element gradient. Superconvergence for functionals (1) is not enough to recover the finite element gradient. We need an additional recovery process, i.e., constructing a higher interpolation. This was done by [29] and is presented as follows.

Let T_h be the rectangular mesh of size h, p_i the nodal points of T_h and φ_i the bilinear basis functions. Then, the natural interpolation of u_h reads:

$$u_h = \sum u_h(p_i)\varphi_i,$$

i.e., the bilinear finite element data multiplied by the bilinear basis. Such an interpolation process only has a first order estimate in H_1-norm:

$$||u_h - u||_1 = O(h).$$

In order to recover the above estimate we will use the biquadratic interpolation. Suppose that T_h is obtained from a coarse mesh T_{2h} by subdividing each element into four equal rectangles and that $\tilde{\varphi}_i$ are the biquadratic basis functions over coarse mesh T_{2h}. Introducing the biquadratic interpolation

$$\tilde{u}_h = \sum u_h(p_i)\tilde{\varphi}_i,$$

i.e., the bilinear finite element data multiplied by the biquadratic basis, we get a second order estimate (see Section 4.3 for a proof):

$$||\tilde{u}_h - u||_1 = O(h^2).$$

We list below, in the bilinear case, the global recovery of H_1-estimate by using the higher interpolation $\tilde{u}_h$ (or $\tilde{u}^h$ for eigensolution).

Mesh	Lame equs	$\Delta u = \lambda u$		Δ^2	Stokes
	$\tilde{u}_h - u$	$\tilde{u}^h - u$	$\frac{\lVert \tilde{u}^h \rVert_1^2}{\lVert \tilde{u}^h \rVert_0^2} - \lambda$	$\tilde{u}_h - u$	$\tilde{u}_{h/2} - u$
0)	h^2	h^2	h^4		
0) & a. u.				h^2	h^2
1)	h^2	h^2	h^4		
1) & a. u.				$h^{1.5}$	h^2
2)	$h^{1.5}$	$h^{1.5}$	h^3		
2) & a. u.	$h^2\lvert\ln h\rvert^{1/2}$	$h^2\lvert\ln h\rvert^{1/2}$	$h^4\lvert\ln h\rvert$	$h^{1.5}$	h^2
3)	$h^{1.5}$	$h^{1.5}$	h^3	$h^{0.75}$	h^2
Arbitrary	h	h	h^2	O(1)	h

The proof of eigenvalue problem will be given in Section 6.

Above table answers the second problem, i.e., the higher interpolation of lower finite element data achieves a higher order convergence than the natural interpolation.

1.4. New concept of superconvergence. Superconvergence results based on the standard concept have been surveyed by Krizek and Neittaanmaki in [21]. There are many works in China in this direction after Chen and Zhu. We refer to books [4] and [72] for the history and references, in particularly to Zhu's work on quadratic elements and Chen's on biquadratic elements. The earlier results on superconvergence were presented only on the uniform (or almost uniform) mesh. An essential relaxation on the mesh structure is due to the "piecewise" concept introduced by Lin and Xu [49], Lin and Lu [41, 40]. After them, most of superconvergence results are presented on the piecewise uniform mesh, see Chen, Huang, Wahlbin, Xie, Zhu, et al. [5, 7, 10, 18, 19, 35, 39, 45, 46, 47, 59, 60, 66, 72]. Xie, Huang, Chen, et al. [69, 48, 14, 19, 11, 25–27] also considered the superconvergence for corner singular problems.

The standard concept of superconvergence concentrated on finding exceptional stress points in interior region. The new concept of superconvergence is based on the higher interpolation stated in Section 1.3, by which the superconvergence is not an exceptional but a global phenomenon. Such a new concept makes the superconvergence theory to be direct and better understood.

2. Integral identities for rectangular mesh

As stated in Section 1.1, integrals (1) achieve the optimal order when the rectangular mesh is imposed on the domain. Such results have been observed in [24, 29, 30, 48] by building certain integral identities. Such integral identities are basic for optimal analysis and it is worth to present in details.

For a fixed *rectangular element*

$$\tau = [x_\tau - h_\tau, x_\tau + h_\tau] \times [y_\tau - k_\tau, y_\tau + k_\tau]$$

define the interpolation error functions

$$E(x) = \frac{1}{2}((x - x_\tau)^2 - h_\tau^2), \ F(y) = \frac{1}{2}((y - y_\tau)^2 - k_\tau^2).$$

Then, exploiting the specialness of error functions and test function, for the bilinear case, we have the following explicit identities (see Section 4.1 for a proof):

$$\int_\tau (u - i_h u)_x \varphi_x = \int_\tau (F\varphi_x - \frac{1}{3}(F^2)_y \varphi_{xy}) u_{xyy}, \tag{2}$$

$$\int_\tau (u - i_h u)_x \varphi = \int_\tau \left(\left[F(\varphi - E_x \varphi_x) - \frac{1}{3}(F^2)_y (\varphi_y - E_x \varphi_{xy}) \right] u_{xyy} - E\varphi_x u_{xx} \right), \tag{3}$$

$$\int_\tau (u - i_h u)_x \varphi_y = \int_\tau (F(\varphi_y - E_x \varphi_{xy}) u_{xyy} - E\varphi_{xy} u_{xx}), \tag{4}$$

i.e., the finite element interpolation $i_h u$ is expressed by error functions E and F. Their orders are respectively

$$\begin{aligned}
(2) &= O(h^2)||\varphi||_{1,\tau}||u||_{3,\tau}, \\
(3) &= O(h^2)(||\varphi||_{0,\tau}||u||_{3,\tau} + ||\varphi||_{1,\tau}||u||_{2,\tau}), \\
(4) &= O(h^2)(||\varphi||_{1,\tau}||u||_{3,\tau} + ||\varphi||_{2,\tau}||u||_{2,\tau}).
\end{aligned}$$

The real role of explicit identities (2)–(4), however, is to reduce the norm $||\varphi||_1$ to $||\varphi||_0$. This can be achieved by integration by parts.

Integrating (2) by parts and summing for τ lead to the further identity with the boundary integrals defined on the boundary parts L_i of the domain Ω:

$$\begin{aligned}
\int_\Omega (u - i_h u)_x \varphi_x = &\left(\int_{L_1} - \int_{L_2} \right) \left(F\varphi - \frac{1}{3}(F^2)_y \varphi_y \right) u_{xyy} dy \\
&- \int_\Omega \left(F\varphi - \frac{1}{3}(F^2)_y \varphi_y \right) u_{xxyy}.
\end{aligned} \tag{5}$$

Integrating (3) by parts and summing for τ lead to the further identity with the Abel summation of the edge integrals defined on the common edge l of two adjacent elements τ and τ' :

$$\begin{aligned}\int_\Omega (u-i_hu)_x\varphi = &\left(\int_{L_1}-\int_{L_2}\right)\frac{1}{3}h_\tau^2\varphi u_{xx}dy\\ &+\sum_l\int_l\frac{1}{3}(h_\tau^2-h_{\tau'}^2)\varphi u_{xx}dy\\ &+\int_\Omega\left[\left(\frac{1}{6}(E^2)_x\varphi_x-\frac{1}{3}(h_\tau)^2\varphi\right)u_{xxx}+[F(\varphi-E_x\varphi_x)\right.\\ &\left.-\frac{1}{3}(F^2)_y(\varphi_y-E_x\varphi_{xy})\Big]u_{xyy}\right],\end{aligned} \tag{6}$$

where

$$E=\frac{1}{6}(E^2)_{xx}-\frac{1}{3}h_\tau^2,\quad F=\frac{1}{6}(F^2)_{yy}-\frac{1}{3}k_\tau^2.$$

Same process for (4) leads to a similar but much longer identity:

$$\begin{aligned}\int_\Omega(u-i_hu)_x\varphi_y = &(\int_{L_3}-\int_{L_4})E\varphi_xu_{xx}dx\\ &+\int_\Omega(E\varphi_xu_{xxy}+F(\varphi_y-E_x\varphi_{xy})u_{xyy})\\ &=-(\int_{L_3}-\int_{L_4})E\varphi_xu_{xx}dx\\ &-(\int_{L_1}-\int_{L_2})\frac{1}{3}h_\tau^2\varphi u_{xxy}dy-(\int_{L_3}-\int_{L_4})\frac{1}{3}k_\tau^2\varphi u_{xyy}dx\\ &-\sum_l\int_l\frac{1}{3}\left(h_\tau^2-h_{\tau'}^2\right)\varphi u_{xxy}dy-\sum_l\int_l\frac{1}{3}\left(k_\tau^2-k_{\tau'}^2\right)\varphi u_{xyy}dx\\ &-\int_\Omega[(\frac{1}{6}(E^2)_x\varphi_x-\frac{1}{3}(h_\tau)^2\varphi)u_{xxxy}\\ &+(\frac{1}{6}(F^2)_y\varphi_y-\frac{1}{3}(k_\tau)^2\varphi)u_{xyyy}-FE\varphi_{xy}u_{xxyy}]\end{aligned} \tag{7}$$

with the same meaning of E and F as above. One may integrate by parts again the first boundary integral in the right hand side. This leads to the Abel summation of the function values at the nodal points p on the boundary of the domain, e.g.,

$$(\sum_{p\in L_3}-\sum_{p\in L_4})\frac{1}{3}\left(h_\tau^2-h_{\tau'}^2\right)\varphi(p)u_{xx}(p)\ ,$$

and the boundary integrals $(\int_{L_3}-\int_{L_4})(\frac{1}{6}(E^2)_x\varphi_x-\frac{1}{3}(h_\tau)^2\varphi)u_{xxx}dx$, etc.

Explicit identities (5)–(7) are the basis for achieving optimal order.

Boundary integrals contained in identities (5)–(7) vanish by using boundary conditions. For the Abel summation of the edge integrals defined on the common edge of two adjacent elements, we need the assumption of almost uniformity, i.e.,

$$|h_\tau-h_{\tau'}|+|k_\tau-k_{\tau'}|=O(h^2),$$

to achieve the optimal order:

$$\int_\Omega (u - i_h u)_x \varphi_x = O(h^2)||\varphi||_0||u||_4 \text{ if } \tfrac{\partial u}{\partial n} = 0 \text{ (or } \varphi = 0) \text{ on } \partial\Omega;$$
$$\int_\Omega (u - i_h u)_x \varphi = O(h^2)||\varphi||_0||u||_3$$

if mesh is almost uniform and $\varphi = 0$ on $\partial\Omega$;

$$\int_\Omega (u - i_h u)_x \varphi_y = O(h^2)||\varphi||_0||u||_4$$

if mesh is almost uniform and $\varphi = 0$ on $\partial\Omega$;

$$= O(h^2)||\varphi||_1||u||_3 \text{ if } \varphi = 0 \text{ on } \partial\Omega,$$

here, the edge integral is directly estimated by the element integral:

$$||\varphi||_l \leq ch^{-1/2}||\varphi||_\tau \quad \forall\, \varphi \in S_h, \quad ||u||_l \leq ch^{-1/2}||u||_{1,\tau}.$$

Even without the boundary condition we also have the sharper estimates:

$$\begin{aligned}
\int_\Omega (u - i_h u)_x \varphi_x &= O(h^{1.5})||\varphi||_0||u||_4 \\
&= O(h^2)||\varphi||_1||u||_3, \\
\int_\Omega (u - i_h u)_x \varphi &= O(h^{1.5})||\varphi||_0||u||_3 \text{ if mesh is almost uniform}, \\
&= O(h^2)||\varphi||_1||u||_3, \\
\int_\Omega (u - i_h u)_x \varphi_y &= O(h)||\varphi||_0||u||_4 \text{ if mesh is almost uniform}, \\
&= O(h^2|lnh|^{1/2})||\varphi||_1||u||_4 \text{ if mesh is almost uniform}, \\
&= O(h^{1.5})||\varphi||_1||u||_3 .
\end{aligned}$$

3. From rectangular mesh to optimal meshes of different shapes

The rectangular mesh can not adapt to a general domain. For the domains with different shapes we need the optimal meshes of different shapes constructed in [31, 28 or 23].

1) *Deformed rectangular mesh.*

Suppose we have an arbitrary (convex) quadrilateral domain D. Since the basis function is defined by the bilinear function on the standard element it is natural to use the transformation from the map D into a rectangular domain Ω. Consider a deformed rectangular mesh on D, i.e., the quadrilateral mesh constructed by connecting the equi-proportional points of two opposite boundaries (see [23]). Then each quadrilateral element e in D is mapped into a rectangular element τ in Ω and the associated integral on e is mapped into the integral on τ with the variable coefficient $r_e(x,y)$ consisted of Jacobi and the derivative:

$$\begin{aligned}\int_e (U - i_h U)_\xi \Phi_\eta &= \int_\tau r_e(x,y)(u - i_h u)_x \varphi_y + \cdots \\ &= \int_\tau r_e^*(u - i_h u)_x \varphi_y + \cdots + O(h^2)||\varphi||_{1,\tau}||u||_{2,\tau}\end{aligned}$$

where the function $r_e(x,y)$ is approximated by the constant r_e^* characterized by the slopes (a_e, b_e) of two mid-lines of e, and the remainder can be considered as the variable coefficient pollution. We obtain from (7)

$$\int_\tau r_e^*(u - i_h u)_x \varphi_y = -(\int_{l_3} - \int_{l_4}) r_e^* E \varphi_x u_{xx} dx + O(h^2)||\varphi||_{1,\tau}||u||_{3,\tau}.$$

Summing for τ leads to the Abel summation of the edge integrals defined on the common edge of two adjacent elements:

$$\begin{aligned}|\sum_l \int_l (r_e^* - r_{e'}^*) E \varphi_x u_{xx} dx| &\le C \max_e (|a_e - a_{e'})| + |b_e - b_{e'}|) h ||\varphi||_1 ||u||_3 \\ &= O(h^2)||\varphi||_1||u||_3. \end{aligned} \tag{8}$$

Then we have the sharper estimates

$$\begin{aligned}\int_D (U - i_h U)_\xi \Phi_\eta &= O(h)||\Phi||_0||U||_4 \text{ if mesh is almost uniform,} \\ &= O(h^2)||\Phi||_1||U||_3 \text{ if } \Phi = 0 \text{ on } \partial D, \\ &= O(h^2|\ln h|^{1/2})||\Phi||_1||U||_4 \text{ if mesh is almost uniform,} \\ &= O(h^{1.5})||\Phi||_1||U||_3, \end{aligned} \tag{9}$$

and, with the same argument, the sharper estimates

$$\int_D (U - i_h U)_\xi \Phi = O(h^2)||\Phi||_0||U||_3$$

if mesh is almost uniform and $\Phi = 0$ on ∂D,

$$\begin{aligned}&= O(h^{1.5})||\Phi||_0||U||_3 \text{ if mesh is almost uniform,} \\ &= O(h^2)||\Phi||_1||U||_3. \end{aligned} \tag{10}$$

Also from (8) we can see that, for achieving an optimal order, the mid-lines of two adjacent elements must be parallel almost, i.e., the element is an almost *parallelogram.*

This answers the first problem, i.e., the almost parallelogram elements are the main shape for optimal meshes.

An essential relaxation on the mesh structure is the following

2) *Piecewise deformed rectangular mesh.*

Suppose we have a polygonal domain D. We first divide D into several convex quadrilateral subdomains D_i, and each of them is then divided into the deformed rectangular mesh (see [23]). Since there is no boundary condition imposed on the boundary of D_i, for the biharmonic equation with Herrmann-Miyoshi scheme, for the first order hyperbolic equation with standard Galerkin scheme and for the Stokes equation with $Q_1 - Q_1$ scheme, we obtain by (9) and (10) that the almost uniform piecewise deformed rectangular mesh leads to an error of a half (even one) order higher than that for the mesh of arbitrary shape.

For a curved domain D, one may use the mesh of Babuska type discussed by Lin and Xie [47], Chen and Lin [10].

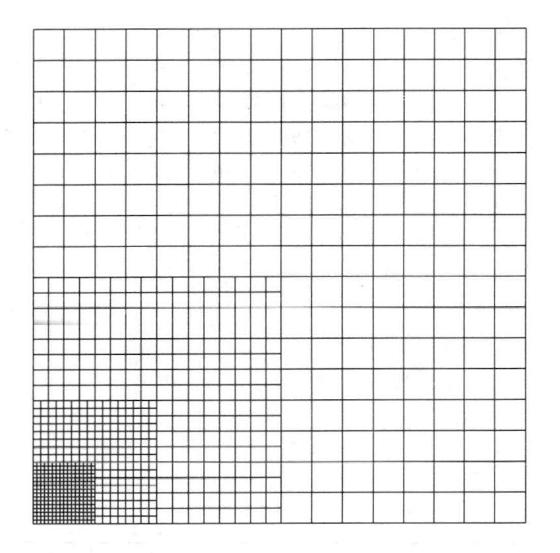

The further generalization of the mesh structure is the (piecewise) rectangular meshes inserting a few boundary layers of ir-rectangular elements (between adjacent pieces).

3) *Mostly rectangular mesh.*

For a complicated domain D which can not realize the piecewise deformed rectangular mesh as in 2), we can make a rectangle $\Omega \supset D$ and use a rectangular mesh to cover D and make some modification on the boundary of D, we thus get the mostly rectangular mesh. Denote by Ω_0 the collection of those rectangular elements contained completely in D, then, by the ir-rectangular area pollution, the error loses a half order:

$$\begin{aligned}\int_D (u - i_h u)_x \varphi_x &= \int_{\Omega_0} (u - i_h u)_x \varphi_x + \int_{D\backslash\Omega_0} (u - i_h u)_x \varphi_x \\ &= O(h^2)||\varphi||_{1,\Omega_0}||u||_{3,\Omega_0} + O(h^{1.5})||\varphi||_{1,D\backslash\Omega_0}||u||_{2,\infty,D\backslash\Omega_0} \\ &= O(h^{1.5})||\varphi||_1(||u||_3 + ||u||_{2,\infty}), \qquad (11)\end{aligned}$$

$$\int_D (u - i_h u)_x \varphi = \int_{\Omega_0} (u - i_h u)_x \varphi + \int_{D\backslash\Omega_0} (u - i_h u)_x \varphi$$
$$= O(h^{1.5}) ||\varphi||_0 (||u||_3 + ||u||_{2,\infty}), \tag{12}$$
$$\int_D (u - i_h u)_x \varphi_y = \int_{\Omega_0} (u - i_h u)_x \varphi_y + \int_{D\backslash\Omega_0} (u - i_h u)_x \varphi_y$$
$$= O(h^{1.5}) ||\varphi||_{1,\Omega_0} ||u||_{3,\Omega_0} + O(h^{1.5}) ||\varphi||_{1,D\backslash\Omega_0} ||u||_{2,\infty,D\backslash\Omega_0}$$
$$= O(h^{1.5}) ||\varphi||_1 (||u||_3 + ||u||_{2,\infty}) \tag{13}$$

but they still have the superconvergence.

Our mostly rectangular mesh includes also the local refinement mesh of optimal size-shape structure, such as that indicated in the figure above, for the domain with a reentrant corner *without knowing the singular exponent.*

Initially, the rectangular domain Ω with size L is covered by a uniform mesh of size h. Consider the subdomain Ω_0 (the quarter of Ω) with size $L/2$ and the reentrant corner of Ω. The element in Ω_0 is then refined into four elements. Repeating the procedure we obtain the subdomain Ω_1 (the quarter of Ω_0) with size $L/2^2$ and a refinement mesh of size $h/2^2, \cdots$, until Ω_{k-1} with size $L/2^k$ and a refinement mesh of size $h/2^k$. Thus, we get a piecewise rectangular mesh, each piece of which is an L-shape subdomain. Although each piece is the quarter of the adjacent piece they have the same number of rectangular elements. Some irregular points appear on the boundaries of two adjacent pieces but they can be avoided by, e.g., connecting each irregular point with vertices of the adjacent element. This causes boundary layers of triangular elements between two adjacent pieces but all of them only capture a small region: let D_0 denote all triangular elements, then

$$\text{means}\,(D_0) \leq 2 \times h \times \frac{L}{2} + 2 \times \frac{h}{2} \times \frac{L}{2^2} + \cdots + 2 \times \frac{h}{2^{k-1}} \times \frac{L}{2^k}$$
$$= hL \sum_{l=0}^{k-1} \frac{1}{4^l} = O(h)\ .$$

Thus, we get a mostly rectangular mesh. We then have the same superconvergent estimates as (11)-(13), under the regular solution.

For the corner singular solution we refer to Lin and Yan [50].

For the triangular mesh (of piecewise almost uniform or mostly uniform structure) we refer to Lin and Xie [47], Blum [2], Lin and Liu [35].

4. About the proof

4.1. Proof of basic integral identities (2)-(4). Such identities have been proved in [24, 30] and in [48] for variable coefficients.

Denote

$$W = (u - i_h u)_x\ .$$

Then

$$\int_{l_i} W\,dx = 0.$$

We have, through the integration by parts and exploiting the specialness of F and E,

$$\begin{aligned}
\int_\tau W &= \int_\tau F_{yy}W = -\int_\tau F_yW_y + \left(\int_{l_1} - \int_{l_2}\right) F_yW\,dx \\
&= \int_\tau FW_{yy} - \left(\int_{l_1} - \int_{l_2}\right) FW_y\,dx = \int_\tau FW_{yy}\,, \\
\int_\tau (y-y_\tau)W &= \frac{1}{6}\int_\tau (F^2)_{yyy}W \\
&= \frac{1}{6}\int_\tau (F^2)_yW_{yy} + \frac{1}{3}\left(\int_{l_1} - \int_{l_2}\right) FF_yW_y\,dx \\
&= \frac{1}{6}\int_\tau (F^2)_yW_{yy}\,, \\
\int_\tau (x-x_\tau)W &= \int_\tau E_xW = -\int_\tau EW_x\,, \\
\int_\tau (x-x_\tau)(y-y_\tau)W &= \int_\tau E_xF_yW = \int_\tau EFW_{xy}\,.
\end{aligned}$$

Exploiting the bilinearity of φ, we have

$$\begin{aligned}
\int_\tau W\varphi_x &= \int_\tau W\varphi_x(x,y_\tau) + \int_\tau (y-y_\tau)W\varphi_{xy} \\
&= \int_\tau \left(F\varphi_x - \frac{1}{3}(F^2)_y\varphi_{xy}\right) W_{yy}\,, \\
\int_\tau W\varphi &= \int_\tau W\varphi(x_\tau,y_\tau) + \int_\tau (x-x_\tau)W\varphi_x(x,y_\tau) \\
&\quad + \int_\tau (y-y_\tau)W\varphi_y(x_\tau,y) + \int_\tau (x-x_\tau)(y-y_\tau)W\varphi_{xy} \\
&= \int_\tau \left(\left[F(\varphi - E_x\varphi_x) - \frac{1}{3}(F^2)_y(\varphi_y - E_x\varphi_{xy})\right] W_{yy} - E\varphi_xW_x\right)\,, \\
\int_\tau W\varphi_y &= \int_\tau W\varphi_y(x_\tau,y) + \int_\tau (x-x_\tau)W\varphi_{xy} \\
&= \int_\tau (FW_{yy}(\varphi_y - E_x\varphi_{xy}) - EW_x\varphi_{xy})\,.
\end{aligned}$$

4.2. Proof of Section 1.2. Now we explain how the superconvergence for functionals (1) can be used to sharpen the global error of various finite element schemes listed in the tables in Sections 1.2 and 1.3.

The first example is the standard Galerkin scheme for Lame equations. The argument is direct and has been done by Liu and Lin [39].

Another example is the mixed scheme of Ciarlet-Raviart for the biharmonic equation. Let $(v_h, u_h) \in S_h \times S_h^0$ be the approximation pair:

$$
\begin{aligned}
&(\nabla(v - v_h), \nabla\varphi) = 0 && \forall\varphi \in S_h^0,\\
&(\nabla(u - u_h), \nabla\psi) + (v - v_h, \psi) = 0 && \forall\psi \in S_h,
\end{aligned}
$$

where S_h^0 denotes the bilinear element space with zero boundary condition. We restrict ourselves to, for definiteness, the rectangular mesh. The error between the finite element interpolation and the finite element solution is then reduced to functionals (1) and estimates in Section 2:

$$
\begin{aligned}
||i_h v - v_h||^2 =&(i_h v - v_h, i_h v - v) - (\nabla(u - i_h u), \nabla(i_h v - v_h))\\
&- (\nabla(i_h v - v), \nabla(i_h u - u_h))\\
\leq& ch^2(||i_h v - v_h||_0 + ||i_h u - u_h||_1)||u||_5,\\
(\nabla(i_h u - u_h), \nabla\varphi) =&(\nabla(i_h u - u), \nabla\varphi) - (v - i_h v + i_h v - v_h, \varphi)\\
\leq& c(h^2||u||_4 + ||i_h v - v_h||_0)||\varphi||_0,
\end{aligned}
$$

$$
\begin{aligned}
&||i_h u - u_h||_1 \leq ch^2||u||_4 + c||i_h v - v_h||_0,\\
&||i_h v - v_h||_0^2 \leq ch^4||u||_5^2 + \varepsilon||i_h u - u_h||_1^2, \quad \varepsilon \ll 1,
\end{aligned}
$$

thus

$$
||i_h v - v_h||_0 + ||i_h u - u_h||_1 \leq ch^2||u||_5,
$$

especially we have $||v - v_h||_0 \leq ch^2||u||_5$.

The above argument was done by Lin, Li and Zhou [32]. For superconvergence estimates of u_h in H_1- and L_2-norms we refer to Lin and Liu [34].

The same argument also applies to the Herrmann-Miyoshi scheme, see Lin, Li and Zhou [32], to the $Q_1 - Q_1$, $Q_{1,2} \times Q_{2,1} - Q_0$ and $Q_2 - P_1$ schemes of Stokes problem, see Lin, Li and Zhou [33], Lin and Pan [36], and to the standard Galerkin scheme of first order hyperbolic equation, see Lin and Zhou [55, 57].

It is not difficult to work on the meshes of shapes 1)-3) but they lead, as shown in Sections 1.2 and 1.3, to different convergent orders. See [30, 28]. For the triangular mesh with piecewise almost uniform or mostly uniform structure we refer to Lin and Liu [35].

4.3. Proof of Section 1.3. Note that the higher interpolation of lower finite element data is equivalent to the interpolated finite elements defined in Lin and Yang [53, 54], Lin, Yan and Zhou [52]:

$$
\tilde{u}_h = I_{2h}^2 u_h
$$

where u_h is the bilinear finite element solution, I_{2h}^2 the biquadratic interpolation on coarse mesh T_{2h}. For the bilinear interpolation i_h on fine mesh T_h we have

$$
I_{2h}^2 i_h = I_{2h}^2,
$$

and hence, in H_1-norm,

$$\tilde{u}_h - u = I_{2h}^2(u_h - i_h u) + I_{2h}^2 u - u = O(h^2).$$

5. Global superconvergence

As a by-product of Sections 1.3 and 1.4, we obtain an error estimator (Zhu and Lin [73]):

$$||u - u^h||_{1,\Omega_0}^2 \approx ||\tilde{u}^h - u^h||_{1,\Omega_0}^2$$

where $\Omega_0 \subset \Omega$.

6. Eigenvalue superconvergence

The global superconvergence of the higher interpolation stated in Section 1.3 is particularly useful for the eigenvalue problem.

Let (λ, u) and (λ^h, u^h) denote the exact eigenpair of Laplace operator and the associated approximate eigenpair in the finite element space S_h^p over a rectangular mesh. In the particular case $p = 1$, (λ^h, u^h) is the bilinear finite element eigenpair. Let $R_h u \in S_h^p$ denote the Ritz projection of Laplace operator.

PROPOSITION 1. *There holds, for $p \geq 2$,*

$$\|u^h - R_h u\|_{1,2} \leq ch^{p+2},$$

i.e., the finite element eigensolution is more close to the Ritz projection than that to the exact eigensolution.

PROPOSITION 2. *For $w \in H_0^1(\Omega)$ and $w \neq 0$, there holds*

$$\left| \frac{\| \nabla w\|_{0,2}^2}{\|w\|_{0,2}^2} - \lambda \right| \leq c \frac{\| \nabla (w - u)\|_{0,2}^2}{\|w\|_{0,2}^2}.$$

In Proposition 2 setting the higher interpolation of finite element eigensolution

$$w = \tilde{u}^h,$$

we can obtain, similar to Section 6, an order of $O(h^{2p+4})$ for the eigenvalue approximation, i.e., the higher interpolation of finite element eigensolution makes the eigenvalue approximation of four order higher than the standard finite element eigensolution itself. In particular, when $p = 2$, there holds

$$\frac{\| \tilde{u}^h \|_1^2}{\| \tilde{u}^h \|_0^2} - \lambda = O(h^8).$$

The details of proofs were given in Lin and Yang [53, 54], Lin, Yan and Zhou [52].

7. Global extrapolation and defect correction

The extrapolation technique (ET) in FEM has been surveyed by Rannacher in a summer school at Helsinki in 1987, see [62] for the history and the references before 1988. Only a few remarks are added in this report. In 1978, numerical results of ET applied to Poisson equation on the rectangular domain were reported in Huang and Liu [15]. And, in 1980, the first finite element analysis of ET was done by Lin and Liu [37, 38] for 1D and for a triangle domain. The essential progress was done in 1983 by Lin, Lu and Shen [42, 43], Lin and Lu [40, 41], Lin and Wang [44] in examining the analogous finite element analysis of ET in a general quadrilateral domain and in a general polygonal domain where the "piecewise" concept of mesh structure was introduced for the first time.

Most of extrapolation analysis were presented on nodal points of the triangular mesh in 2D. However, by using further integral expansion and higher interpolation, a global extrapolation has been presented on the rectangular (or cube) mesh by Lin [24], Lin and Yan [51], including the 3D and the biquadratic FEM. See also the work of Shaidurov [64] on the global extrapolation.

The defect correction technique in FEM was initiated in 1987 by Rannacher [62, 63] and developed by Blum [2] in Germany and by Yang in China. They and Shaidurov [64] have widely used, as an auxiliary technique, the interpolated finite elements. Again, by using integral identities and higher interpolation, a direct and global analysis of the defect correction has been given in Lin and Zhou [56], Lin, Yan and Zhou [52].

Both of the above techniques can be also considered as tools for generating superconvergence phenomena, see Ciarlet [12], for both of the solution and the gradient. For example, the two techniques started from bilinear elements can yield fourth order convergence for both of solution and gradient. See Lin and Zhu [59, 60], Lin and Xie [45–47], Lin [25], Lin and Zhou [56].

8. Extension

So far we only considered the bilinear elements in 2D. Some extensions of explicit integral identities to elements of degree $p \geq 2$, to 3D and to further expansions have been done by Lin, Yan and Zhou [52], Lin and Yan [51]. These results are basic but the representations will be too long. We omit the details.

References

1. I. Babuska and B. Szabo, *Lecture notes on finite element analysis*, Springer-Verlag, 1991.
2. H. Blum, *Defect correction schemes for mixed and nonconforming finite elements*, Bonn. Math. Schr., **228**(1991), 3–20.
3. H. Blum, Q. Lin and R. Rannacher, *Asymptotic error expansion and Richardson extrapolation for linear finite elements*, Numer. Math., **49**(1986), 11–37.
4. C. Chen, *Finite element method and its analysis in improving accuracy*, Hunan Science Press, 1982.
5. C. Chen and Y. Huang, *Extrapolation of triangular linear element in general domain*, Numer. Math. J. Chinese Univ., **1**(1989), 1–15.

6. C. Chen and J. Jin, *Extrapolation of biquadratic finite element approximation in rectangle domain*, J. Xiangtan Univ., **3**(1987), 16–28.
7. C. Chen and J. Liu, *Superconvergence of gradient of triangular linear element in general domain*, J. Xiangtan Univ., **1**(1987), 114–126.
8. C. Chen and Q. Lin, *Extrapolation of finite element approximation in a rectangular domain*, J. Comput. Math., **3**(1989), 227–233.
9. H. Chen, *Analysis of rectangular finite element*, J. Xiangtan Univ., **4**(1989).
10. H. Chen and Q. Lin, *Extrapolation for isoparametric bilinear finite element approximation on smooth bounded domains*, Syst. Sci. Math. Scis., 1992.
11. H. Chen and Q. Lin, *Finite element approximation using a graded mesh on domains with reentrant corners*, Syst. Sci. Math. Scis., 1992.
12. P. Ciarlet, *Basic error estimates for elliptic problems*, Handbook of Numer. Anal., Vol. II, North-Holland, Amsterdam, 1991, 17–351.
13. K. Feng, *Difference scheme based on variational principle*, J. Appl. Comp. Math., **2**(1965), 237–261.
14. I. Graham, Q. Lin and R. Xie, *Extrapolation of Nystrom solution of boundary integral equations on non-smooth domains*, J. Comput. Math., 1992.
15. H. Huang and G. Liu, *Solving elliptic boundary value problems by successive mesh subdivisions*, Math. Numer. Sinica, **2**, **3**(1978), 41–52, 28–35.
16. H. Huang, M. Mu and W. Han, *Asymptotic expansion for numerical solution of a less regular problem in a rectangle*, Math. Numer. Sinica, **4**(1986).
17. H. Huang, W. E and M. Mu, *Extrapolation combined with MG method for solving finite element equations*, J. Comput. Math., **4**(1986).
18. Y. Huang and Q. Lin, *A finite element method on polygonal domains with extrapolation*, Math. Numer. Sinica, **3**(1990), 239–249.
19. Y. Huang and Q. Lin, *Finite element method on polygonal domains*, Parts I-II, J. Sys. Sci. Math. Scis., 1992.
20. M. Krizek, Q. Lin and Y. Huang, *A nodal superconvergence arising from combination of linear and bilinear elements*, Syst. Sci. Math. Scis. **2**(1988).
21. M. Krizek and P. Neittaanmaki, *On superconvergence techniques*, Acta Appl. Math., **9**(1987), 175–198.
22. Q. Lin, *High accuracy from the linear elements*, Proc. of 1984 Beijing Symp. on DGDE-Comp. of PDE, ed. Feng Kang, Science Press, 258–262.
23. Q. Lin, *Finite element error expansion for non-uniform quadrilateral meshes*, Syst. Sci. Math. Scis., **3**(1989), 275–282.
24. Q. Lin, *An integral identity and interpolated postprocess in superconvergence*, Research Report 90-07, Institute of Systems Science, Academia Sinica, Beijing, 1990.
25. Q. Lin, *Extrapolation of finite element gradients on nonconvex domains*, Bonn. Math. Schr., **228**(1991), 21–30.
26. Q. Lin, *Fourth order eigenvalue approximation by extrapolation on domains with reentrant corners*, Numer. Math., **58**(1991), 631–640.
27. Q. Lin, *Superconvergence of the finite element method for singular solution*, J. Comput. Math., **1**(1991), 111–114.
28. Q. Lin, *Global error expansion and superconvergence for higher order interpolation of finite elements*, J. Comput. Math., Suppl. Issue (1992), 286–289.
29. Q. Lin, *Best mesh and elevated basis expression for finite element method*, Proceedings of Systems Science & Systems Engineering, Great Wall (H.K.) Culture Publish Co., 1991, 169–173.
30. Q. Lin, *A rectangle test for finite element analysis*, ibid, 213–216.
31. Q. Lin, *From rectangle test to optimal adaptive meshes*, ibid, 247–251.
32. Q. Lin, J. Li and A. Zhou, *A rectangle test for Ciarlet-Raviart and Herrmann-Miyoshi schemes*, ibid, 230–233.
33. Q. Lin, J. Li and A. Zhou, *A rectangle test for Stokes equations*, ibid, 240–241.
34. Q. Lin and M. Liu, *A rectangle test for biharmonic problem*, ibid, 238–239.
35. Q. Lin and M. Liu, *A rectangle test for nonrectangular domains*, ibid, 359–361.

36. Q. Lin and J. Pan, *Global superconvergence for rectangular elements in Stokes problem*, ibid, 371–376.
37. Q. Lin and J. Liu, *A discussion for extrapolation method for finite elements*, Tech. Rep. Inst. Math., Academia Sinica, Beijing, 1980.
38. Q. Lin and J. Liu, *Extrapolations and corrections for the two-dimensional integral and differential equation*, A lecture given at die Mathematischen Institute der Freien Universitat, Berlin, December 1981.
39. Q. Lin and J. Liu, *Defect corrections of finite element approximation for singular problem*, Proc. China-France Sympos. on FEM Beijing, 1982.
40. Q. Lin and T. Lu, *Asymptotic expansions for finite element approximation of elliptic problem on polygonal domains*, Comp. Math. on Appl. Sci. Eng. (Proc. Sixth Int. Conf. Versailles, 1983), LN in Comp. Sci., North-Holland, INRIA, 1984, 317–321.
41. Q. Lin and T. Lu, *Asymptotic expansions for finite element eigenvalues and finite element solution*, (Proc. Int. Conf., Bonn, 1983), Math. Schr., **158**(1984), 1–10.
42. Q. Lin, T. Lu and S. Shen, *Asymptotic expansions for finite element approximations*, Research Report IMS-11, Chengdu Branch of Academia Sinica, 1983.
43. Q. Lin, T. Lu and S. Shen, *Maximum norm estimate, extrapolation and optimal point of stresses for finite element methods on strongly regular triangulation*, J. Comput. Math., **4**(1983), 376–383.
44. Q. Lin and J. Wang, *Some expansions of finite element approximation*, Research Report IMS-15, Chengdu Branch of Academia Sinica, 1984.
45. Q. Lin and R. Xie, *Some advances in the study of error expansion for finite elements*, J. Comput. Math., **4**(1986), 368–382.
46. Q. Lin and R. Xie, *Error expansions for finite element approximation and its applications*, LN in Math., **1297**(1987).
47. Q. Lin and R. Xie, *How to recover the convergence rate for Richardson extrapolation on bounded domain*, J. Comput. Math., **4**(1988), 68–79.
48. Q. Lin and R. Xie, *A rectangle test for finite element problem with variable coefficients*, Proc. Syst. Sci. Syst. Eng., Great Wall (H.K.) Culture Publish Co., 1991, 368–370.
49. Q. Lin and J. Xu, *Linear finite elements with high accuracy*, J. Comput. Math., **3**(1985), 115–133.
50. Q. Lin and N. Yan, *A rectangle test for singular solution with irregular meshes*, Proceedings of Systems Science & Systems Engineering, Great Wall (H. K.) Culture Publish Co., 1991, 236–237.
51. Q. Lin and N. Yan, *A rectangle test in R^3*, ibid, 242–246.
52. Q. Lin, N. Yan and A. Zhou, *A rectangle test for interpolated finite elements*, ibid, 217–229.
53. Q. Lin and Y. Yang, *The finite element interpolated correction method for elliptic eigenvalue problems*, Math. Numer. Sinica, **3**(1992), 334–338.
54. Q. Lin and Y. Yang, *Interpolation and correction of finite element methods*, Math. in practice and theory, **3**(1991), 29–35.
55. Q. Lin and A. Zhou, *Some arguments for recovering the finite element error of hyperbolic problems*, Acta Math. Sci., **3**(1991), 291–298.
56. Q. Lin and A. Zhou, *Defect correction for finite element gradient*, Syst. Sci. Math. Scis., **4**(1991).
57. Q. Lin and A. Zhou, *A rectangle test for first order hyperbolic equation*, Proceedings of Systems Science & Systems Engineering, Great Wall (H. K.) Culture Publish Co., 1991, 234–235.
58. Q. Lin and Q. Zhu, *Unidirectional extrapolation of finite difference and finite elements*, J. Eng. Math., **2**(1984), 1–12.
59. Q. Lin and Q. Zhu, *Asymptotic expansions for the derivative of finite elements*, J. Comput. Math., **4**(1984), 361–363.
60. Q. Lin and Q. Zhu, *Local asymptotic expansion and extrapolation for finite elements*, J. Comput. Math., **3**(1986), 263–265.
61. T. Lu, *Asymptotic expansion and extrapolation for finite element approximation of nonlinear elliptic equations*, J. Comput. Math., **2**(1987), 194–199.

62. R. Rannacher, *Extrapolation techniques in the finite element method (A survey)*, Summer School on Numerical Analysis, Helsinki, Univ. of Tech., MAT-C 7, 1988, 80–113.
63. R. Rannacher, *Defect correction techniques in the finite element method*, Proc. Metz Days on Numerical Analysis, Univ. Metz, June 1990, Pitman Research Notes in Math.
64. V. Shaidurov, *Multigrid Methods of Finite Elements*, Moscow, 1989.
65. G. Strang and G. Fix, *An analysis of the finite element method*, Prentice-Hall, 1973.
66. L. Wahlbin, *Local behavior in finite element methods*, Handbook of Numerical Analysis, Vol. II, North-Holland, Amsterdam, 1991, 352–522.
67. J. Wang, *Asymptotic expansion and L^∞-error estimates for mixed finite element methods*, Numer. Math., **55**(1989), 401–430.
68. J. Wang and Q. Lin, *Expansion and extrapolation for the finite element method*, J. Sys. Sci. & Math. Scis., **2**(1985), 114–120.
69. R. Xie, *Error estimates for finite element Green function on nonconvex domains*, Math. Numer. Sinica, **4**(1988).
70. Y. Yang, *L_p-estimates and superconvergence of finite element approximations for eigenvalue problems*, Numer. Math. J. Chinese Univ., **9**(1987).
71. Y. Yang, *A theorem on superconvergence of shifts of finite element approximations for eigenvalue problems*, J. Math., **10**(1990), 229–234.
72. Q. Zhu and Q. Lin, *Superconvergence Theory of Finite Element Methods*, Hunan Science Press, 1989.
73. Q. Zhu and Q. Lin, *Asymptotically exact a posteriori error*, Submitted to Math. Numer. Sinica, 1991.

INSTITUTE OF SYSTEMS SCIENCE, ACADEMIA SINICA, BEIJING 100080, CHINA

Contemporary Mathematics
Volume 163, 1994

Mathematical Theory of Some Non-standard Finite Element Methods

SHI ZHONG-CI AND WANG MING

1. Introduction

The main idea of finite element method is to use discrete solutions on finite element spaces V_h, usually consisting of piecewise polynomials of a certain degree, to approximate the continuous solutions on infinite dimension space V according to a variational principle. For the mathematical foundation of the finite element method, there is a well-known result: the convergence of finite element solutions u_h to the real solution u is dependent on the approximation of finite element spaces V_h to the space V, when V_h are subspaces of V. The approximability of a finite element space is guaranteed if it contains all piecewise linear or quadratic polynomials in accordance with second or fourth order elliptic problems, i.e., the constant strain condition, in mechanical terminology, holds.

When finite element spaces V_h are subspaces of V, the method is called conforming or a standard finite element method. In general, the conforming property of finite element spaces implies, for a $2m$th order elliptic boundary value problem, that V_h are subspaces of $C^{m-1}(\overline{\Omega})$, where Ω is the domain of the given problem. This condition is very strong for 4th order problems. For example, a quintic polynomial with at least 18 parameters is required when elements are triangles. This suffers serious computational difficulty, either the dimension of finite element spaces is fairly large or their structure is very complicated. It is interested how to decrease the number of nodal parameters and the degree of polynomials.

A successful approach is to relax the C^{m-1} continuity on Ω. The elements obtained in this way are called nonconforming. To overcome the difficulty with C^{m-1} continuity, there are several new methods suggested by mechanists, such as, quasi-conforming method proposed by Tang Limin et al. [**42, 11, 15, 20**], the generalized conforming method by Long Yuqiu [**17**], the free formation method

1991 *Mathematics Subject Classification.* **65N30, 73V05.**

by Bergan et al. [**4, 5, 6**], the energy-orthogonal method by Bergan et al [**13**]. Many elements constructed by these methods have very good numerical performance.

The finite element spaces V_h, constructed by nonconforming method and the other methods, named above, are all not contained in V. Hence these methods are called non-standard finite element methods. For the mathematical theory of these methods, Shi, Zhang and Wang et al. have done a series of works [**9**], [**10**], [**21**]-[**37**], [**44**]-[**67**]. In this paper, a brief description of the results is summarized.

Let Ω be a polyhedroid domain in R^n. Denote the Sobolev spaces, Sobolev norm and semi-norm by $H^m(\Omega)$, $H_0^m(\Omega)$, $\|\cdot\|_m$ and $|\cdot|_m$ respectively. Define, for $u, v \in H^1(\Omega)$,

$$a(u,v) = \int_\Omega \sum_{i=1}^n \frac{\partial u}{\partial x_i}\frac{\partial v}{\partial x_i}\,dx. \tag{1.1}$$

Let $f \in L^2(\Omega)$. The variational formula of Dirichlet boundary value problem of Poisson equation is finding $u \in H_0^1(\Omega)$, such that,

$$a(u,v) = \int_\Omega fv\,dx, \quad \forall v \in H_0^1(\Omega), \tag{1.2}$$

When $n = 2$, define, for $u, v \in H^2(\Omega)$,

$$\begin{aligned}\bar{a}(u,v) = \int_\Omega \Big\{ \triangle u \triangle v + (1-\sigma)\big(2\frac{\partial^2 u}{\partial x_1 \partial x_2}\frac{\partial^2 v}{\partial x_1 \partial x_2} \\ - \frac{\partial^2 u}{\partial x_1^2}\frac{\partial^2 v}{\partial x_2^2} - \frac{\partial^2 u}{\partial x_2^2}\frac{\partial^2 v}{\partial x_1^2}\big)\Big\} dx,\end{aligned} \tag{1.3}$$

where $\sigma \in (0, \frac{1}{2})$ is the Poisson ratio. The variational formula of a clamped plate bending problem is to find $u \in H_0^2(\Omega)$, such that,

$$\bar{a}(u,v) = \int_\Omega fv\,dx, \quad \forall v \in H_0^2(\Omega). \tag{1.4}$$

We will discuss the non-standard finite element methods for solving problem (1.2) and (1.4).

For each h with $h \to 0$, let T_h be a finite subdivision of Ω, and suppose T_h satisfy the following conditions: 1) For each $T \in T_h$, T is a closed convex polyhedroid, and for each h, $\overline{\Omega} = \cup_{T \in T_h} T$. 2) For any two distinct $T_1, T_2 \in T_h$, $T_1 \cap T_2$ is either an empty set or a common face. 3) There exists a constant η independent of h, such that, for arbitrary $T \in T_h$, $\eta h \leq \rho_T < h_T \leq h$, here ρ_T is the diameter of the largest ball contained in T and h_T is the one of the smallest ball containing T.

Corresponding to T_h, let $V_h \subset L^2(\Omega)$ be a finite element space, such that, $\forall v \in V_h$, $\forall T \in T_h$, $v|_T \in P_r(T)$, while $P_r(T)$ is the space of all polynomials on T with the degree not greater than r.

V_h is a subspace of $H^m(\Omega)$ if and only if $V_h \subset C^{m-1}(\overline{\Omega})$. When V_h is not a subset of $H^m(\Omega)$, we define mesh dependent semi-norm $|\cdot|_{m,h}$ as follows,

$$|v|_{m,h} = \Big(\sum_{T \in T_h} |v|^2_{m,T} \Big)^{1/2}, \quad \forall v \in V_h.$$

2. The nonconforming finite elements

Since the nonconforming finite element spaces V_h are not subspaces of $H^m(\Omega)$, the bilinear forms (1.1) and (1.3) must be extended. We define,

$$a_h(u,v) = \sum_{T \in T_h} \int_T \sum_{i=1}^{n} \frac{\partial u}{\partial x_i} \frac{\partial v}{\partial x_i}\, dx, \quad \forall u, v \in H^1(\Omega) + V_h, \tag{2.1}$$

$$\begin{aligned} \bar{a}_h(u,v) = \sum_{T \in T_h} \int_T \Big\{ \triangle u \triangle v + (1-\sigma)\Big(2 \frac{\partial^2 u}{\partial x_1 \partial x_2} \frac{\partial^2 v}{\partial x_1 \partial x_2} - \frac{\partial^2 u}{\partial x_1^2} \frac{\partial^2 v}{\partial x_2^2} \\ - \frac{\partial^2 u}{\partial x_2^2} \frac{\partial^2 v}{\partial x_1^2}\Big) \Big\} dx, \quad \forall u, v \in H^2(\Omega) + V_h. \end{aligned} \tag{2.2}$$

The nonconforming method for problem (1.2) is to find $u_h \in V_h$, such that,

$$a_h(u_h, v_h) = \int_\Omega f v_h\, dx, \quad \forall v_h \in V_h, \tag{2.3}$$

while the one for problem (1.4) is to find $u_h \in V_h$, such that,

$$\bar{a}_h(u_h, v_h) = \int_\Omega f v_h\, dx, \quad \forall v_h \in V_h. \tag{2.4}$$

What are the conditions ensuring the convergence of nonconforming finite elements? Like conforming elements, the approximability condition is apparently needed for nonconforming methods, that is, the finite element space must contain all piecewise linear or quadratic polynomials in accordance with problem (1.2) or problem (1.4). However, in nonconforming cases, the approximability condition alone does not guarantee the convergence. Some further conditions are required. Engineers proposed a condition, so called 'patch test', based on mechanical considerations and numerical experience, and they believe that it is a necessary and sufficient condition for convergence, provided the approximability condition holds. Unfortunately, rigorous mathematical analyses established that the patch test in the original version [**14**] or its modification [**43**] is neither necessary nor sufficient [**21, 22, 23, 30, 40**]. Papers [**21, 30, 40**] give the examples which pass the patch test but do not converge. A necessary and sufficient condition–the generalized patch test was suggested by Stummel [**39**] from a mathematical point of view.

By means of the generalized patch test, Shi has analyzed several nonconforming elements, violating the patch test.

In paper [**22**], the quadrilateral Wilson element, without the modifications of the variational formulation, is shown to be convergent provided the following condition (A) is true.

Condition (A): *The distance d_T between the midpoints of the diagonals of a quadrilateral $T \in T_h$ is of order $o(h_T)$ uniformly for all T as $h \to 0$.*

For the quadrilateral elements of Sander and Beckers in [**19**], Shi shown, in paper [**23**], that the elements of 8 degrees of freedom and 12 degrees of freedom, which do not pass the patch test, pass the generalized patch test and hence converge if condition (A) is satisfied, while the element with 16 degrees of freedom does not pass the generalized patch test and diverges.

Further, a strange convergence behaviour of nonconforming approximation has been successfully observed by Shi and Cai [**21, 25, 7**]. Several examples show that the nonconforming element solution may converge not to the desired solution of the given problem, but to a wrong limit. Paper [**21**] presents a constrictive approach for analysis of Stummel's counter-examples to the patch test, using the explicit solutions of the finite element equations to establish the exact limits of the nonconforming approximations involved in these examples. Moreover, a simple device for remedy these defects is introduced [**21**] by a combination of hybrid and penalty methods, which may be used to obtain a convergent procedure for a divergent element.

For the plate bending problem (1.4), it is known that Zienkiewicz incomplete cubic triangular element [**3**] is nonconforming and it converges if all triangles are generated by three sets of parallel lines [**16**]. In paper [**24**], Shi proves that Zienkiewicz element also passes the generalized patch test under the condition of parallel lines and it does not pass the test, and thus does not converge, for the union-jack mesh and the fish-bone mesh. Paper [**7**] establishes the wrong limit of Zienkiewicz approximate solution when the union-jack mesh is used.

For Morley element, paper [**32**] obtains the optimal error estimates directly from the nonconforming finite element technique different from Arnold-Brezzi [**2**] approach via a mixed method. If the solution u of problem (1.4) is in $H^3(\Omega) \cap H_0^2(\Omega)$, then

$$|u - u_h|_{2,h} \leq Ch(|u|_3 + h|f|_0) \tag{2.5}$$

and

$$|u - u_h|_{1,h} \leq Ch^2(|u|_3 + h|f|_0) \tag{2.6}$$

when Ω is convex.

3. F-E-M-test

Based on the work of convergence studies of nonconforming finite elements, Shi proposed in paper [**28**] a sufficient condition, named F-E-M-Test, for nonconforming finite elements . The F-E-M-Test is simple to apply, it checks only the local properties of shape functions along each interface or on each element. Furthermore, it seems to be easier for engineers in practical situations than the generalized patch test does.

F_1-Test: The finite element space V_h is said to pass the F_1-Test for $2m$ order boundary value problems, if for every function $v_h \in V_h$ the jump of v_h, denoted by $[v_h]$, across each interface F of two adjacent elements T_1 and T_2 satisfies the condition

$$|\int_F [v_h]\, ds| \leq o(h_T^{n/2})\|v_h\|_{m,T_1\cup T_2}, \quad h_T = \max\{h_{T_1}, h_{T_2}\}. \tag{3.1}$$

For every outer boundary $F \subset T \cap \partial\Omega$ with Dirichlet boundary conditions, the jump $[v_h]|_F \equiv v_h|_F$ and the condition (3.1) is understood as

$$|\int_F v_h ds| \leq o(h_T^{n/2})\|v_h\|_{m,T}. \tag{3.2}$$

F_2-Test: For fourth order problems, the F_2-Test requires that the jumps $[\frac{\partial}{\partial x_k} v_h]$ across the interface F satisfy the condition

$$|\int_F [\frac{\partial}{\partial x_k} v_h] ds| \leq o(h_T^{n/2})\|v_h\|_{2,T_1\cup T_2}, \quad k = 1, 2, \cdots, n. \tag{3.3}$$

For every outer boundary $F \subset T \cap \partial\Omega$ with Dirichlet boundary conditions, $[\frac{\partial}{\partial x_k} v_h]_F \equiv \frac{\partial}{\partial x_k} v_h|_F$ and condition (3.3) reads

$$|\int_F \frac{\partial}{\partial x_k} v_h ds| \leq o(h_T^{n/2})\|v_h\|_{2,T}, \quad k = 1, 2, \cdots, n. \tag{3.4}$$

In particular, if in the condition (3.1) or (3.3) the equality

$$\int_F [v_h] ds = 0$$

or

$$\int_F [\frac{\partial}{\partial x_k} v_h] ds = 0, \quad k = 1, 2, \cdots, n$$

holds for all $F \subset \partial T$, respectively, the F-Test are called the strong F_1-Test or F_2-Test, respectively.

E_1-M_1-Test: The finite element space V_h is said to pass the E_1-M_1-Test for $2m$th order problems, if every function $v_h \in V_h$ can be decomposed into two parts, a continuous part $C_1(v_h)$ and a discontinuous part $d_1(v_h)$:

$$v_h = C_1(v_h) + d_1(v_h), \tag{3.5}$$

such that, on each element T the discontinuous part $d_1(v_h)$ satisfies the two conditions

$$\Big|\int_{\partial T} d_1(v_h)N_i ds\Big| \le o(h_T^{n/2})\|v_h\|_{m,T}, \quad i = 1, 2, \cdots, n, \tag{3.6}$$

$$\Big|\int_{\partial T} d_1^2(v_h)ds\Big| \le o(h_T^{-1})\|v_h\|^2_{m,T}. \tag{3.7}$$

E_2-M_2-Test: For fourth order problems, the E_2-M_2-Test requires that the first derivatives $\frac{\partial}{\partial x_k}v_h$ can be decomposed into two parts,

$$\frac{\partial}{\partial x_k}v_h = C_2\Big(\frac{\partial}{\partial x_k}v_h\Big) + d_2\Big(\frac{\partial}{\partial x_k}v_h\Big), \quad k = 1, 2, \cdots, n, \tag{3.8}$$

where $C_2(\frac{\partial}{\partial x_k}v_h)$ are continuous functions over all elements and $d_2(\frac{\partial}{\partial x_k}v_h)$ are the associated remainder terms such that on each element T the discontinuous parts $d_2(\frac{\partial}{\partial x_k}v_h)$ satisfy conditions

$$\Big|\int_{\partial T} d_2\Big(\frac{\partial}{\partial x_k}v_h\Big)N_i ds\Big| \le o(h_T^{n/2})\|v_h\|_{2,T}, \quad i, k = 1, 2, \cdots, n, \tag{3.9}$$

$$\Big|\int_{\partial T} d_2^2\Big(\frac{\partial}{\partial x_k}v_h\Big)ds\Big| \le o(h_T^{-1})\|v_h\|^2_{2,T}, \quad k = 1, 2, \cdots, n. \tag{3.10}$$

Similar to the strong F-Tests, if the equalities

$$\int_{\partial T} d_1(v_h)N_i ds = 0, \quad i = 1, 2, \cdots, n$$

or

$$\int_{\partial T} d_2\Big(\frac{\partial}{\partial x_k}v_h\Big)N_i ds = 0, \quad k, i = 1, 2, \cdots, n$$

hold for every element T, respectively, the tests are called the strong E_1-M_1-Test or E_2-M_2-Test, respectively.

Let V_h possess the approximability and the bilinear form (2.1) or (2.2) be uniformly V_h-elliptic.

THEOREM 3.1. *For problem (1.2) ($m = 1$), the F_1-Test or the E_1-M_1-Test implies the convergence. For problem (1.4) ($m = 2$), the F_1-Test or the E_1-M_1-Test together with the F_2-Test or the E_2-M_2-Test imply the convergence.*

In paper [**28**], the applications of F-E-M-Test to the convergence of various nonconforming elements are given.

4. The double set parameter method

In the discussion of mathematical foundation of some non-standard finite elements, such as 9 parameter generalized conforming element [**17**] and 9-parameter quasi-conforming element [**11, 42**], Shi and Chen found that the elements actually use two sets of parameters, one of which consists of usual nodal parameters, like function values and their derivatives at nodes of elements. The second set may be chosen some special type of linear functionals on the given finite element space, such as integrals of functions and their derivatives along edges of elements in order to meet certain convergence requirements. From this fact, they suggested a double set parameter method, DSP method for short, in paper [**10**], and the corresponding mathematical theory was established.

For element T, let the shape function space be

$$\bar{P}(T) = \text{ span } \{\phi_1, \cdots, \phi_m\}, \tag{4.1}$$

where $\phi_1, \cdots, \phi_m$ are linearly independent polynomials, and the degrees of freedom be

$$D(v) = (d_1(v), \cdots, d_m(v))^\top, \tag{4.2}$$

with $d_1, \cdots, d_m$ the linear functionals on $H^k(T)$ $(k \geq 1)$.

For each $p \in \bar{P}(T)$, it can be represented by the linear combination of the basis functions ϕ_i, i.e.,

$$p = \beta_1\phi_1 + \cdots + \beta_m\phi_m. \tag{4.3}$$

Substituting this formula into the functionals d_i, we get a relation between β_i and $D(p)$,

$$Cb = D(p) \tag{4.4}$$

where $b = (\beta_1, \cdots, \beta_m)^\top$ and $C = (d_i(\phi_j))_{m\times m}$ is a matrix. To determine b by $D(p)$, we assume the well posed condition

$$\det\ C \neq 0 \tag{4.5}$$

holds.

In usual finite element calculations, the degrees of freedom $D(p)$ are taken as the unknowns of a discrete system and the shape functions are, therefore, represented by these unknowns throughout (4.3) and (4.4). Substituting p in the bilinear form of the given variational problem, one gets the stiffness matrix. In this case, the degrees should be chosen simple and that the size of resulting discrete system is small. On the other hand, they must make the shape functions have certain continuity across interelement boundaries to ensure the convergence. These two goals are, sometimes, not easy to be satisfied simultaneously.

Trying to meet these two requirements, Shi and Chen suggested the DSP method. The method takes usual nodal parameters as another set of unknowns. Let the nodal parameters are

$$Q(v) = (q_1(v), \cdots, q_l(v))^\top. \tag{4.6}$$

Then one can use linear combinations of nodal parameters to discretize the degrees of freedom (4.2), which yields

$$D(v) = GQ(v) + \epsilon(v), \tag{4.7}$$

where $\epsilon(v)$ is the remainder term of discretization, G is the discretizing matrix dependent on nodal parameters and independent of v. Neglecting $\epsilon(v)$ in (4.7), we define the real shape function space on T as follows

$$P(T) = \{p \in \bar{P}(T) \mid p = \sum_{i=1}^{m} \beta_i \phi_i, b = C^{-1}GQ, \forall Q \in R^l\}. \tag{4.8}$$

Then the finite element space is $V_h = \{v \in L^2(\Omega) \mid v|_T \in P(T), \forall T \in T_h\}$ and the final set of unknowns is $Q(v)$. The finite element so constructed is called the DSP element.

The convergence of DSP elements is given in the following theorem.

THEOREM 4.1. *Let* $s = 1$ *or* 2, *and let the DSP element satisfy:* 1) *The well-posed condition* (4.5) *holds;* 2) $|\cdot|_{s,h}$ *is a norm on finite element space* V_h; 3) *Polynomial space* $P_s(T) \subset \bar{P}(T)$ *and* $\epsilon(v)$ *in* (4.7) *vanishes for all* $v \in P_s(T)$; 4) V_h *passes the F-E-M-Test. Then the DSP element is convergent for problem* (1.2) *when* $s = 1$ *and for problem* (1.4) *when* $s = 2$.

DSP element method has some advantages. The nodal parameters $Q(v)$ in (4.6) and the degrees of freedom $D(v)$ in (4.2) can be chosen independently. The nodal parameters should be simple and that the total numbers of discrete system are small, while the degrees of freedom and discrete formula (4.7) are chosen to satisfy the generalized patch test or F-E-M-Test. DSP method can also better solve the matching problem between the shape functions and nodal parameters. Moreover, DSP method can be put into the general framework of finite element packages.

Based on the theory of DSP method, Shi and Chen have proved the convergence of 9-parameter quasi-conforming element and the generalized conforming element in papers [**34, 35**], and constructed some new elements [**10**].

5. The quasi-conforming finite elements

Let V_h be nonconforming finite element spaces. For $v_h \in V_h$ define $v_h^i \in L^2(\Omega)$ by $v_h^i|_T = \frac{\partial}{\partial x_i}(v_h|_T), \forall T \in T_h$. Comparing (2.1) with (1.1), we see that v_h^i takes the position corresponding to the one of derivative. On the other hand, v_h^i and v_h have, in general, no derivative relation on whole domain Ω, but they have on each elements. From the point that the derivative relations may not be satisfied on each elements, Tang et al. suggested the quasi-conforming method, in mechanical terminology. The mathematical description and a general theory of quasi-conforming finite element method was given by Zhang and Wang. By the multiple set function method, MSF method for short, they give a unified convergence theory of nonconforming and quasi-conforming methods. Their results are summarized in book [**67**]. Now we take problem (1.4) as an example to view the main idea.

Firstly, let us see how quasi-conforming finite element method relaxes the derivative relations. Let T be an element. By Green formula, for $v \in H^1(T)$, $i = 1, 2$, we have

$$\int_T w\frac{\partial}{\partial x_i}v\,dx = \int_{\partial T} vwN_i ds - \int_T v\frac{\partial}{\partial x_i}w\,dx, \quad \forall w \in H^1(T). \tag{5.1}$$

$\frac{\partial}{\partial x_i}v$ is uniquely determined by (5.1) as soon as v is given, that is, $\frac{\partial}{\partial x_i}v$ satisfies variational formula (4.1) of Galerkin type. Hence we can choose a finite dimensional subspace W_i of $H^1(T)$ to get an approximate function $\Pi_T^i v \in W_i$ of $\frac{\partial}{\partial x_i}v$ by

$$\int_T w\Pi_T^i v\,dx = \int_{\partial T} vwN_i ds - \int_T v\frac{\partial}{\partial x_i}w\,dx, \quad \forall w \in W_i. \tag{5.2}$$

Since the continuity of shape functions on the boundary ∂T plays an important role for nonconforming approximations, we can choose an approximate function $\Pi_{\partial T}v$ of v on ∂T to replace $v|_{\partial T}$, which has better continuity than $v|_{\partial T}$ does. Instead of (5.2), $\Pi_T^i v \in W_i$ is now determined by

$$\int_T w\Pi_T^i v\,dx = \int_{\partial T} w\Pi_{\partial T}vN_i ds - \int_T v\frac{\partial}{\partial x_i}w\,dx, \quad \forall w \in W_i. \tag{5.3}$$

We denote the 2-index by $\alpha = (\alpha_1, \alpha_2)$ with $|\alpha| = \alpha_1 + \alpha_2$. Generally, let V_h be a finite element space associated with T_h. For each element T, denote the shape function space by P_T, and choose 6 polynomial subspaces $N_T^\alpha, |\alpha| \le 2$. Giving linear operators $\Pi_T^0 : P_T \to N_T^0$ and $\Pi_{\partial T}, \Pi_{\partial T}^N, \Pi_{\partial T}^s : P_T \to L^\infty(\partial T)$, we define

$\Pi_T^\alpha : P_T \to N_T^\alpha, 0 < |\alpha| \le 2$, as follows. For $p \in P_T$, let

$$\begin{cases} \int_T q\Pi_T^{(1,0)} p\,dx = \int_{\partial T} q\Pi_{\partial T} pN_1 ds - \int_T \frac{\partial q}{\partial x_1}\Pi_T^0 p\,dx, & \forall q \in N_T^{(1,0)}, \\ \int_T q\Pi_T^{(0,1)} p\,dx = \int_{\partial T} q\Pi_{\partial T} pN_2 ds - \int_T \frac{\partial q}{\partial x_2}\Pi_T^0 p\,dx, & \forall q \in N_T^{(0,1)}, \\ \int_T q\Pi_T^{(2,0)} p\,dx = \int_{\partial T} q(N_1^2\Pi_{\partial T}^N p - N_1N_2\Pi_{\partial T}^s p)ds & \\ \qquad\qquad - \int_T \frac{\partial q}{\partial x_1}\Pi_T^{(1,0)} p\,dx, & \forall q \in N_T^{(2,0)}, \\ 2\int_T q\Pi_T^{(1,1)} p\,dx = \int_{\partial T} q(2N_1N_2\Pi_{\partial T}^N p + (N_1^2 - N_2^2)\Pi_{\partial T}^s p)ds & \\ \qquad\qquad - \int_T (\frac{\partial q}{\partial x_1}\Pi_T^{(0,1)} p + \frac{\partial q}{\partial x_2}\Pi_T^{(1,0)} p)dx, & \forall q \in N_T^{(1,1)}, \\ \int_T q\Pi_T^{(0,2)} p\,dx = \int_{\partial T} q(N_2^2\Pi_{\partial T}^N p + N_1N_2\Pi_{\partial T}^s p)ds & \\ \qquad\qquad - \int_T \frac{\partial q}{\partial x_2}\Pi_T^{(0,1)} p\,dx, & \forall q \in N_T^{(0,2)}. \end{cases} \tag{5.4}$$

From (5.4), we see that $\Pi_T^\alpha p$ is the approximation of $D^\alpha p$, and $\Pi_{\partial T} p, \Pi_{\partial T}^N p$ and $\Pi_{\partial T}^s p$ are the ones of $p|_{\partial T}$, $\frac{\partial}{\partial N} p|_{\partial T}$ and $\frac{\partial}{\partial s} p|_{\partial T}$ respectively.

For $v, w \in V_h$, define

$$\begin{aligned} \bar{a}_h(v,w) = \sum_{T\in T_h} \int_T \Big\{ &(\Pi_T^{(2,0)} v + \Pi_T^{(0,2)} v)(\Pi_T^{(2,0)} w + \Pi_T^{(0,2)} w) \\ &+ (1-\sigma)\big(2\Pi_T^{(1,1)} v\Pi_T^{(1,1)} w - \Pi_T^{(2,0)} v\Pi_T^{(0,2)} w \\ &- \Pi_T^{(0,2)} v\Pi_T^{(2,0)} w\big)\Big\} dx. \end{aligned} \tag{5.5}$$

The quasi-conforming finite element method for problem (1.4) is to find $u_h \in V_h$, such that,

$$\bar{a}_h(u_h, v_h) = \int_\Omega f v_h\,dx, \quad \forall v_h \in V_h. \tag{5.6}$$

If we take N_T^α containing $P_r(T)$ with r sufficiently large and $\Pi_T^0 p = p, \Pi_{\partial T} p = p|_{\partial T}, \Pi_{\partial T}^N p = \frac{\partial}{\partial N} p|_{\partial T}$ and $\Pi_{\partial T}^s p = \frac{\partial}{\partial s} p|_{\partial T}$ for all $p \in P_T$, then the method becomes the nonconforming one.

Under certain assumptions, Zhang and Wang have proved that the quasi-conforming method is convergent and discussed the basic properties, such as imbedding and compact properties, of finite element spaces (see [**53, 61, 66, 67**]). As a result, they showed the convergence of 9-parameter, 12-parameter and 15-parameter quasi-conforming triangular elements as well as 12-parameter quasi-conforming rectangular element.

6. The TRUNC plate element

In paper [**1**], Argyris et al. developed a triangular unconventional plate element, named TRUNC. Numerical experiences show that the element gives a very good results. The mathematical formulation and the convergence analysis of TRUNC element are given by Shi in paper [**27**].

Given a triangle T, we denote by λ_i the area coordinates for T. The shape function has the form

$$\begin{aligned} v = a_1 \lambda_1 + a_2\lambda_2 + a_3\lambda_3 + a_4\lambda_1\lambda_2 + a_5\lambda_2\lambda_3 + a_6\lambda_3\lambda_1 \\ + a_7(\lambda_1^2\lambda_2 - \lambda_1\lambda_2^2) + a_8(\lambda_2^2\lambda_3 - \lambda_2\lambda_3^2) + a_9(\lambda_3^2\lambda_1 - \lambda_3\lambda_1^2), \end{aligned} \tag{6.1}$$

which is uniquely determined by the nine nodal parameters: the function values and two first order derivatives at the three vertices of T. The finite element space V_h corresponding to shape function (6.1) is just the Zienkiewicz element, which converges only for very special mesh subdivisions. But now the bilinear form (2.2) is slightly changed as follows.

Each function $v_h \in V_h$ can be split into two parts,

$$v_h = \bar{v}_h + v_h', \tag{6.2}$$

with

$$\bar{v}_h|_T = a_1\lambda_1 + a_2\lambda_2 + a_3\lambda_3 + a_4\lambda_1\lambda_2 + a_5\lambda_2\lambda_3 + a_6\lambda_3\lambda_1, \tag{6.3}$$

$$v_h'|_T = a_7(\lambda_1^2\lambda_2 - \lambda_1\lambda_2^2) + a_8(\lambda_2^2\lambda_3 - \lambda_2\lambda_3^2) + a_9(\lambda_3^2\lambda_1 - \lambda_3\lambda_1^2) \tag{6.4}$$

on each $T \in T_h$. Then the bilinear form in (2.2) may be written as

$$\bar{a}_h(v_h, w_h) = \bar{a}_h(\bar{v}_h, \bar{w}_h) + \bar{a}_h(\bar{v}_h, w_h') + \bar{a}_h(v_h', \bar{w}_h) + \bar{a}_h(v_h', w_h'). \tag{6.5}$$

It is proved in [**27**] that the application of the trapezoidal rule to the integrals involved in the intermediate terms $\bar{a}_h(\bar{v}_h, w_h')$ and $\bar{a}_h(v_h', \bar{w}_h)$ in (6.5) yields zero value of these two terms. Therefore, neglecting these terms, the bilinear form $\bar{a}_h(v_h, w_h)$ may be modified as

$$b_h(v_h, w_h) = \bar{a}_h(\bar{v}_h, \bar{w}_h) + \bar{a}_h(v_h', w_h'), \quad \forall v_h, w_h \in V_h. \tag{6.6}$$

The associated variational equation

$$b_h(u_h, v_h) = \int_\Omega f v_h dx, \quad \forall v_h \in V_h \tag{6.7}$$

has exactly the same solution as that produced by the TRUNC element

It is seen that TRUNC element is much simpler than the original Zienkiewicz one. There are no coupling terms between $\bar{v}_h, w_h'$ and $v_h', \bar{w}_h$. Moreover, it is shown that the solution of (6.7) converges to the true solution of (1.4) for an arbitrary mesh geometry.

THEOREM 6.1. *Let* $u \in H^3(\Omega) \cap H_0^2(\Omega)$ *be the solution of problem* (1.4) *and* $u_h \in V_h$ *be the solution of problem* (6.7). *Then*

$$|u - u_h|_{2,h} \leq Ch|u|_3, \tag{6.8}$$

and

$$|u - u_h|_{1,h} \leq Ch^2|u|_3, \tag{6.9}$$

when Ω *is convex.*

7. The free formation scheme and the energy-orthogonal elements

Bergan et al. [**4**] proposed the so called Individual Element Test for constructing a stiffness matrix. This technique was subsequently improved [**5**] and recent development is the free formation scheme [**6**]. By this new scheme, a stiffness matrix can be separated into two parts: the first one is only related to the constant strain modes of a given problem, while the second one is determined by higher order modes of shape functions using a standard finite element technique. There is no coupling term in the stiffness matrix. Bergan's free formation has a very simple structure. Using this scheme the TRUNC element and an energy-orthogonal element have been constructed. They have fairly good performance in applications. Bergan et al. have derived their free formation purely from mechanical considerations.

In paper [**9**], the mathematical aspects of the scheme are discussed. The paper analyzes the essential step in the scheme, i.e., the establishment of the lumping matrix and clearly explains the construction of the main term in the stiffness matrix. It is proved that the free formation is actually equivalent to a special kind of nonconforming finite elements.

Paper [**37**] gives a detailed mathematical analysis for Bergan's energy-orthogonal element. Its convergence together with error estimates are derived. It is interesting to mention that if in finite element calculations of Bergan's energy-orthogonal element, the standard right-hand side $\int_\Omega f v_h dx$ of (2.4) is replaced by a simple one $\int_\Omega f \bar{\bar{v}}_h dx$, where $\bar{\bar{v}}_h$ is the linear part of the shape function v_h, then this slight change gives better accuracy with less computational costs.

In paper [**36**], Shi and Zhang present a modified scheme of free formation in accordance with a simple convergence requirement of nonconforming elements. The element stiffness matrix is also consisting of two separate parts, one corresponds to constant strain modes and the other to high order modes of shape functions. However, the main term of stiffness matrix now is simply derived from the convergence requirement. It seems a more direct way of its derivation than that in Bergan's scheme. The treatment of high order modes is the same as in Bergan's scheme, using the conventional method. Starting from the shape function space of Bergan's energy-orthogonal element, the modified scheme provides a new energy-orthogonal element. Numerical experiments show that this new element gives more accurate results than Bergan's. The convergence proof as well as the error estimates are also given.

REFERENCES

1. J. H. Argyris, M. Haase and H. P. Mlejnek, *On an unconventional but natural formation of a stiffness matrix*, Comput. Meths. Appl. Mech. Eng., **22**(1980), 1–22.
2. D. N. Arnold and F. Brezzi, *Mixed and nonconforming finite element methods: Implementation, postprocessing and error estimates*, M^2AN, **19**(1985), 7–32.
3. G. P. Bazeley, Y. K. Cheung, B. M. Irons and O. C. Zienkiewicz, *Triangular elements in bending – conforming and nonconforming solutions*, in Proc. Conf. Matrix Methods in Structural Mechanics, Air Force Ins. Tech., Wright-Patterson A. F. Base, Ohio, 1965.
4. P. G. Bergan and L. Hansen, *A new approach for deriving "good" finite element stiffness matrices*, in J. R. Whiteman ed: The Mathematics of Finite Elements and Applications, Vol II, 1976, 483–498.
5. P. G. Bergan, *Finite elements based on energy orthogonal functions*, Int. J. Numer. Meths. Eng., **15**(1980), 1541–1555.
6. P. G. Bergan and M. K. Nygard, *Finite elements with increased freedom in choosing shape functions*, Int. J. Numer. Meths. Eng., **20**(1984), 634–664.
7. Cai Wei, *The limit problems for Zienkiewicz triangle elements*, Numerica Mathematica Sinica, **8**:4(1986), 345–353.
8. G. F. Carey, *An analysis of finite element equations and mesh subdivisions*, Comput. Meths. Appl. Mech. Engrg. **9**(1976), 165–179.
9. Chen Shaochun and Shi Zhong-ci, *On the free formulation scheme for construction of stiffness matrices*, in Proc. International Conference on Scientific Computation, (T. Chan & Z-C, Shi eds), World Scientific, 1992, 18–27.
10. Chen Shaochun and Shi Zhong-ci, *Double set parameter method of constructing stiffness matrices*, Numerica Mathematica Sinica, **13**:3(1991), 286–296.
11. Chen Wanji, Liu Yingxi and Tang Limin, *The formulation of quasi-conforming elements*, Journal of Dalian Institute of Technology, **19**:2(1980), 37–49.
12. P. C. Ciarlet, *The Finite Element Method for Elliptic Problems*, North-Holland, Amsterdam, New York, 1978.
13. C. A. Felippa and P. G. Bergan, *A triangular bending element based on energy-orthogonal free formulation*, Comput. Meths. Appl. Mech. Eng., **61**(1987), 129–160.
14. B. M. Irons and A. Razzaque, *Experience with the patch test*, in Proc. Symp. on Mathematical Foundations of the Finite Element Method, (ed. A. R. Aziz), Academic Press, 1972, 557–587.
15. Jiang Heyang, *Derivation of higher precision triangular plate element by quasi-conforming element method*, Journal of Dalian Institute of Technology, **20**, Suppl. 2 (1981), 21–28.
16. P. Lascaux, P. Lesaint, *Some nonconforming finite elements for the plate bending problem*, RAIRO, Anal. Numer., **R-1**(1985), 9–53.
17. Long Yu-qiu and Xin Ke-gui, *Generalized conforming elements*, J. Civil Engineering, **1**(1987), 1–14.
18. L. S. D. Morley, *The triangular equilibrium element in the solution of plate bending problems*, Aero. Quart., **19**(1968), 149–169.
19. G. Sander and P. Beckers, *The influence of the choice of connectors in the finite element methods*, Internat. J. Numer. Methods Engrg., **11**(1977), 1491–1505.
20. Shi Guangyu, *12-parameter rectangle quasi-conforming element for plate bending*, Master Thesis, Dalian Institute of Technology, 1980.
21. Shi Zhong-ci, *An explicit analysis of Stummel's patch test examples*, Int. J. Numer. Meth. Eng., **20**(1984), 1233–1246.
22. Shi Zhong-ci, *A convergence condition for the quadrilateral Wilson element*, Numer. Math., **44**(1984), 349–361.

23. Shi Zhong-ci, *On the convergence properties of quadrilateral elements of Sander and Beckers*, Math. Comp., **42**(1984), 493–504.
24. Shi Zhong-ci, *The generalized patch test for Zienkiewicz's triangles*, J. Computational Mathematics, **2**(1984), 279–286.
25. Shi Zhong-ci, *Convergence properties of two nonconforming finite elements*, Comput. Meths. Appl. Mech. Eng., **48**(1985), 123–137.
26. Shi Zhong-ci, *A united formulation of shape functions for two kinds of nonconforming plate bending elements*, Numerica Mathematica Sinica, **8**(1986), 428–434.
27. Shi Zhong-ci, *Convergence of the TRUNC plate element*, Comput. Meths. Appl. Mech. Eng., **62**(1987), 71–88.
28. Shi Zhong-ci, *The F-E-M-Test for nonconforming finite elements*, Math. Comp., **49**(1987), 391–405.
29. Shi Zhong-ci, *On the nine degree quasi-conforming plate element*, Numerica Mathematica Sinica, **10**(1988), 100–106.
30. Shi Zhong-ci, *On Stummel's examples to the patch test*, Computational Mechanics, **5**(1989), 81–87.
31. Shi Zhong-ci, *On the accuracy of the quasi-conforming and generalized conforming finite elements*, Chin. Ann. Math., **11B**:2(1990), 148–156.
32. Shi Zhong-ci, *On the error estimates of Morley element*, Numerica Mathematica Sinica, **12**:2(1990), 113–118.
33. Shi Zhong-ci and Chen Shaochun, *An analysis of a nine parameter plate element of Specht*, Numerica Mathematica Sinica, **11**:3(1989), 312–318.
34. Shi Zhong-ci and Chen Shaochun, *Direct analysis of a nine parameter quasi-conforming plate element*, Numerica Mathematica Sinica, **12**:1(1990), 76–84.
35. Shi Zhong-ci and Chen Shaochun, *Convergence of a nine degree generalized conforming element*, Numerica Mathematica Sinica, **13**:2(1991), 193–203.
36. Shi Zhong-ci and Zhang Fei, *Construction and analysis of a new energy-orthogonal unconventional plate element*, JCM, **8**(1990), 75–91.
37. Shi Zhong-ci et al, *Convergence analysis of Bergan's energy-orthogonal plate element based on free formulation scheme and its modification*, Mathematical Model and Methods in Applied Sciences, World Scientific, 1993.
38. G. Strang and G. J. Fix, *An Analysis of the Finite Element Method*, Prentice-Hall, 1973.
39. F. Stummel, *The generalized patch test*, SIAM J. Numer. Analysis, **16**(1979), 449–471.
40. F. Stummel, *The limitation of the patch test*, Int. J. Numer. Meth. Eng., **15**(1980), 177–188.
41. F. Stummel, *Basic compactness properties of nonconforming and hybrid finite element spaces*, RAIRO, Anal. Numer., **4**:1(1980), 81–115.
42. Tang Limin, Chen Wanji and Liu Yingxi, *Quasi-conforming elements in finite element analysis*, J. Dalian Inst. of Technology, **19**:2(1980), 19–35.
43. R. L. Taylor, T. C. Simo, O. C. Zienkiewicz and A. H. C. Chan, *The patch test — a condition for assessing FEM convergence*, Int. J. Numer. Meth. Eng., **22**(1986), 39–62.
44. Wang Ming, *A note on the generalized patch test*, Journal of Dalian Institute of Technology, **23**:3(1984), 127–129.
45. Wang Ming, *Finite element method for a class of nonlinear coupled problems of complex Schrödinger equation and real Klein-Gordon equation*, Journal of Dalian Institute of Technology, **25**:1(1986), 101–105.
46. Wang Ming, *On the penalty finite element methods for the stationary Navier-Stokes equations*, Journal of Dalian Institute of Technology, **25**, Suppl. (1986), 7–13.
47. Wang Ming, *A new approach to the upwind finite element*, Journal of Math. Res. & Exposition, **7**:1(1987), 124.
48. Wang Ming, *On the penalty finite element methods for stationary Stokesian problem*, Mathematica Numerica Sinica, **9**:3(1987), 309–318.
49. Wang Ming, *Finite element method for a class of nonlinear problems I- Abstract results*, Journal of Math. Res. & Exposition, **7**:4(1987), 671–680.
50. Wang Ming, *Finite element method for a class of nonlinear problems II- Applications*,

Journal of Math. Res. & Exposition, **8**:3(1988), 427–438.

51. Wang Ming, *A comment on patch test*, Computational Structural Mechanics and Applications, **5**:4(1988), 115–117.
52. Wang Ming, *On the inequalities for the maximum norm of nonconforming finite element spaces*, Mathematica Numerica Sinica, **12**:1(1990), 104–107.
53. Wang Ming, *The trace embedding and compact properties of finite element spaces*, Journal of Math. Res. & Exposition, **10**:2(1990), 187–194.
54. Wang Ming, *A comment on the quasi-conforming finite element methods*, Mathematica Numerica Sinica, **12**:2(1990), 206–207.
55. Wang Ming, *The multigrid method for finite elements solving the biharmonic equation*, in Proc. of Second Conference of Numerical Methods for partial Differential Equations, World Scientific, Singapore, 1992, 107–125.
56. Wang Ming, *The multigrid method for TRUNC plate element*, JCM, accepted.
57. Wang Ming, *The W-Cycle multigrid method for finite elements with nonnested spaces*, Research Report No.50, 1991, Institute of Mathematics and Department of Mathematics, Peking University.
58. Wang Ming and Zhang Hongqing, *On the convergence of quasi-conforming elements for linear elasticity problem*, JCM, **4**:2(1986), 131–145.
59. Wang Ming and Zhang Hongqing, *A note on some finite element methods*, Mathematica Numerica Sinica, **8**:3(1986), 303–313.
60. Wang Ming and Zhang Hongqing, *Finite element methods for the stationary Navier-Stokes equations in the stream function formulation*, Journal of Dalian Institute of Technology, **25**, Suppl. (1986), 1–6.
61. Wang Ming and Zhang Hongqing, *On the embedding and compact properties of finite element spaces*, Appl. Math. Mech. (English edition), **9**:2(1988), 135–142.
62. Zhang Hongqing, *The generalized patch test and 9-parameter quasi-conforming element*, Proc. the Sino-France Symposium on Finite Element Methods, (ed. K. Feng), 566–583, Science Press, Gordan and Breach, 1983.
63. Zhang Hongqing, *The generalized patch test of multiple sets of functions and 12-parameter quasi-conforming element*, Journal of Dalian Institute of Technology, **21**:3(1982), 11–19.
64. Zhang Hongqing and Wang Ming, *Finite element approximations with multiple sets of functions and quasi-conforming elements*, in Proc. the 1984 Beijing Symp. on Diff. Geometry and Diff. Equations (ed. K. Feng), 354–365, Science Press, Beijing, 1985.
65. Zhang Hongqing and Wang Ming, *Finite element approximation with multiple set of functions and quasi-conforming elements for plate bending problems*, Appl. Math. Mech. (English edition), **6**:1(1985), 41–52.
66. Zhang Hongqing and Wang Ming, *On the compactness of quasi-conforming element spaces and the convergence of quasi-conforming element method*, Appl. Math. Mech. (English edition), **7**:5(1986), 443–459.
67. Zhang Hongqing and Wang Ming, *The Mathematical Theory of Finite Element Methods*, Science Press, Beijing, 1991.

Computing Center, Academia Sinica, Beijing 100080, China

Department of Mathematics, Peking University, Beijing 100871, China

Contemporary Mathematics
Volume **163**, 1994

Some results on the field of spline theory and its applications*

SUN JIA-CHANG

1. Spline function in local coordinates

For practical point of view, the usual spline method has to be modified if there are some data with steep gradients. In this case the model of piecewise polynomial in a global coordinate is not available. Based on the principle of "Geometry Invariant", a method so called "Spline function in local coordinates" developed by the author were published in 1977-1980 ([3]-[6]). Some corresponding variational properties have been studied in [4].

Let $\{P_j\}$ be a sequence of ordered points in a plane. Connecting each two neighboring interpolating points with a straight line, we select such a local coordinate that the abscissa is the chord through the two points and the ordinate is a line perpendicular to the abscissa. This is a natural simulation to the so-called moving frame in Differential Geometry. An analog for space curves was also investigated in [3]. It has been pointed that the corresponding system, published by the D.H. Thomes in the Journal "Mathematics of Computation" in 1976, was too complicated and partly wrong. Our nonlinear system was rewritten as semi-linear form. For example, to find a cubic spline in local coordinates reduces to solve the following system of equations in terms of curvature ρ at knots

$$\mu_j \rho_{j-1} + 2\mu_j + \lambda_j \rho_{j+1} = 3d_j + h_j(\rho) \tag{1}$$

where μ_j and $\lambda_j = 1 - \mu_j$ are mesh ratio, and d_j is an approximation curvature evaluation of the circle passing through the three points P_{j-1}, P_j and P_{j+1}

$$d_j = -\frac{2}{l_j + l_{j+1}} \operatorname{tg} \phi_j,$$

1991 *Mathematics Subject Classification.* 65D07.

* This work was supported by National Natural Science Foundation of China.

and

$$\begin{aligned}h_j(\rho) = &-\frac{6\,\mathrm{tg}\,\phi_j}{l_j+l_{j+1}}m_j\tilde{m}_j + \mu_j\rho_{j-1}\{1-[1+m_{j-1}^2]^{3/2}\}\\ &+\lambda_j\rho_{j+1}\{1-[1+\tilde{m}_{j+1}^2]^{3/2}\}\\ &+2\rho_j\{\mu_j[1-(1+\tilde{m}_j^2)^{3/2}]+\lambda_j[1-(1+m_j^2)^{3/2}]\}\end{aligned}$$

The corresponding matrix form reduces to

$$Au = f + G(u) \tag{2}$$

where A is a tridiagonal matrix which is the same to the usual cubic spline in the form and u is the curvature vector or the first derivatives in the local coordinates at knots. The simple iteration is a natural solver

$$Au^{(0)} = f, \quad Au^{(k)} = f + G(u^{(k-1)}). \tag{3}$$

The convergence of the procedure with second order has been proved in [3] if the maximum angle between the neighboring coordinates is reasonable small. In fact only a few extra amount of working time is needed for most practical interpolations.

Furthermore, it was proved that the approximation order of the local splines is the same as the usual splines. However, the coefficients before the main order now become to be related geometry variables instead of algebraic variables, such as curvature [3].

2. Circular splines

The length of arc is the most natural parameter for geometry curves. Based on the idea we have constructed some geometry splines. So called circular spline is useful tools for the aim of computer graphics and numerical control. Requiring a circle around each data point with common tangent to the two neighboring circles leads to a nonlinear system

$$\lambda_j\cos\frac{\alpha_{j-1}+\beta_j}{4}\sin\frac{-\alpha_{j-1}+3\beta_j}{4}+\mu_j\cos\frac{\alpha_j+\beta_{j+1}}{4}\sin\frac{3\alpha_j-\beta_{j+1}}{4}=0, \tag{4}$$

where

$$\alpha_j+\beta_j=\phi_j.$$

As local splines, the above system can also be reduced to the following form if the angles ϕ are small

$$\mu_j\rho_{j-1}+3\mu_j+\lambda_j\rho_{j+1}=3b_j+G_j(\rho) \tag{5}$$

where

$$\begin{aligned} b_j &= -\frac{8\phi_j}{l_j + l_{j+1}}, \\ G_j &= \frac{1}{l_j + l_{j+1}}(g_{j-1} + 3\tilde{g}_j + 3g_j + \tilde{g}_{j+1}), \\ g_j &= 3\alpha_j - \beta_{j+1} - 4\cos\frac{\alpha_j + \beta_{j+1}}{4}\sin\frac{3\alpha_j - \beta_{j+1}}{4}, \\ \tilde{g}_j &= \alpha_{j-1} - 3\beta_j - 4\cos\frac{\alpha_{j-1} + \beta_j}{4}\sin\frac{\alpha_{j-1} - 3\beta_j}{4}. \end{aligned}$$

The approximation order is the same as quadratic splines which can not be used for curves with multiple values. The circular spline method and resulting software package has widely been used in many Chinese factories in the purpose of numerical control and CAD ([3], [7], [20]). Furthermore, we have developed a class of so-called cubic geometric splines. The continuous relation among three knots is the following:

$$\arcsin[\frac{l_j}{6}(2\rho_j + \rho_{j-1})] + \arcsin[\frac{l_{j+1}}{6}(2\rho_j + \rho_{j+1})] = -\phi_j - O(l^4). \tag{6}$$

3. Semi-linear Means and its applications to O.D.E. problems

The definition of B-splines originally comes from divided difference of a truncated power function at some points. A relationship between divided difference and Mean values leads us to develop a class of Generalized Means and corresponding inequalities [8]. For a given positive sequence $a_1, \ldots, a_n$, we define $S(r,t)$ to be the means of (a) with two parameters as follows

$$S(r,t) := S(r,t;a) = \{\frac{(n-1)!\Gamma(t+1)}{\Gamma(t+n)}[a_1^r, \ldots, a_n^r]y^{n-1+t}\}^{\frac{1}{rt}}. \tag{7}$$

Moreover, we have shown two main inequalities in t and r directions:

THEOREM 1.

$$\frac{\partial}{\partial t}S(r,t)\begin{cases} \leq 0 & \text{if } r > 0 \\ \geq 0 & \text{if } r < 0 \end{cases} \tag{8}$$

and

$$\frac{\partial}{\partial r}S(r,t)\begin{cases} \geq 0 & \text{if } t+n > 0 \\ \leq 0 & \text{if } t+n < 0 \end{cases} \tag{9}$$

where the both equalities hold if and only if $\min\{a_j\} = \max\{a_j\}$.

Being its semi-linearity, we have found that the Generalized Means are very useful to make finite difference scheme [9].

For instance, we may have some semi-linear numerical differentiation formulas

$$u'(x_j) = \frac{2r}{(1+r)(x_{j+1}-x_{j-1})}\{\frac{u_{j+1}^{1+r}-u_j^{1+r}}{u_{j+1}^r-u_j^r} - \frac{u_j^{1+r}-u_{j-1}^{1+r}}{u_j^r-u_{j-1}^r}\}$$
$$- \frac{h^2}{12}\frac{d}{dx}\{2u'' - (1-r)\frac{u'^{\,2}}{u}\}|_{x=x_j} + O(h^4) \tag{10}$$

and

$$u'(x_j) = \frac{1}{h}\{\frac{u(x_{j+1})-u(x_j)}{\log u(x_{j+1}) - \log u(x_j)} - \frac{u(x_j)-u(x_{j-1})}{\log u(x_j) - \log u(x_{j-1})}\}$$
$$- \frac{h^2}{12}\frac{d}{dx}\{2u'' - \frac{u'^{\,2}}{u}\}|_{x=x_j} + O(h^4). \tag{11}$$

The above scheme can be used to solve some ordinary and partial differential equations with steep gradients ([2], [19]). The corresponding proof of convergent order has been obtained. Moreover, for solving the following initial problem

$$y' = f(x,y), \quad y(a) = y_a$$

we have introduced a class of nonlinear implicit one-step scheme

$$Y_{n+1} = Y_n + hS(f_n, f_{n+1})$$

with

$$S(f,g) = \frac{r}{1+r}\frac{f^{1+r}-g^{1+r}}{f^r-g^r}. \tag{12}$$

We have proved the following results on convergence and stability about the above scheme.

THEOREM 2. *The above scheme with any parameter r is second order with truncation estimation*

$$L(f;r) = -\frac{h^3}{12}\{f'' - (1-r)\frac{f'^2}{f}\} + O(h^5) \tag{13}$$

and the scheme will be fourth order if r is chosen as

$$r = 1 - \frac{ff''}{f'^2}\Big|_{x=x_n+\frac{h}{2}}$$

or

$$r = \frac{1}{h}(\frac{f_{n+1}}{f'_{n+1}} - \frac{f_n}{f'_n}).$$

THEOREM 3. *The above scheme is A-stable if $r \geq 0$, otherwise it is conditionally A-stable.*

Nearly at the same time, we have introduced so called semi-linear finite element [10], by using subsets instead of usual linear subspace. As an example, we have considered the following perturbation problem:

$$Lu = -\epsilon u'' + p(x)u' + q(x)u = f(x), \quad u(0) = u(1) = 0.$$

Associated with a partition Δ_h we have set up two subsets in the solution space $H_1^0[0,1]$. One is the usual piecewise linear space, the other subset SP is defined as follows: $for\ x_{j-1} \le x \le x_j,\ t = \frac{x-x_{j-1}}{h}$

$$u_s^h = \begin{cases} u_{j-1}(1-t) + u_j t & \text{if } \frac{|u_j - u_{j-1}|}{h} < d \\ (u_{j-1}+c)^{1-t}(u_j+c)^t - c & \text{otherwise} \end{cases} \tag{14}$$

where c is a parameter to make the formula to be well defined, d is a control constant. It has been shown that the following convergent rate holds

$$||u_s^h - u||_{1,s} = O(h). \tag{15}$$

4. Multivariate Bernstein polynomials

4.1. Today multivariate Bernstein polynomial is one of the main tool for multivariate approximation theory, and is now considered as a main mathematical basis for so-called Computer Aided Geometric Design. It was proved that each Bernstein polynomial can be approximated to any degree by its Bézier nets – a sequence of piecewise linear functions consisted from the original B-net. The following facts were explored by the author and his student Kang Zhao. The distance between two neighboring Bézier nets is $O(m^{-2})$ and it has been shown how the piecewise linear surface uniformly tends to the Bernstein surface belonged to C^∞ with a rate of $O(m^{-1})$ [22].

4.2. The dual bases theory of univariate B-splines developed by Carl de Boor played a fundamental role in 1-D case. We have derived a dual bases for the multivariate Bernstein polynomials on simplex domain as follows. Let Bernstein polynomial of order n on simplex S_m is defined by

$$\begin{aligned} B[f](u) &= \sum_{|i|=n} f_i B_i(u), \\ B_i(u) &= \frac{n!}{i!} u^i \ \forall i \in Z_+^m,\ |i| = n \text{ and } u \in R^m \end{aligned} \tag{16}$$

and in [23] we set the following linear functional defined by

$$\lambda_{r,s}[f] := \sum_{|\alpha| \le r} \binom{r}{\alpha} \frac{(n-|\alpha|)!}{n!} D_{\alpha,s} f(e_s), \tag{17}$$

where

$$D_{\alpha,s} := \prod_{t \ne s} (D_t - D_s)^{\alpha_t}, \quad \forall 1 \le s \le m.$$

THEOREM 4. *$\{\lambda_{r,s} : |r| \leq n\}$ is a dual basis of these Bernstein polynomials in the sense*

$$\lambda_{r,s}(B_i) = \delta_{r,i}. \tag{18}$$

By using the above dual basis, we can derive most of conclusions about multivariate Bernstein polynomials immediately, such as bases properties, degree raising formulas, domain transformation relations and subdivision algorithms, etc.

5. B-net approach to multivariate B-splines and B-spline FEM

5.1. Our research on the B-net approach to multivariate B-splines has been begun independently since 1985. First, we reformulated univariate B-splines in terms of B-nets [13]. Then more attention has been paid to explore B-net structure of bivariate B-splines with three direction mesh [12], [14]-[15].

Denote triangle shifting operator in each of the three directions by

$$E_1(\Delta)\Omega_{ijk} = \Omega_{i+1,j,k},$$
$$E_2(\Delta)\Omega_{ijk} = \Omega_{i,j+1,k},$$
$$E_3(\Delta)\Omega_{ijk} = \Omega_{i,j,k+1}$$

and the third order difference operator in the three direction

$$D_3 = (I - E_1E_2^{-1})(I - E_2E_3^{-1})(I - E_3E_1^{-1}).$$

THEOREM 5. *Suppose a piecewise polynomial surface $B^n(P,Q)$ to be a B-spline in space $S^{\nu}_{n,\Delta}$ and another piecewise polynomial surface $B^{n+3}(P,Q)$ is defined by the following derivative-difference relation in three directions for each $\Omega_{\lambda\mu\nu}$*

$$D^3_{e_1e_2e_3}B^{n+3}_{rst}(\Omega_{\lambda\mu\nu}) = D_3(\Delta)B^n_{r-1,s-1,t-1}(\Omega_{\lambda\mu\nu}) \tag{19}$$

if $rst \neq 0$

$$D^3_{e_1e_2e_3}B^{n+3}_{rst} = 0$$

if $rst = 0$, where $r+s+t = n+3$, then $B^{n+3}(P,Q)$ must be a B-spline with the same parameter Q in the space $S^{\nu+2}_{n+3,\Delta}$. Furthermore, there is a difference recurrence in terms of B-nets for each pair of piecewise polynomial surfaces $B^n(P,Q)$ and $B^{n+3}(P,Q)$ [14].

By using the above recurrence we may easily obtain B-net representation for some usual B-spline spaces, such as S^1_3, S^2_4 (see Fig. 1-2) and S^3_6 etc.

Moreover, we may derive dimension formulas and bases for some usual spline spaces. Instead of some classical and rather complicated algebraic relations, we may obtain some results in terms of geometry relations as follows:

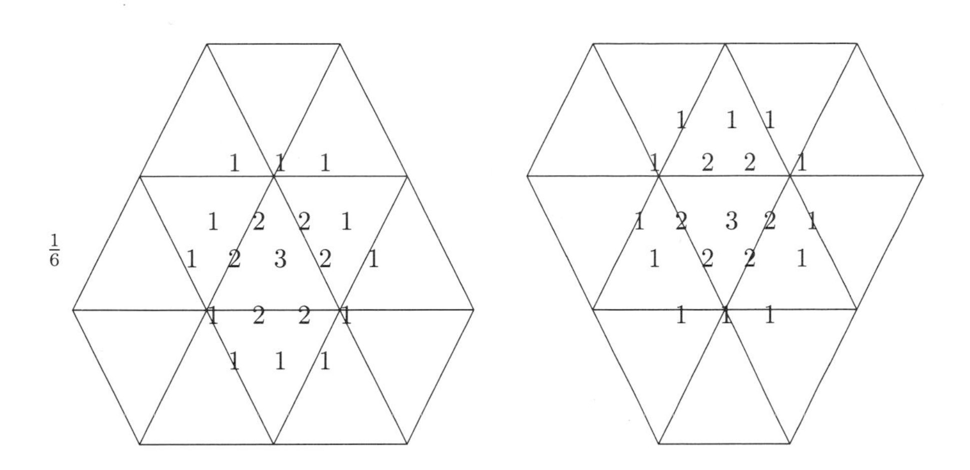

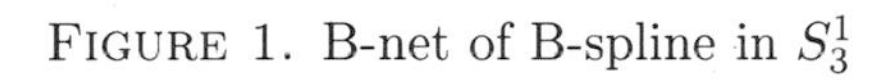
FIGURE 1. B-net of B-spline in S_3^1

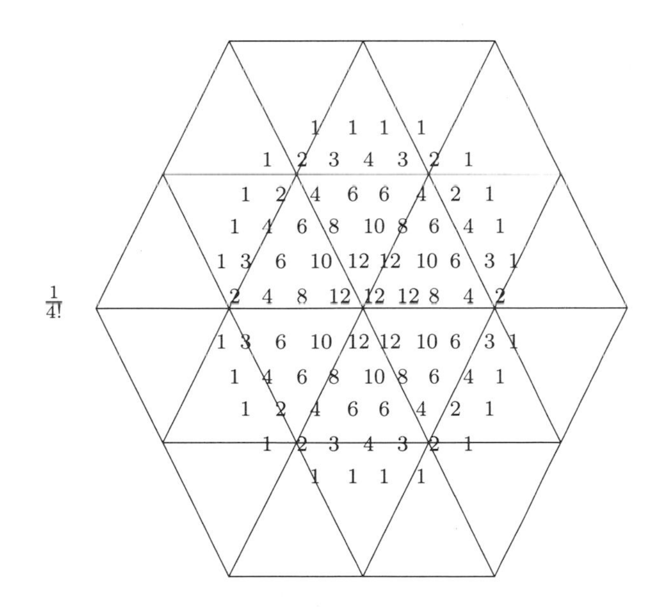

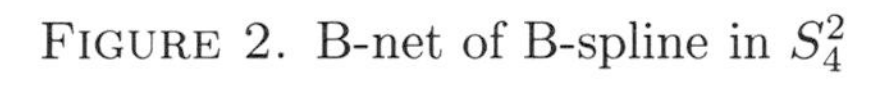
FIGURE 2. B-net of B-spline in S_4^2

THEOREM 6.

$$\dim S_3^1(\Omega) = T_1 - 3,$$
$$\dim S_{3\nu}^{2\nu-1,2\nu-1}(\Omega) = T_{-\nu}, \tag{20}$$
$$\dim S_{3\nu+1}^{2\nu,2\nu}(\Omega) = V_{-\nu}.$$

$$S_{3\nu}^{2\nu-1,2\nu-1}(\Omega) = Span_{Q\in\Omega_{-\nu}}\{B_{3\nu}^{2\nu-1}(P,Q)\},$$

$$S_{3\nu+1}^{2\nu,2\nu}(\Omega) = Span_{Q\in\Omega_{-\nu}}\{B_{3\nu+1}^{2\nu}(P,Q)\}, \tag{21}$$

where domains Ω_ν and $\Omega_{-\nu}$ are called to be n-order expanding domain and contracting domain, respectively. T and V denote the number of internal subtriangles and vertices, respectively.

Based on the above discussion, a class of quasi-Hermitian interpolation has been studied.

5.2. A fast parallel algorithm of bivariate spline surfaces. The bivariate B-spline is a very useful tool for designing surface modeling. One main difficulty in practice, however, is the lack of an efficient algorithm for evaluating and displaying the resulting surface. In fact, for a given partition Ω a bivariate spline in the space $S_k^\mu(\Omega)$ is a piecewise bivariate polynomial of total degree k with global continuity degree μ. It means that in each subdomain the surface can be represented as a Bernstein-Bézier form. By using well-known subdivision technique one may give an algorithm for B-B surface in each subtriangles. However, the working amount along this way would still be very large. Because in three direction case there is no analogy of efficient recurrence like so called de Boor-Cox algorithm in univariate case, we have to find another way by using some B-spline properties.

At first, we have found a recurrence between two level bases with mesh size h and $2h$ in the space $S_{3\nu}^{2\nu-1}$ and in the space $S_{3\nu+1}^{2\nu}$, respectively.

THEOREM 7.

$$B_{3\nu}^{2\nu-1}(P,Q;\Delta_{2h}) = S_\nu(\Delta_h)B_{3\nu}^{2\nu-1}(P,Q;\Delta_h)$$

where projector

$$S_\nu(\Delta_h) = \{G_3(\Delta_h)\}^\nu(I + E_1 + E_2 + E_3),$$

$$G_3 = (I + E_1E_2^{-1})(I + E_2E_3^{-1})(I + E_3E_1^{-1}).$$

THEOREM 8.

$$B_{3\nu+1}^{2\nu}(P,Q;\Delta_{2h}) = \hat{S}_\nu(\Delta_h)B_{3\nu+1}^{2\nu}(P,Q;\Delta_h)$$

where

$$\hat{S}_\nu(\Delta_h) = \frac{1}{2}\{G_3(\Delta_h)\}^{\nu+1}$$

or

$$\hat{S}_\nu = G_3 \hat{S}_{\nu-1}.$$

Then we have the following fast algorithm

$$S(P) = \sum_Q C(Q)B(P,Q;\Delta_h) = \sum_Q C^{[1]}(Q)B(P,Q;\Delta_{\frac{h}{2}}) = \cdots$$

$$= \sum_Q C^{[m]}(Q)B(P,Q;\Delta_{2^{-m}h}).$$

For instance, in the spline space S_3^1 a fast algorithm is as follows

$$S(P) = \sum_Q C(Q)S(\Delta_{\frac{h}{2}})B(P,Q;\Delta_{\frac{h}{2}}) = \sum_Q C^{[1]}(Q)B(P,Q;\Delta_{\frac{h}{2}}).$$

The above formulas evaluating coefficients $C^{[1]}$ are divided into two parts:

$$C_I^{[1]}(Q) = \frac{1}{4}(I + E_1^{-1} + E_2^{-1} + E_3^{-1})C_I(Q),$$

$$C_{II,1}^{[1]}(Q) = \frac{1}{8}(4I + E_1E_2^{-1} + E_1E_3^{-1} + E_2^{-1} + E_3^{-1})C_I(Q),$$

$$C_{II,2}^{[1]}(Q) = \frac{1}{8}(4I + E_2E_3^{-1} + E_2E_1^{-1} + E_3^{-1} + E_1^{-1})C_I(Q),$$

$$C_{II,3}^{[1]}(Q) = \frac{1}{8}(4I + E_3E_1^{-1} + E_3E_2^{-1} + E_1^{-1} + E_2^{-1})C_I(Q).$$

Moreover, in the spline space S_4^2 the fast algorithm also consists of two parts: At integer points

$$C_I^{[1]} = \frac{1}{16}(10I + E_1E_2^{-1} + E_1E_3^{-1} + E_2E_3^{-1} + E_2E_1^{-1} + E_3E_1^{-1} + E_3E_2^{-1})C(Q).$$

At mid-points

$$C_{II,1}^{[1]} = \frac{1}{8}(3I + 3E_2E_3^{-1} + E_1E_3^{-1} + E_1^{-1}E_2)C(Q),$$

$$C_{II,2}^{[1]} = \frac{1}{8}(3I + 3E_1E_3^{-1} + E_2E_3^{-1} + E_2^{-1}E_1)C(Q),$$

$$C_{II,3}^{[1]} = \frac{1}{8}(3I + 3E_1E_2^{-1} + E_3E_2^{-1} + E_3^{-1}E_1)C(Q),$$

$$C_{II,4}^{[1]} = \frac{1}{8}(3I + 3E_3E_2^{-1} + E_1E_2^{-1} + E_1^{-1}E_3)C(Q),$$

$$C_{II,5}^{[1]} = \frac{1}{8}(3I + 3E_3E_1^{-1} + E_2E_1^{-1} + E_2^{-1}E_3)C(Q),$$

$$C_{II,6}^{[1]} = \frac{1}{8}(3I + 3E_2E_1^{-1} + E_3E_1^{-1} + E_3^{-1}E_2)C(Q).$$

In general we may construct a corresponding subdivision algorithm for evaluating and displaying spline surface in the space $S_{3\nu}^{2\nu-1}$ and $S_{3\nu+1}^{2\nu}$ for any integer

ν. As another example we may present the following fast algorithm for bivariate spline in the space S_6^3 as follows. At the center

$$\begin{aligned} C_I^{[1]}(Q) =& \frac{1}{64}\{10I + 14(E_1^{-1} + E_2^{-1} + E_3^{-1})\} \\ &+ (E_1E_2^{-1} + E_1E_3^{-1} + E_2E_3^{-1} + E_2E_1^{-1} + E_3E_1^{-1} + E_3E_2^{-1}) \\ &+ 2(E_1E_2^{-1}E_3^{-1} + E_2E_3^{-1}E_1^{-1} + E_3E_1^{-1}E_2^{-1})\}C(Q). \end{aligned}$$

At the neighbors

$$\begin{aligned} C_{II}^{[1]}(Q) =& \frac{1}{64}\{22I + 6(E_2^{-1} + E_3^{-1}) \\ &+ 9(E_1E_3^{-1} + E_1E_2^{-1}) + 3(E_3E_2^{-1} + E_2E_3^{-1}) \\ &+ 2(E_1E_2^{-1}E_3^{-1} + E_1^{-1}) + (E_2E_1^{-1} + E_3E_1^{-1})\}C(Q), \\ C_{II}^{[2]}(Q) =& \frac{1}{64}\{22I + 6(E_3^{-1} + E_1^{-1}) \\ &+ 9(E_2E_1^{-1} + E_2E_3^{-1}) + 3(E_3E_1^{-1} + E_1E_3^{-1}) \\ &+ 2(E_2E_3^{-1}E_1^{-1} + E_2^{-1}) + (E_3E_2^{-1} + E_1E_2^{-1})\}C(Q), \\ C_{II}^{[3]}(Q) =& \frac{1}{64}\{22I + 6(E_1^{-1} + E_2^{-1}) \\ &+ 9(E_3E_2^{-1} + E_3E_1^{-1}) + 3(E_1E_2^{-1} + E_2E_1^{-1}) \\ &+ 2(E_3E_1^{-1}E_2^{-1} + E_3^{-1}) + (E_1E_3^{-1} + E_2E_3^{-1})\}C(Q). \end{aligned}$$

5.3. Bibariate B-spline finite element. As tensor product of univariate B-spline, a B-spline element was investigated by Shi Zhong-ci in 1980. We have developed a conforming B-spline finite element over domains which can be completely divided by a three direction mesh.

Consider the partial differential equation

$$\Delta^2 u = f \quad \text{in } \Omega$$

with boundary conditions

$$u = \frac{\partial u}{\partial n} = 0 \quad \text{on } \partial\Omega$$

or

$$u = \Delta u = 0 \quad \text{on } \partial\Omega,$$

where Ω is a convex polygon with a partition of a three direction mesh. The approximation has been shown to have order $O(h)$ in the space $H_2(\Omega)$. Numerical tests also verify the convergence for eigenvalue problem over some parallel hexagon domains, including triangles, trapezoids and rectangles [21].

In the particular case $h_1 = h_2 = h_3 = h$, we obtain a 37-point difference scheme for biharmonic operator in the following figure.

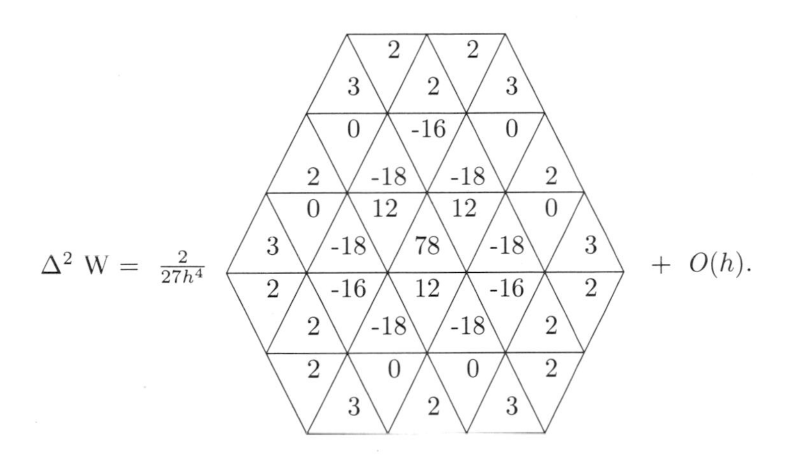

The number in each triangle is the weight coefficient of the function value related.

Recently, the above B-spline finite element has been applied to domain decomposition methods for fourth order problems [1], [16]-[18], [21].

References

1. T. F. Chan, W. E and Jiachang Sun, *Domain decomposition interface preconditioner for fourth order elliptic problems*, Applied Numerical Mathematics, **8**(1991), 317–331.
2. Jian Li and Jiachang Sun, *An application of generalized means difference schemes to convection-diffusion equations*, Chinese J. Numer. Math. & Appl., **13**(1991), 48–59.
3. Jiachang Sun, *Spline function in local coordinates*, Acta Mathematicae Sinica, **20**(1977), 28–40. (in Chinese).
4. Jiachang Sun, *Variational properties of cubic splines in local coordinates*, Acta Mathematicae Appligate Sinica, **1**(1978), 115–122 (in Chinese).
5. Jiachang Sun, *The spline interpolation for space curves in local coordinates*, Acta Mathematicae Appligate Sinica, **2**(1979), 340–343 (in Chinese).
6. Jiachang Sun, *Generalized splines in local coordinates*, Mathematicae Numerica Sinica, **2**(1980), 142–145 (in Chinese).
7. Jiachang Sun, *Spline Functions and Computational Geometry*, Academic Press, Beijing, 362pp. 1982 (in Chinese).
8. Jiachang Sun, *Generalization of the mean values and their inequalities*, Chinese Annals of Mathematics, **4b**(1983), 493–500.
9. Jiachang Sun, *Semi-linear difference schemes*, J. Comput. Math., **2**(1984), 93–111.
10. Jiachang Sun, *A Semi-linear finite element method*, J. Comput. Math., **3**(1985), 97–114.
11. Jiachang Sun, *A note on local dimension of bivariate spline space with multicells*, J. Numer. Methods and Computer Appl., **8**(1987), 249–252.
12. Jiachang Sun, *The B-net approach and application of bivariate B-splines*, UMSI (University of Minnesota, Supercomputer Institute) Research Report 88/139, December, 1988.
13. Jiachang Sun, *The B-net approach to B-spline in one dimension*, Chinese J. Numer. Math. & Appl., **11**(1989), 69–81.
14. Jiachang Sun, *The B-net structure and recurrence algorithms for B-splines with three directions*, Chinese J. Numer. Math. & Appl., **13**(1989), 48–59.
15. Jiachang Sun, *Dual bases and quasi-interpolation of B-spline in space S_3^1*, Acta Mathe-

maticae Appligate Sinica, **17**(1991), 470–477 (in Chinese).

16. Jiachang Sun, *Domain decomposition and multilevel PCG method for solving fourth order problems in 3-D*, Proceeding of Sixth International Conference on Domain Decomposition Methods. Como, Italy, June, 1992.
17. Jiachang Sun and Binkun Li, *Bivariate B-spline finite element method on type-1 triangulations*, Chinese J. Numer. Math. & Appl., **13**(1991), 42–54.
18. Jiachang Sun and Binkun Li, *Parallel multilevel B-spline Preconditioners for the biharmonic problems*, Proceeding of International Conference on Scientific Computation, Hangzhou, China, 1991.
19. Jiachang Sun and Ken Jackson, *Nonlinear implicit one-step schemes for solving I.V.P. of O.D.E. with steep gradients*, Journal of Computational Mathematics, **2**(1984), 264–281.
20. Jiachang Sun and Wei-lin Zeng, *On arc and bi-arc approximation*, Mathematica Numerica Sinica, **3**(1981), 97–112 (in Chinese).
21. Jiachang Sun and Jun Zou, *DDM preconditioner for 4-th order problems by using B-spline finite element methods*, The 4-th International Conf. on DDM, Moscow, 1990.
22. Jiachang Sun and Kang Zhao, *On the structure of Bézier nets*, J. Comput. Math., **5**(1987), 376–383.
23. Kang Zhao and Jiachang Sun, *Dual bases of multivariate Bernstein-Bézier polynomials*, Computer Aided Geometry Design, **5**(1988), 119–125.

COMPUTING CENTER, ACADEMIA SINICA, P.O.BOX 2719, BEIJING 100080, CHINA

Contemporary Mathematics
Volume **163**, 1994

Approximation Theory and Spline Function

WANG REN-HONG

First of all, we would like to deeply cherish our memory of Professors K.K. Chen and F.T. Wang, two Chinese pioneers in the field of Approximation Theory. In particular, K.K. Chen not only had organized the first mathematical seminar on Approximation Theory in the early 1950's, but also had written a lot of papers, an excellent book, and a survey on Chinese mathematicians' contributions in the theory of trigonometric series before 1964 ([**11, 13, 12**]). Even now, his ideas and results are still strong influences on many contemporary young Chinese mathematicians.

1. Approximation theory and harmonic analysis

Denote by $\mathcal{G} = \{T(\xi)\}$, $(\xi > 0)$, a one-parameter semi-group of linear operators on a complex Banach space X to itself such that $T(\xi_1+\xi_2)x = T(\xi_1)[T(\xi_2)x]$ for all ξ_1, $\xi_2 > 0$, and all $x \in X$. Assume $\|T(\xi)\| \leq M < +\infty$, $0 < \alpha \leq \xi \leq \max(\alpha + 1, 2\alpha)$, where M depends on $T(\xi)$. Denote $A_\eta = [T(\eta) - I]/\eta$, of which the strong limit $A = \lim_{\eta\to 0} A_\eta$ (whenever it exists) is known as the infinitesimal generator of $\mathcal{G}$. Moreover, for a given interval $[\alpha, \beta] \subset (0, \infty)$, denote $\mu(\delta, x) = \sup_{\substack{\alpha\leq\xi_1,\xi_2\leq\beta \\ |\xi_1-\xi_2|\leq\delta}} \|T(\xi_1)x - T(\xi_2)x\|$, which is the rectified modulus of continuity of $T(\xi)x$ in $[\alpha, \beta]$. Similarly, we denote $\mu(\delta) = \sup \|T(\zeta_1) - T(\zeta_2)\|$ in case $T(\xi)$ is uniformly continuous for $\xi > 0$.

Hille's theorem on the "first exponential formula" [**44**] is one of the most fundamental formulae in the theory of semi-group of linear operators. It was sharpened by L.C. Hsu [**47**] to the following quantitative form: If $T(\xi)$ is strongly continuous for $\xi > 0$, then for every $x \in X$, every $\xi (0 < \alpha \leq \xi \leq \beta)$, and for $\eta > 0$ being small, we have $\| \exp[(\xi - \alpha)A_\eta]T(\alpha)x - T(\xi)x\| \leq \mu(\eta^{1/3}, x) + K \cdot \eta^{1/3} \cdot \|x\|$,

1991 *Mathematics Subject Classification.* 41A10, 41A35, 41A36, 41A05, 41A15, 42A10, 60D07.

where $K = K(\beta, M)$ is a positive constant independent of η. Moreover, if $T(\xi)$ is uniformly continuous for $\xi > 0$, then $\|\exp[(\xi - \alpha)A_\eta]T(\alpha) - T(\xi)\| \leq \mu(\eta^{1/3}) + K \cdot \eta^{1/3}$.

C.L. Mao ([**84**]) discussed the uniform approximation of continuous functions, and generalized Machado's theorem to the case of an arbitrary set of continuous functions on a Hausdorff compact space X. W.X. Zheng ([**144**]) introduced and discussed in detail about the approximation properties of Bohman-Zheng operator $B_\sigma(f;x)$ (cf. Lecture Notes in Maths, No.556, 1976).

The method of growing multiplier has been known as a convenient process for dealing with the approximation of unbounded functions by linear positive operators ([**46, 48, 56, 118, 121, 122, 123**]). Denote by E a Banach space. Let $f(t)$ be a functional defined on E, and let $L_n[f(t);x](n = 1, 2, \dots)$ be a sequence of linear operators transforming $f(t)$ to functionals of $x(x \in E)$, where each $L_n[f;x]$ is positive on the sphere $\mathcal{G}(\|x\| \leq 1)$. Denote by $\Omega(\|x\|)$ a "bounding function" satisfying $\Omega(\|x\|) \in C(E)$, $\Omega(\|x\|) \geq 1$, and $\Omega(\|x\|) \uparrow \infty$ $(\|x\| \to \infty)$. L.C. Hsu and R.H. Wang ([**56**]) obtained the following result: Let $\alpha_n \uparrow \infty$ $(n \to \infty)$, and let the conditions (i) $L_n[1, \alpha_n^{-1}x] \to 1 (n \to \infty)$; (ii) $L_n[\|\alpha_n t - x\|^2 \cdot \Omega(\|\alpha_n t\|); \alpha_n^{-1}x] \to 0 (n \to \infty)$ hold uniformly on every closed sphere of E. Then for every $f(x) \in C(E)$, $f(x) = O(\Omega(\|x\|))(\|x\| \to \infty)$, we have $\lim_{n\to\infty} L_n[f(\alpha_n t); \alpha_n^{-1}x] = f(x), (x \in E)$. Moreover, the limit relation holds uniformly on any bounded set in a finite dimensional subspace E^* of E. This is a quite general principle on the method of growing multiplier. Since then, R.H. Wang ([**123**]) introduced the quasi-local positive operator, and established the corresponding principle about the method of growing multiplier. Further, it was proved ([**121, 122, 123**]) that any unbounded continuous function defined on an arbitrary given infinite domain can be approximated by explicit polynomials. In fact, for any given unbounded continuous function $f(x)$, we can find "matched bounded function" $\Omega_f(|x|) \uparrow \infty(|x| \to \infty)$ such that $f(x) = O(\Omega_f(|x|))$, and $\Omega_f(|x|) \succ \exp|x|$, $(x \to \infty)$. Denote by $\bar{B}_n[f(t);x]$ the Bernstein operator defined on $[-1, 1]$, Wang ([**121, 122, 123**]) proved that for any given $f(x) \in C(-\infty, +\infty)$, the limit relation $\lim_{n\to\infty} \bar{B}_n[f(\Omega_f^{-1}(\log n) \cdot t); x/\Omega_f^{-1}(\log n)] = f(x)$, $-\infty < x < +\infty$ holds, where $\Omega_f^{-1}(\cdot)$ is an inverse function of $\Omega_f(\cdot)$. Moreover, the above limit holds uniformly on any bounded interval in $(-\infty, +\infty)$. Of course, the limit as mentioned above also holds for Landau, Hermit-Fejér, and other operators ([**121, 122, 123**]).

S.L. Wang ([**132**]) discussed the approximation by Hermit-Fejér operator $H_n(f;\ x)$, and showed that $\|f - H_n(f;x)\|_c = O(\log n/n)(n \to \infty)$ holds if and only if $\omega(g, \delta) + \omega(\tilde{g}, \delta) = O(\delta \log \delta^{-1})(\delta \to 0^+)$, where $g(\theta) = f(\cos\theta)$, and $\tilde{g}$ is the conjugate function of g.

T.F. Xie ([**137**]) investigated the $(0, 2)$ interpolation polynomial. Particularly, he showed that the condition $\int_0^1 \frac{\omega(f', \delta)}{\delta} d\delta < \infty$ in the convergence theorem of Balazs and Turan (1958) can be waived.

It is well known that for any positive integer k, there is a $p_n \in P_n$, such that $|f-p_n| \leq c_k \cdot \omega_k(f, \Delta_n(x))$, where $\Delta_n(x) = \sqrt{1-x^2}/n + 1/n^2$, (A. Timan, and J. Brydnyi). X.M. Yu ([**142**]) showed precisely that $\Delta_n(x)$ can not be changed to $\sqrt{1-x^2}/n + \varepsilon_n/n^2$, for $k \geq 2$, and $\varepsilon_n \downarrow 0$. Thus answer to the famous Lorentz's question (1963) is negative.

R.A. DeVore and X.M. Yu ([**39**]) proved the following theorems: (i) For any given increasing function $f \in C[-1,1]$, there exists an increasing polynomial $p_n(x)$ of degree n, such that $|f - p_n| < C \cdot \omega_2(f, \sqrt{1-x^2}/n)$, $-1 \leq x \leq 1$. (ii) Let $0 < \alpha < 2$. Then $f(x)$ is increasing and belongs to $\mathrm{Lip}^* \alpha$, if and only if there exist increasing polynomials p_n such that $|f - p_n| \leq C(\sqrt{1-x^2}/n)^\alpha$, $-1 \leq x \leq 1$, $n = 1, 2, \cdots$.

C.Y. Yu ([**139**]) discussed the problem of approximating functions on the positive real axis by generalized polynomials, and extended some results of S. Mandelbrot and S. Agmon on uniqueness of the solution of the moment problem. Yu also extended some results of J.F. Ritt, S. Mandelbrot and G. Valison on the Dirichlet series which converge everywhere to those which converge only in a half-plane ([**140**]).

Denote by X the class $C_{2\pi}$ or $L^1_{2\pi}$. For $f \in X$, define $L_n(f) = K_n * f$, where $K_n(t) = \frac{1}{2} + \sum_1^n \lambda_k^{(n)} \cos kt$. According to Jackson's estimate, we have $C_n^*(X) = \sup_{f \in X, f \neq \mathrm{const}} \|f - L_n(f)\|_X / \omega_f(\frac{\pi}{n+1})_X$. X.H. Wang ([**134**]) showed that for $K_n(t) \geq 0, C_n^*(C_{2\pi}) = 1 + \frac{2}{\pi}\int_0^\pi [\frac{n+1}{\pi} \cdot t] \cdot K_n(t) dt$.

Let $X = L_{2\pi}$ or $L_{2\pi}^\infty$, and

$$W_X^r := \Big\{ f \in X : f(x) = \frac{a_0}{2} + \int_0^{2\pi} \sum_{k=1}^\infty k^{-r} \cos(k(x-t) - \frac{\pi r}{2}) \varphi(t) dt,$$

$$\int_0^{2\pi} \varphi(t) dt = 0, \|\varphi\|_X \leq 1 \Big\}.$$

Y.S. Sun extended the estimate of $\varepsilon_n(W_X^r)_X$ established by S.B. Stechkin to the case of $r > 1$. He [**115**] showed

$$\varepsilon_n(W_X^r)_X = \frac{4}{\pi n^r} \left| \sum_{\nu=0}^\infty \frac{\sin[(2\nu+1)\beta\pi - (\pi r)/2]}{(2\nu+1)^{r+1}} \right|,$$

where $\beta\pi$ is the zero of $\sum_{\nu=0}^\infty \cos[(2\nu+1)t - (\pi r)/2]/(2\nu+1)^r$.

Y.G. Shi ([**96**]) proved the following theorem: Let V_1 and V_2 be Chebyshev subspaces with dimension m and $n (m < n)$ respectively. Let $V_1 \subset V_2$, and $v_j \in V_j (j = 1, 2)$. If there exists an $f \in C[-1,1]$ such that v_j is a best $L_p(1 < p < \infty)$ approximation to f from $V_j (j = 1, 2)$, then the function $v_2 - v_1$ changes sign at least m times in $[-1,1]$. The above theorem partially answered a problem of Rivlin on $L_p(1 < p < \infty)$ approximation: Characterize those n-tuples of algebraic polynomials $\{p_0, \cdots, p_{n-1}\}$ with deg $p_j = j$, for which there exists an $f \in C[-1,1]$ such that the polynomial of best uniform of degree j to f is $p_j (j = 0, 1, \cdots, n-1)$ (*Abstract spaces and approximation*, Berkhäuser-Verlag-Basel/Stuttgart, 1969). Y.G. Shi ([**97**]) also discussed the limits of a Chebyshev-type theory of restricted range approximation. Shi noted that for restricted range approximation in $\tilde{C}(X) = \{f \in C(X) : l(x) \leq f(x) \leq u(x) \text{ for all } x \in X = [0,1]\}$ a fairly complete Chebyshev theory exists and now Shi proceeds to study the limit of this theory. That is, under what sort of circumstances will best approximations always be characterized by alternation, or by alternation or the occurrence of a simultaneous positive and negative extreme point; when will uniqueness in $C(X)$ or $\tilde{C}(X)$ always hold; and what about combinations of the above?

X.H. Sun ([**110**]) used the generalized definition of Λ-bounded variation of a function given by D. Waterman ([**136**]) to study Lagrange interpolation of functions of generalized bounded variation.

X.P. Sun ([**111**]) discussed a problem on monotone approximation, and showed that if $f \in C[-1,1]$ satisfies $(f(x_1) - f(x_2))/(x_1 - x_2) \geq \delta$, $x_1, x_2 \in [-1,1]$, $x_1 \neq x_2$, and $E_n(f) = o(n^{-2})$, then the best approximation polynomial p_n is increasing for all n sufficiently large.

Denote by M any subspace of a normed linear space Y. The metric projection P_M from Y onto M is defined as

$$P_M(y) = \{\tilde{g} \in M : \|y - \tilde{g}\| = \inf\{\|y - g\| : g \in M\}\}(y \in Y).$$

M is called proximinal in Y if $P_M(y) \neq \phi$ for all $y \in Y$. A continuous mapping s from Y to M is called a continuous selection for P_M if $s(y) \in P_M(y)$ for all $y \in Y$. For a normed space X and a compact Hausdorff space T, Wu Li ([**70**]) studied the space $C(T, X)$ of all continuous maps from T to X endowed with the norm $\|f\| := \sup\{\|f(t)\|_X : t \in T\}(f \in C(T, X))$, where $\|\cdot\|_X$ is the norm on X. In quite general cases Li answered a problem posed by E.W. Cheney and C. Franchetti ([**17**]): Assume that T is a compact metric space with uncountably many points, and G denotes a finite dimensional subspace of X. Then $C(T, G)$ is proximinal in $C(T, X)$ if and only if P_G has a continuous selection.

X.C. Shen discussed the problem on the basis of rational functions in a certain class of domains, and exhibited some the efficient solution to the problem of multiple interpolation in H^p over the upper half plane ([**92**]).

G.L. Xu ([**138**]) gave a sufficient and necessary condition for the existence of rational interpolation, which also corrected Macon-Dupree's theorem (The Amer. Math. Monthly, 69(1962), 751–759).

An example given by C.B. Dunham ([**40**]) shows that the simultaneous exchange algorithm for Chebyshev approximation by interpolating rationals may fail to produce an interpolating rational approximant. Y.Q. Huang and J. Williams ([**60**]) showed that the example given by Dunham cannot occur as a step of the exchange algorithm. Huang and Williams also gave an example for which Dunham's objection to the algorithm is valid.

W.X. Zheng ([**145**]) introduced a kind of approximate identity kernels on Walsh system, and showed a criteria for determining whether a kernel is an approximate identity kernel. Zheng ([**146**]) also defined the derivative of functions on local fields K, and obtained the formula $\chi_\lambda^{(1)}(x) = |\lambda|\chi_\lambda(x)$ for characters $\chi_\lambda, \lambda \in K$. Let $f \in L^r(K)$, $1 \le r < \infty$, and consider the linear operator

$$L(f,x,\lambda) = \int_K f(t)|\lambda| \cdot w(\lambda(x-t))dt, \quad \lambda \in K,$$

where the kernel w is generated by some $\omega \in L^1(K), W = \omega$. By using the above derivative, Zheng proved several lemmas including the Bernstein inequality and established some inverse approximation theorems for the class $W[L^r, |x|^\alpha]$ and $\mathrm{Lip}_r\alpha$.

W.Y. Su ([**108**]) investigated boundedness of pseudo-differential operators $T_\sigma f$ over local fields in Lebesgue and Besov spaces, and obtained the L^p and $B(\alpha, r, s)$ estimates for them.

The earliest results on the uniqueness of multiple trigonometrical series were given by M.T. Cheng ([**20**]). He defined firstly the Gibbs phenomenon on a sequence of bivariate functions, and showed that if $f(x,y) = f_1(x)\cdot f_2(y)$, f_1 and f_2 are bounded variation functions, then Bochner-Riesz means of the Fourier series of f, $\{S_R^\delta(x,y;f)\}(\forall\delta > 0)$ have the Gibbs phenomenon at all discontinuity points of f([**21, 22**]). M.T. Cheng also established some Tauberian theorems on Bochner-Riesz summation ([**19**]).

Multivariate approximation seems to be one of the main trends in the future research of approximation theory. Under the leadership of M.T. Cheng, ever since 1956, Chinese mathematicians have started the investigation of the approximation of multivariate functions. M.T. Cheng and Y.H. Chen first introduced a kind of very useful trigonometric polynomial operators

$$S_N^{(\delta,k)}(f;x,y) = \sum_{\nu \le N^2} (1 - (\frac{\sqrt{\nu}}{N})^k)^\delta \cdot A_\nu(x,y),$$

where $A_\nu(x,y) = \sum_{m^2+n^2=\nu} C_{m,n} \cdot \exp\{i(mx+ny)\}$, and $f(x,y) \sim \sum C_{m,n} \cdot \exp\{i(mx+ny)\}$. It is just the Riesz's means when $k = 2$. The typical results are: $f - S_N^{(\delta,2)} = O(\omega(\frac{1}{N};f))$ provided $f \in C, \delta > \frac{1}{2}$; $f - S_N^{(\delta,2)} = O(\frac{1}{N}\cdot\omega_1(\frac{1}{N};f))$ provided $f \in C^1$, where $\omega_1(\delta) = \max(\omega(\delta;f'_x), \omega(\delta;f'_y))$. In particular, Cheng and Chen showed that $f - S_N^{(\delta,2)} = O(N^{-2})$ holds, only if $f \equiv$ const. So, the Riesz's means can not be utilized to improve the approximation order of f ([**24, 25, 26**]).

D.G. Deng ([**38**]) introduced a generalized area integral and tent integral as follows. Let $F(y,t)$ be defined on R_+^2. Denote

$$A_q(F)(x) = (\int_{\Gamma(x)} \frac{|F(y,t)|^q}{t} dydt)^{1/q},$$

$$C_q(G)(x) = \sup_{X \in I}\left(\frac{1}{|I|}\int_{\hat{I}} |G(y,t)|^q dydt\right)^{1/q},$$

where $\Gamma(x) = \{(y,t) : |y-x| < t\}$, a cone in R^2_+ vertexed at $x \in R$, $\hat{I} = \{(y,t) : y \in I, (y-t, y+t) \subset I\}$, a tent in R^2_+ based on $I \in R$.

Deng obtained a generalized Carleson inequality

$$\left|\int_{R^2_+} F(x,t)G(x,t)dxdt\right| \leq C\int_R A_q(F)C_{q'}(G)dx,$$

where $1 \leq q \leq \infty$, $\frac{1}{q} + \frac{1}{q'} = 1$.

Let $b(x) = b_0(|x|), b_0(t)$ be a bounded function with the Λ-bounded mean variation and $1 < p < \infty$, and $H(x) = K(x)b(x)$. X.L. Shi ([**94**]) discussed boundedness of the convolution operator $Tf(x) = p.v.H * f(x)$ in the $L^p(R^n)$ sense. He showed that for any $n = 1, 2, \cdots$, and arbitrary $K(x)$ satisfying conditions

$$\int_{R_1<|x|<R_2} K(x)dx = 0 \quad (0 < R_1 < R_2 < \infty),$$

$$\int_{|x|>2|y|} |K(x+y) - K(x)|dx \leq B \quad (|y| \neq 0)(\text{Hörmander condition}),$$

and $K(x) = O(|x|^{-n})$, the operators $Tf = H * f$ are bounded in $L^p(R^n)$ if and only if

$$m^{1/\Lambda_m} = O(1) \quad (m \to \infty),$$

where $\{\lambda_k\}$ is a sequence of positive number satisfying $\lambda_k \uparrow$, and $\Lambda_m = \sum_{j=1}^{m} 1/\lambda_j \uparrow \infty$.

Denote by P the non-constant spherical harmonic defined on the surface $\sum_{n-1}$ of the unit sphere in R^n, let $K(x) = |x|^{-n} \cdot P(x/|x|)$, for $x \neq 0$, and $T^n = \{x = (x_1, \cdots, x_n) : -\pi < x_k \leq \pi, 1 \leq k \leq n\}$. The conjugate Fourier series of $f \in L^1(T^n)$ associated with the Calderon-Zygmund kernel K is $\sum_{m\neq 0, m\in Z^n} \hat{K}(m)\hat{f}(m)e^{im\cdot x}$. S.Z. Lu ([**79**]) studied the Bochner-Riesz means

$$\tilde{S}_R^{\frac{n-1}{2}} f(x) = \sum_{0<|m|<R} \hat{f}(m)\hat{K}(m)\left(1 - \frac{|m|^2}{R^2}\right)^{\frac{n-1}{2}} \cdot e^{im\cdot x},$$

where the index $\frac{n-1}{2}$ is the "critical index" for $p = 1$. Lu showed that if the entropy of a function f satisfies

$$J(f) = \int_0^\infty E(x : |f(x)| > s)ds < \infty,$$

then $\lim_{R\to\infty} \tilde{S}_R^{\frac{n-1}{2}} f(x)$ exists a.e.. He also proved that there exists a function f, such that $J(f) < \infty$ with $J(R_j f) = \infty (1 \leq j \leq n)$, where $R_j f$ is the j-th Riesz transform of f ([**80**]).

S.Z. Lu, M.H. Taibleson and G. Weiss ([**81, 82**]) proved the almost everywhere convergence of Bochner-Riesz's means of multiple Fourier series, and of multiple conjugate Fourier series, respectively.

X.Z. Liang found the relationship between the bivariate polynomial interpolation and the theory of algebraic curves. He also gave several criteria for finding suitable knots of bivariate interpolation by polynomials ([**91, 127**]).

G.Z. Chang and P.J. Davis ([**3**]) showed that if the Bézier net associate with the n-th Bernstein polynomial $B_n(f)$ of f over triangle T on R^2 is convex on T, then $B_n(f)$ is also convex on T. Chang and Y.Y. Feng proposed some conditions for preserving convexity, and showed that it is the necessary and sufficient condition in the case of $n = 2$ and 3. They also discussed the iterative limit of multivariate Bernstein operator ([**4**]).

2. Spline functions and their applications

The spline function is one of the most useful tools in numerical approximation. It has attracted much attention of Chinese mathematicians. First of all, Bu-Chin Su created a new outlook, a new view of the Computational Geometry. In fact, Su originated with success the studies of geometric invariants in computational geometry ([**104**]). By using geometric invariants, Su investigated and solved the problems concerning the distribution of singular points and inflection points on algebraic parametric curve of degree n. For instance, when $n = 3$, he considered parametric curve (C) of degree 3:

$$\text{(C)} \qquad \begin{aligned} x &= a_0 + a_1 t + \frac{1}{2}a_2 t^2 + \frac{1}{6}a_3 t^3, \\ y &= b_0 + b_1 t + \frac{1}{2}b_2 t^2 + \frac{1}{6}b_3 t^3, \end{aligned}$$

and found a relative affine invariant of the above curve,

$$I = (\frac{q}{p})^2 - \frac{2r}{p},$$

where $p = a_2b_3 - a_3b_2$, $q = a_3b_1 - a_1b_3$, and $r = a_1b_2 - a_2b_1$.

Su also obtained the following important result ([**100**]):

THEOREM. *If $I > 0$, then the curve* (C) *has two real inflection points, and no real singular point; If $I = 0$, then the curve* (C) *has one double point, and no real inflection.*

Under the leadership of Su, his younger colleagues D.Y. Liu, X.J. Hua, Y.L. Hsin, and himself have obtained a lot of results concerning the affine invariant of higher degree, higher dimension, and other related problems. The detail of results have been presented in the excellent book *Computational Geometry* written by B.C. Su and D.Y. Liu ([**106**]).

It is well known that the main tools for fitting curves and surfaces are Bézier method and B-spline method. According to the geometric point of view, however, these methods are just the representations of algebraic parametric curve in different bases. So, most important problems in the fitting curve and surface become the problems in control of singular points, inflection points, and convexity etc. By using the geometric invariant, Su and his colleagues also systematically investigated Bézier curve (surface), parametric spline, non-linear spline, the convexity of curves and surfaces, smoothing curve and surface, fitting curve (surface) and so on ([**100**]–[**107**]). The approach of Geometric invariant proposed by Su has become one of the most important tools for studying Computational Geometry.

The cubic spline has been widely used in solving various kinds of technical problems. Z.C. Shi ([**98**]) constructed a kind of spline finite element methods. The method based on cubic B-spline was used to obtain approximate solution for the equilibrium problems of elastic composite structures on regular regions. By using eigenvalue analysis techniques, Shi also found several excellent conditions on nonsingularity of the interpolating matrix of the cubic spline ([**99**]).

X.H. Wang worked out a useful formula on the remainder. His formula provided various precise estimates for the error terms involved in spline function ([**135**]).

According to the principle of Geometric invariant, J.C. Sun independently proposed and constructed quadratic and cubic spline in local coordinates to fit data with steep gradient or multivalues ([**109**]).

Z.R. Guo ([**42**]) investigated the saturation properties of the lacunary interpolating splines, and obtained the saturation theorems which are better than the original conjecture of B.K. Swartz and R.S. Varga. Z.R. Guo and M.D. Ye ([**43**]) discussed the lacunary interpolation by quintic splines. For several interpolation conditions, they obtained some necessary and sufficient conditions, and error estimates.

It is well known that when interpolation points coincide with knots, the knot sequence must obey some restriction in order to guarantee the existence and boundedness of the interpolation projector. However, when the interpolation points are chosen to be the knot averages, the corresponding quadratic or cubic spline interpolation projectors are bounded independently of the knot sequence. Based on this fact, de Boor (1975) made a conjecture that the interpolation by splines of order k at knot averages is bounded for any k. R.Q. Jia ([**69**]) disproved de Boor's conjecture for $k \geq 20$.

Let k be a positive integer, and assume that $x = (x_i)_{i \in Z}$ is a real nondecreasing knot sequence with $x_i < x_{i+k}$ for all i. Set $I = (x_{-\infty}, x_\infty)$ where $x_{-\infty} = \lim_{i \to -\infty} x_i$, and $x_\infty = \lim_{i \to \infty} x_i$. Denote by S the normed linear space of all bounded splines of order k with the knot sequence x endowed with the L_∞-norm $\|\cdot\|_\infty$. R.Q. Jia considered the orthogonal projection P_s from $L_\infty(I)$ onto S with respect to the ordinary inner product $(f, g) = \int_I f(x)g(x)dx$. Jia ([**68**]) showed that $\sup_x \|P_s\| \le \cos t_k < \infty$, where x runs through all the multiple geometric knot sequences $x = (x_i)_{i \in Z}$ (i.e., $x_{li} = x_{li+1} = \cdots = x_{li+l-1} = q^i$, for all i, $0 < q < \infty$, $l \in N$). This result extends a result of K. Höllig (J. Approx. Th. **33**(1981), 318–333).

R.Q. Jia ([**61**]–[**63**]) answered a conjecture on cubic spline interpolation at a bi-infinite knot sequence proposed by C.A. Micchelli. Jia ([**61**]–[**65**]) also solved two conjectures proposed in the book *Total positivity* written by S. Karlin. C. de Boor, Jia and Pinkus ([**1**]) discussed the structure of invertible bi-infinite totally positive matrix.

J.Z. Wang, J.Z. Wang and D.R. Huang investigated the optimal error bounds concerning various interpolating splines, and obtained several results ([**58, 59, 117**]). Particularly, Wang ([**117**]) showed the optimal error bounds for Hermit interpolating splines, and verified a conjecture proposed by P.R. Lipow and I.J. Schoenberg (1973).

T.P. Chen ([**15**]) established a very useful error estimate on general Hermit interpolating polynomial. By using the estimate, Chen gave an asymptotic expression for the difference between $f^{(q)}(x)$ and $S_{\Delta_n}^{(q)}(x)$, where $f \in C^r[0,1]$ $(m + 4 \le r)$, $S_{\Delta_n}(x)$ is the $(m+3)$-th interpolating spline, and Δ_n is a uniform partition of $[0, 1]$ with step length $1/n$ ([**16**]).

H.L. Chen obtained a fundamental theorem of algebra for certain class of g-splines, and gave a necessary and sufficient condition for the existence and uniqueness of the interpolating g-spline with mixed boundary conditions ([**8, 9**]).

C.K. Lu considered the interpolating complex cubic splines with deficiency 2 for a function defined on a closed or open plane curve ([**76**]). The error analysis for these approximations and their derivatives was also made. Lu also discussed the approximation problem of Cauchy type integrals by interpolating splines of degree up to 3 ([**76**]). The error bounds between Cauchy type integrals with the function and the spline as kernel density were given by Lu ([**77**]). Moreover, these results were extended to quartic splines ([**78**]). When the curve is the unit circle, H.L. Chen ([**6, 7**]) showed that the existence of interpolating complex quadratic spline depends on the odevity of knots. Chen also showed that the interpolating complex cubic spline always exists uniquely. Chen and T. Hvaring ([**10**]) established a boundary approximation of the conformal mappings from unit circle to Jordan domain by solving certain integral equation. They also constructed complex harmonic splines by using the above boundary values. Chen investigated the problem on zeros of complex splines on the unit circle, and estimated exactly the number of zeros of a certain class of splines of degree n ([**6, 7**]).

Denote by $S_k^{\mu}(\Delta)$ the space of piecewise polynomials of degree k and smoothness μ over partition Δ consisting of irreducible algebraic curves. R.H. Wang ([**119**]) showed the existence of bivariate spline. That is, $s \in S_k^{\mu}(\Delta)$ if and only if for any given interior edge of Δ, there exists a "smoothing cofactor" and satisfies the "conformality condition" at each interior vertex. The theorem has been extended to the case of higher dimension ([**121, 122, 124, 127**]). According to "smoothing cofactor-conformality condition" method stated as above, the multivariate spline is equivalent to a corresponding algebraic problem. So, in principle one can use the algebraic methods to study any problem on the multivariate spline. Using this method, Wang gave the formulae on the dimension of $S_k^\mu(\Delta)$ for any given partition consisting of irreducible algebraic curves ([**119, 120, 124**]). For the so-called cross-cut, or even quasi cross-cut partition Δ, C.K. Chui and Wang obtained the explicit formula on dimension of the space $S_k^\mu(\Delta)$. Make use of the formula, a criterion for testing the existence of non-trivial spline with compact support was presented by Chui and Wang, independently ([**33**]). For any given triangulation Δ, Morgan and Scott (Math. Comp. 1975) obtained the dimension of $S_k^1(\Delta), k \geq 5$. For the cases of $k = 3$ and 4, dimension of $S_k^1(\Delta)$ was computed by Chou, Su and Wang ([**27**]). Wang and X.G. Lu obtained the dimensional formulae of $S_k^\mu(T)$, for all $\mu \geq 1, k \geq 4\mu + 1$, and arbitrary triangulation T ([**128**]). Recently, D. Hong ([**45**]) gave an explicit formula for the dimension of $S_k^\mu(k \geq 3\mu + 2)$ over arbitrary triangulation, and constructed a local basis for them.

It appears that the most useful basis functions are the functions with locally supported. For the uniform crisscross partition C_{mn}, a basis of bivariate quadratic B-spline with smallest support was given and proved by Chui and Wang ([**35**]). Chui and Wang also constructed a bivariate quadratic B-spline for non-uniform crisscross partition $\tilde{C}_{mn}$, and proved these B-splines span the whole $S_2^1(\tilde{C}_{mn})$ ([**34**]). For uniform three direction mesh Δ_{mn}, using Fredrickson's B-spline, with its translations and inversions, Chui and Wang ([**35**]) established the criteria for finding various bases of $S_3^1(\Delta_{mn})$. In the same paper, however, Chui and Wang also showed that the space $S_4^2(\Delta_{mn})$ possesses no basis consisting of only splines with locally supported. The results have been generalized by de Boor and K.Höllig (J.Comp. & Appl. Math. 9(1983)). Wang and T.X. He ([**126**]) showed a basis of the space $S_k^\mu(\Delta_{mn})(k \geq \mu + 1)$, where Δ_{qc} is the quasi cross-cut partition. C.K. Chui, T.X. He and R.H. Wang ([**29**]) discussed the C^2 quartic spline space on a four-directional mesh, and showed the various properties of certain splines in this space. They also discussed ([**28**]) the problem on interpolation by bivariate linear spline, and gave an algorithm for placing the sample points for bivariate linear spline interpolation where the triangulation is arbitrary.

C.K. Chui, L.L. Schumaker and R.H. Wang ([**31**]), R.H. Wang and T.X. He ([**125**]) showed the dimension and basis of space of bivariate splines satisfying certain boundary conditions for the cases of uniform and non-uniform partitions, respectively. Z. Sha ([**89, 90**]) discussed the bivariate interpolation problem on the spaces $S_2^1(\Delta_{mn}^{(2)})$ and $S_3^1(\Delta_{mn}^{(1)})$.

The problem on linear independent of the box splines has been solved by Jia ([**64**]–[**67**]). Moreover, Jia discussed the approximation order of bivariate spline space, and obtained a fine result on the controlled approximation order of the box spline spaces over the three direction mesh. Specially, Jia ([**67**]) gave a counterexample to a result of G. Strang and G.Fix *Constructive aspects of functional analysis* (G. Geymonat, ed (C.I.M.E.II, Ciclo Erice, 1971), 1973, 793–844): If the order of controlled approximation from a collection of locally supported elements is k, then there is a linear combination Ω of those elements and their translates such that any polynomial of degree less than k can be reproduced by Ω and its translates.

Y.S. Li discussed multivariate optimal interpolation to scattered data throughout a rectangle. Li ([**71, 72**]) also gave the characterization, existence, uniqueness theorems, and construction of the optimal interpolation. C.K. Chui and Y.S. Hu discussed the geometric properties of certain bivariate splines ([**30**]).

K. Zhao and J.C. Sun ([**143**]) constructed the explicit expressions for dual bases of multivariate Bernstein-Bézier polynomials on simplices, which are shown to be useful in proving some known results, such as the convergence of the degree raising, domain transformation. Based on the dual bases, Zhao and Sun also formulated generalized Hermit interpolations on simplices which gives the general 9-parameter interpolants on triangles discussed by G. Farin.

It is well known that the sequence of Bézier nets $(\hat{f})_n(x)$ associated with a Bernstein-Bézier surface over a triangle uniformly converges to the surface as n goes to infinity. Y.Y. Feng ([**41**]) gave the precise rates of convergence, and presented the pointwise convergence result and saturation theorem. Y. Cao and X.J. Hua ([**2**]) discussed the convexity of quadratic parametric triangular Bernstein-Bézier surface in the light of the property of Gaussian curvature which keeps the sign invariant under affine transformation and poses a sufficient condition for convex surfaces of explicit geometric meaning.

The GC^1 conditions in several special cases might be useful for CAD surface modeling. D.Y. Liu and J. Hoschek ([**74**]) presented the necessary and sufficient conditions for geometric C^1 continuity (GC^1) that cover all the four combinations between rectangular and triangular Bézier surface patches. Further, they also showed some more practical sufficient conditions in [**74**]. D.Y. Liu ([**73**]) discussed the GC^1 continuity conditions between two adjacent rational Bézier surface patches. He also showed that there are many weights in the GC^1 conditions which are useful to control the shape of patches and to compose a GC^1 smooth surface.

R.H. Wang and X.Q. Shi ([**130**]) gave a practical method for constructing $S^{\mu}_{\mu+1}$ surface interpolation over any given triangulation, where μ is any positive integer. If $\mu = 1$, the method is Powell-Sabin's six-triangles method ([**88**]). This kind of surface interpolations can be realized on parallel computers. R.H. Wang and X.Q. Shi ([**129**]) also presented a kind of cubic C^1-interpolations over n-dimensional simplices. This technique is similar to n-dimensional Clough-Tocher scheme, and can be used to solve the problem of n-dimensional finite element method. X.Q. Shi ([**95**]) proposed "interpolation-conformality" method on higher-dimensional spline.

By using the duality principle for Kolmogorov width and Gelfand width, Y.S. Sun ([**112**]–[**114**]) established some fundamental relations between the two widths for several important function classes. In fact, Sun obtained some results on the extremum property of the L-spline, and extremum problem on best approximation. Sun and D.R. Huang showed that generalized Bernoulli splines and Euler splines are fundamental for establishing Landau- Kolmogorov type inequalities as well as in the exact estimation of N-width and the construction of extremal subspaces ([**116, 115**]). Some of their results have generalized or improved the related results of Korneichuk, Karlin, Schoenberg, Micchelli and Pinkus. The Landau-Kolmogorov type inequality given by them generalized Sharma-Tzimbalarko's related results. Using Korneichuk's rearrangement method, Sun gave a Taikov type inequality which has become a basic tool for establishing the estimates of N-width in L_p-norm ([**115**]). The Kolmogorov type inequality on self-conjugate ordinary differential operator were studied by D.R. Huang and J.Z. Wang ([**59**]).

3. Numerical integration

L.K. Hua and Y. Wang obtained a series of useful results on Numerical integration by the Number-Theoretic method. Particularly, the famous "Hua-Wang" method has completely solved the problem of computing extreme coefficients. As a consequence, a certain method of numerical integration of multivariate periodic functions has been constructed ([**57**]).

By means of different tools, S.H. Min ([**85, 86**]), C.D. Pan ([**87**]), and L.C. Hsu ([**46**]–[**51**]) also obtained, respectively, some interesting results on the Number-Theoretic method of approximating integration of higher dimension.

L.C. Hsu was delighted with his early finding of a general asymptotic expansion formula for a wide class of oscillatory integrals in 1958 ([**52**]). The formula with remainder term given by Hsu ([**50**]) is also valid for non-integer parameters, which was proved by X.H. Wang and Y.S. Chou. Using Euler-Maclaurin's formula, Hsu and Chou investigated the expansions for integrals of rapidly oscillating functions. As an extension of various asymptotic expansions for parameter integrals due to Erugin-Sobolev, Krylov, Riekstens and Havie, respectively, Hsu and Chou ([**53**]) established a general asymptotic formula for a type of singular oscillatory integral of the form $\int_0^1 x^{-\alpha} f(x, Nx)dx (0 < \alpha < 1)$ in which $f(x, y)$ is periodic in y with period unity, and N is a large parameter.

L.C. Hsu, R.J. Tomkins and C.L. Wang ([**55**]) showed a constructive process which leads to a class of quadrature formulae with any preassigned compound precision for the numerical integration of rapidly oscillating functions on $(0, \infty)$ of the form $e^x F(x, \lambda)$, where λ is a large parameter and $F(\cdot, y)$ is periodic of period unity in y.

L.C. Hsu found "reducing-dimensionality" expansions for computing high-dimensional integrations ([**49**]). Using this technique, Hsu, Wang(R.H.) Chou and other colleagues (e.g., J.X. Yang, T.X. He, and G.Q. Zhu) ([**52**]) have also established a series of "boundary type" cubature formulae for various special domains in high dimensions.

C.K. Lu gave different mechanical quadrature formulae of singular integrals with Chebyshev weights, among which the quadrature formulae of the transformed weight type of the first kind were first proposed in [**78**].

L.C. Hsu and Y.S. Chou ([**54**]) constructed two kinds of multivariate numerical integration formulae that make use of merely boundary point as evaluation points. They also obtained a kind of symmetric boundary type formula that possesses some homogeneous algebraic precision.

Because there have been several surveys on Approximation Theory (e.g., [**12, 51, 91**]), some results obtained by Chinese mathematicians are not mentioned.

REFERENCES

1. C. de Boor, R.Q. Jia and A. Pinkus, Linear Alge. & its Appl., **47**(1982), 41–55.
2. Y. Cao And X.J. Hua, Comput. Aid. Geo. De., **8**(1991), 1–6.
3. G.Z. Chang and P.J. Davis, J. Approx. Theo., **40**(1984), 11–28.
4. G.Z. Chang and Y.Y. Feng, Comput. Aid. Geo. DE., **1**(1984), 279–283.
5. G.Z. Chang and Hoschek, Multi. Approx. Theo. III, eds. W. Schempp and K. Zeller, 1985, 61–70.
6. H.L. Chen, J. Approx. Theo., **38**:4(1983).
7. ———, J. Approx. Theo., **39**.4(1983).
8. ———, J. Approx. Theo., **43**:2(1985).
9. ———, J. Approx. Theo. Appl., **1**:2(1985), 1–14.
10. H.L. Chen and T. Hvaring, Math. Comp., **42**(1984), 151–164.
11. K.K. Chen, *Theory of Trigonometric Series*, Vol. I, Shanghai Sci. Tech. Press, 1964.
12. ———, Adv. Math. Sinica, **8**(1965), 337–351.
13. ———, *Theory of Trigonometric Series*, Vol. II, Shanghai Sci. Tech. Press, 1979.
14. ———, K.K. Chen's Collectanea, Sci. Press, Beijing, 1981.
15. T.P. Chen, Scientia Sinica, **24**(1981), 606–617.
16. ———, Scientia Sinica, **26**(1983), 125–137; 919–930.
17. E.W. Cheney and C. Franchetti, Boll. Univ. Mat. Ital., B(5) 18(1981), 1003–1015.
18. M.T. Cheng, Ann. of Math., (2) **50**(1949), 356–384.
19. ———, Ann. of Math., (2) **50**(1949), 763–776.
20. ———, Ann. of Math., (2) **52**(1950), 403–416.
21. ———, Duke Math. J, **17**(1950), 83–90.
22. ———, Duke Math. J, **17**(1950), 477–490.
23. ———, Proc. Amer. Math. Soc., **2**(1951), 77–86.
24. M.T. Cheng and Y.H. Chen, J. Peking Univ., **4**(1956), 411–428.
25. ———, Bull. Acad. Poland Sci. C1, III, **4**(1956), 639–641.
26. ———, J. Peking Univ., **3**(1957), 259–282.
27. Y.S. Chou, L.Y. Su and R.H. Wang, Multi. Approx. Theo. III, eds. W. Schempp and K. Zeller, 1985, 71–83.
28. C.K. Chui, T.X. He and R.H. Wang, Coll. Math. Soc. J. Bolyai 49, Budapest, 1985, 247–255.
29. ———, Approx. Theo. Appl. **3**:4(1987), 32–36.
30. C.K. Chui and Y.S. Hu, Approx. Theo. IV, eds. C.K. Chui, L.L. Schumaker, and J. Ward, 1983, 407–412.
31. C.K. Chui, L.L. Schumaker and R.H. Wang, Canadian Math. Soc. Conf. Proc., **3**(1983), 51–66; 67–80.

32. C.K. Chui and R.H. Wang, Math. Comp., **41**(1983), 131–142.
33. ______, J. Math. Anal. Appl., **94**(1983), 197–221.
34. ______, Approx. Theo. IV, eds. C.K. Chui, L.L. Schumaker and J. Ward, 1983, 413–418.
35. ______, J. Approx. Theo. Appl., **1**(1984), 11–18.
36. ______, Scientia Sinica, **27**(1984), 1129–1142.
37. ______, J. Math. Anal. Appl., **101**(1984), 540–554.
38. D.G. Deng, Studia Math., **78**:3(1984).
39. R.A. DeVore and X.M. Yu, Constr. Approx., **1**(1985), 323–331.
40. C.B. Dunham, Math. Comp., **29**(1975), 552–553.
41. Y.Y. Feng, Comput. Aid. Geo. De., **4**(1987), 245–249.
42. Z.R. Guo, Acta Math. Sinica, **18**(1975), 247–253.
43. Z.R. Guo and M.D. Ye, Approx. Theo. IV, eds. C.K. Chui, L.L. Schumaker and J. Ward, 1983, 483–490.
44. E. Hille, *Functional analysis and semi-groups*, Amer. Math. Soc. Coll. Publ. XXXI, 1948.
45. D. Hong, Approx. Theo. Appl., **7**:1(1991), 56–75.
46. L.C. Hsu, Czech. Math. J., **9**(1959), 574–577.
47. ______, Czech. Math. J., **10**(1960), 323–328.
48. ______, Studia Math., **21**(1961), 37–43.
49. ______, Acta Math. Hung., **12**(1962), 387–392.
50. ______, Proc. Camb. Phil. Soc., **59**(1963), 81–88.
51. ______, Approx. Theo. IV, eds. C.K. Chui, L.L. Schumaker and J. Ward, 1983, 123–151.
52. L.C. Hsu and Y.S. Chou, *Numerical integration in higher dimensions*, Sci. Press, Beijing, 1980.
53. ______, Math. Comp., **37**(1981), 503–507.
54. ______, Calcolo, XXIII(1986), 227–248.
55. L.C. Hsu, R.J. Tomkins and C.L. Wang, BIT, **30**(1990), 114–125.
56. L.C. Hsu and R.H. Wang, Doklad Acad. Nauk USSR, **156**(1964), 264–267.
57. L.K. Hua and Y. Wang, *Applications of Numer Theory to Numerical Analysis*, Springer, and Sci. Press, Beijing, 1981.
58. D.R. Huang and J.Z. Wang, Scientia Sinica, **25**(1982), 1130–1141.
59. ______, Acta Math Sinica, **26**(1983), 715–722.
60. Y.Q. Huang and J. Williams, Math. Comp., **42**(1984), 111–113.
61. R.Q. Jia, J. Approx. Theo., **37**(1983), 293–310.
62. ______, J. Approx. Theo., **38**(1983), 284–292.
63. ______, Approx. Theo. IV, eds. C.K. Chui, L.L. Schumaker and J. Ward, 1983, 539–545.
64. ______, J. Approx. Theo., **40**(1984), 158–160.
65. ______, MRC Report No., **2696**(1984).
66. ______, Constr. Approx., **1**(1985), 175–182.
67. ______, Proc. Amer. Math. Soc., **97**(1986), 647–654.
68. ______, Math. Comp., **48**(1987), 675–690.
69. ______, Constr. Approx., **4**(1988), 1–7.
70. W. Li, J. Math. Anal. Appl., **143**(1989), 187–197.
71. Y.S. Li, CAT Report, No. 55, No. 56(1984).
72. ______, J. Approx. Theo., **43**(1985), 359–369.
73. D.Y. Liu, Comput. Aid. Geo. De., **7**:1-4(1990), 151–163.
74. D.Y. Liu and J. Hoschek, Comput. Aid. De., **21**:4(1989), 194–200.
75. R.L. Long, Scientia Sinica, A, **27**(1984), 16–26.
76. C.K. Lu, J. Approx. Theo., **36**(1982), 183–196.
77. ______, J. Approx. Theo., **36**(1982), 197–212.
78. ______, J. Math. Anal. Appl., **100**(1984), 416–435.
79. S.Z. Lu, Lecture Notes Math. No. 908(1982), 319–325.
80. ______, A Monthly J. of Sci., **29**(1984), 722–726.
81. S.Z. Lu, M.H. Taibleson and G. Weiss, Lecture Notes Math. No. 908(1982), 311–318.
82. ______, *Spaces generated by blocks*, Beijing Normal Univ. Press, 1989.
83. S.Z. Lu and K.Y. Wang, *Bochner-Riesz means*, Beijing Normal Univ. Press, 1988.

84. C.L. Mao, C.R. Acad. Sci. Paris, Serie I, Math., **301**(1985), 349–350.
85. S.H. Min, J. Peking Univ., **5**(1959), 127–130.
86. ______, Sci. Record, **3**(1959), 427–429.
87. C.D. Pan, Sci. Record, **11**(1959), 430–432.
88. M.J.D. Powell and M.A. Sabin, ACM Trans. Math. Software, **3**(1977).
89. Z. Sha, J. Approx. Theo. Appl., **1**:2(1985), 71–82.
90. ______, J. Approx. Theo. Appl., **1**:4(1985), 1–18.
91. X.C. Shen, Multi. Approx. Theo. II, eds. W. Schempp and K. Zeller, 1982, 385–406.
92. ______, J. Approx. Theo. Appl., **1**:2(1985), 15–27.
93. X.L. Shi, J. Approx. Theo. Appl., **1**(1984), 81–93.
94. ______, Indiana Univ. Math. J., **35**(1986), 103–116.
95. X.Q. Shi, Ph.D Thesis, Jilin University, 1988.
96. Y.G. Shi, J. Approx. Theo., **40**(1984), 242–245.
97. ______, J. Approx. Theo., **53**(1988), 41–53.
98. C.T. Shih(=Z.C. Shi), Math. Numer. Sinica, **1**(1979), 50–81.
99. ______, Math. Numer. Sinica, **5**(1983), 195–203.
100. B.Q. Su, Acta Math. Appl. Sinica, **1**(1976), 49–58.
101. ______, Acta Math. Appl. Sinica, **1**(1977), 49–54.
102. ______, Acta Math. Appl. Sinica, **2**(1977), 80–89.
103. ______, J. Fudan, **2**(1977), 22–29.
104. ______, J. Fudan, **4**(1979) (with X.G. Hua).
105. ______, Acta Math. Appl. Sinica, **3**(1980), 139–146(with Y.L. Hsin).
106. B.Q. Su and D.Y. Liu, *Computational Geometry*, Shanghai Sci. Tech. Press, Shanghai, 1981.
107. ______, Scientia Sinica, **10**(1982), 876 877.
108. W.Y. Su, J. Approx. Theo. Appl., **4**:2(1988), 119 129.
109. J.C. Sun, *Spline function and computational geometry*, Science Press, Beijing, 1982.
110. X.H. Sun, Acta Math. Hungar., **53**(1989), 75–84.
111. X.P. Sun, J. Approx. Theo., **57**(1989), 239–244.
112. Y.S. Sun, Acta Math. Sinica, **25**(1982), 561–577.
113. ______, Scientia Sinica, **3**(1982), 215–225.
114. ______, Approx. Theo. IV, eds. C.K. Chui, L.L. Schumaker and J. Ward, 1983, 709–713.
115. ______, *Approximation Theory of Functions*, Beijing Normal Univ. Press, 1989.
116. Y.S. Sun and D.R. Huang, J. Approx. Theo. Appl., **1**(1984), 19–35.
117. J.Z. Wang, Scientia Sinica, **25**(1982), 1056–1065.
118. R.H. Wang, Doklad Acad. Nauk USSR, **150**(1963), 1195–1197.
119. ______, Acta Math. Sinica, **18**(1975), 91–106.
120. ______, Scientia Sinica, Math. I, (1979), 215–226.
121. ______, Acta Math. Sinica, **23**(1980), 163–176.
122. ______, Numer. Math. Chin. Univ., **1**(1980), 78–81.
123. ______, *Approximation of unbounded functions*, Sci. Press, Beijing, 1983.
124. ______, J. Comp. Appl. Math., **12** & **13**(1985), 163–177.
125. R.H. Wang and T.X. He, Kexue Tongbao, **30**(1985), 858–861.
126. ______, Scientia Sinica, **1**(1986), 19–25.
127. R.H. Wang and X.Z. Liang, *Approximation of multivariate functions*, Sci. Press, Beijing, 1988.
128. R.H. Wang and X.G. Lu, Scientia Sinica, A. **32**(1989), 674–684.
129. R.H. Wang and X.Q. Shi, Kexue Tongbao, **9**(1988), 716.
130. ______, Approx. Opti. and Computing, eds. A.G. Law and C.L. Wang, North-Holland, IMACS, 1990, 205–208.
131. R.H. Wang, X.Q. Shi, Z.X. Luo and Z.X. Su, *Multivariate spline function and its applications*, Science Press, Beijing, 1993.
132. S.L. Wang, Kexue Tongbao, Spe. Issue (1980), 76–80.
133. ______, Kexue Tongbao, **29**(1984), 1439–1441.
134. X.H. Wang, Acta Math. Sinica, **14**(1964), 231–237.

135. ______, Kexue Tongbao, **24**(1979), 869–872.
136. D. Waterman, Studia Math., **44**(1972), 107–117.
137. T.F. Xie, J. Approx. Theo. Appl., **1**:4(1985), 57–63.
138. G.L. Xu, J. Comp. Math., **2**(1984), 170–179.
139. C.Y. Yu, Acta Math Sinica, **8**(1958), 190–199.
140. ______, C.R. Acad. Sc. Paris, série A, **228**(1979), 891–893.
141. ______, ibid, série I, **296**(1983), 187–190.
142. X.M. Yu, J. Approx. Theo. Appl., **1**:3(1985), 109–114.
143. K. Zhao and J.C. Sun, Comput. Aid. Geo. De., **5**(1988), 119–125.
144. W.X. Zheng, Acta Math. Sinica, **15**(1965), 54–62.
145. ______, J. Approx. Theo. Appl., **1**(1984), 65–76.
146. ______, Rocky Mount. J. Math., **15**:4(1985), 801–815.

Institute of Mathematical Sciences, Dalian University of Technology, Dalian 116024 China

Contemporary Mathematics
Volume **163**, 1994

A Summary on Continuous Complexity Theory

WANG XING-HUA

ABSTRACT. In this paper we present the work of continuous complexity theory, discuss the iterative convergence and give the estimates from data at one point. This summary is mainly based on [**13, 20, 21, 22, 23, 25**].

1. Exact point estimates for Newton's iteration.

"Point estimate" is abbreviation for the estimate from data at one point. First, we give two definitions of approximate zeros (see [**13, 14**]).

Consider Newton's iteration $z_{n+1} = z_n - Df(z_n)^{-1}f(z_n)$ starting from $z_0 = z \in \mathbf{E}$ for an analytic map $f : \mathbf{E} \to \mathbf{F}$, where both $\mathbf{E}$ and $\mathbf{F}$ are real or complex Banach spaces.

DEFINITION 1. *Let $z \in \mathbf{E}$. If Newton's iteration starting from $z_0 = z$ is well defined and satisfies*

$$\|z_{n+1} - z_n\| \leq (\frac{1}{2})^{2^n-1}\|z_1 - z_0\|,$$

then z is called an approximate zero of f for Newton's iteration.

DEFINITION 2. *Let $z \in \mathbf{E}$ and $\zeta \in \mathbf{E}$ such that $f(\zeta) = 0$. If Newton's iteration starting from $z_0 = z$ is well defined and satisfies*

$$\|\zeta - z_n\| \leq (\frac{1}{2})^{2^n-1}\|\zeta - z\|,$$

then z is called an approximate zero of the second kind of f for Newton's iteration.

Comparing with the work of [**12**], above definitions have been reasonably modified. Such modification happens to consider with the concept of an approximate zero of order two introduced by the author in [**26**]. Only using this concept, the superiority of Newton's method can be demonstrated.

1991 *Mathematics Subject Classification.* 65Y20, 68Q25, 65J15.

The subject was supported by China's State Key Project for Basic Research.

The definitions give us a direction to search for an approximate zero. But what criterion can be used so that we can conveniently judge an approximate zero. First, we recall Kontorovich's theory.

Provided that Df is Lipschitz continuous on some convex region $\mathbf{D} \subset \mathbf{E}$, this theory can be applied to judge convergence and convergent speed of Newton's method with the value of h which is defined by

$$h = h(z, f, \mathbf{D}) = \beta \cdot K$$

and

$$\beta = \beta(z, f) = \|Df(z)^{-1} f(z)\|,$$
$$K = K(z, f, \mathbf{D}) = \operatorname{vrai}\sup_{z \in \mathbf{D}} \|Df(z)^{-1} D^2 f(z)\|.$$

But here we must consider the distance from z to the boundary of $\mathbf{D}$ such that it is not less than $2\beta/(1 + \sqrt{1-2h})$.

Obviously, in this case the criterion is complicated because of the information of region $\mathbf{D}$. Smale once proposed to substitute the analyticity of f at z for Lipschitz continuity of Df in $\mathbf{D}$ and use

$$\gamma = \gamma(z, f) = \sup_{n \geq 2} \|Df(z)^{-1} \frac{D^n f(z)}{n!}\|^{\frac{1}{n-1}}$$

to substitute for $K = K(z, f, \mathbf{D})$ such that the criterion becomes

$$\alpha = \alpha(z, f) = \beta \cdot \gamma$$

without region information. This is called the criterion of point estimates.

Now we state two conclusions in [**14**]. (a) There exists an absolute constant $\alpha_0 \approx 0.130707$ such that if $\alpha(z, f) < \alpha_0$, then z is an approximate zero of f. (b) If $\|z - \zeta\|\gamma(z, f) < \frac{3-\sqrt{7}}{2}$, then z is an approximate zero of the second kind of f, where ζ is a zero of f for Newton's iteration.

1.1. Domain estimate and point estimate. Here domain estimate means the estimate obtained from $K = K(z, f, \mathbf{D})$ which connects with some region information $\mathbf{D}$.

A relation between domain estimate and point estimate can be reduced to the following fact.

THEOREM 1. *Suppose $a, k \in \mathbf{R}_+$ such that*

$$a\gamma(z, f) \leq 1 - (k(\sqrt{1 + \frac{8}{27}k} + 1))^{\frac{1}{3}} + (k(\sqrt{1 + \frac{8}{27}k} - 1))^{\frac{1}{3}},$$

then

$$kaK(z, f, \overline{S(z, a)}) \leq 1,$$

where $\overline{S(z, a)}$ denotes a closed ball with the center z and radius a.

Now setting $a = \frac{3}{2}\beta(z, f), k = \frac{3}{2}$ in theorem 1, we have if

$$\alpha(z, f) \leq \frac{3}{2}\{1 - (\frac{\sqrt{13}+3}{2})^{\frac{1}{3}} + (\frac{\sqrt{13}-3}{2})^{\frac{1}{3}}\} \approx 0.121522,$$

then

$$K(z, f, \overline{S(z, \frac{3}{2}\beta(z, f))} \leq \frac{4}{9}.$$

Hence according to Kontorovich's theory improved by [**17, 18**], z is an approximate zero of f for Newton's iteration.

On the other hand, setting $z = \zeta$, $a = \|z - \zeta\|$, $k = 3$ in theorem 1, we have if

$$\|z - \zeta\|\gamma(\zeta, f) \leq 1 - (\sqrt{17}+3)^{\frac{1}{3}} + (\sqrt{17}-3)^{\frac{1}{3}} \approx 0.115378$$

then

$$3\|\zeta - z\|K(\zeta, f, \overline{S(\zeta, \|z - \zeta\|)}) \leq 1.$$

Thus, according to [**19**], z is an approximate zero of the second kind of f for Newton's iteration.

The possibility reducing the point estimate from Kantorovich's theory has been pointed out by [**13, 14**]. But our estimation is optimal.

1.2. A problem of Smale. Smale proved that the inequality $\alpha(r, \varphi_\alpha) \leq 1$ is true for all $r \in (0, 1)$, where $\varphi_d(r) = \sum_{i=0}^{n} r^i, d \in \mathbf{N}\bigcup\{\infty\}$. This inequality is very important, which can be applied to build a generally convergent iteration (see section 3.3 in [**13**]). For an analytic map $f : \mathbf{E} \to \mathbf{F}$ with form $f(z) = \sum_{i=0}^{d} a_i z^i$, the inequality

$$\gamma(z, f) \leq \|f\|\|Df(z)^{-1}\|\frac{\varphi_d'(\|z\|)^2}{\varphi_d(\|z\|)}$$

is true, where $\|f\| = \sup_k \|a_k\|$. This shows that the criterion $\gamma(z, f)$ is easy to estimate.

Smale intends to establish a kind of L^2-version for $\alpha(r, \varphi_d) \leq 1$ in Hilbert space. But no proof is given.

Problem (Smale [**15**]) Is $\alpha(r, \psi_d) \leq 1$ true for all $r \in (0, 1)$? Here $\psi_d(r) = \sqrt{\sum_{i=0}^{d} r^{2i}}$.

For this problem we have

THEOREM 2. *For all $r \in (0, 1)$, $\alpha(r, \psi_d) \geq \frac{1}{2r^2}$. Thus, if $r < 1/\sqrt{2}$, then $\alpha(r, \psi_d) > 1$. That is, the answer is negative.*

1.3. Exact point estimate to Newton's iteration. Let $N_f(z) = z - Df(z)^{-1}f(z)$. Suppose that criterions $\alpha = \alpha(z,f)$, $\beta = \beta(z,f)$ and $\gamma = \gamma(z,f)$ be defined for an analytic map $f : \mathbf{E} \to \mathbf{F}$ and $z \in \mathbf{E}$. Let $h : \mathbf{R} \to \mathbf{R}$ as follows

$$h(t) = \beta - t + \frac{\gamma t^2}{1-\gamma t}, \qquad t \in \mathbf{R}.$$

It is easy to see $\beta(0,h) = \beta$ and $\gamma(0,h) = \gamma$. So $\alpha(0,h) = \alpha$. Function h has two zeros

$$\left.\begin{matrix} t^* \\ t^{**} \end{matrix}\right\} = \frac{1+\alpha \mp \sqrt{(1+\alpha)^2 - 8\alpha}}{4\gamma}.$$

Let $t_n = N_h^n(0)$. Then an expression of t_n is

$$t_{n+1} - t_n = \frac{(1-q^{2^n})\sqrt{(1+\alpha)^2-8\alpha}}{2\alpha(1-\eta q^{2^n-1})(1-\eta q^{2^{n+1}-1})}\eta q^{2^n-1}\beta,$$

where

$$\eta = \frac{1+\alpha-\sqrt{(1+\alpha)^2-8\alpha}}{1+\alpha+\sqrt{(1+\alpha)^2-8\alpha}},$$

$$q = \frac{1-\alpha-\sqrt{(1+\alpha)^2-8\alpha}}{1-\alpha+\sqrt{(1+\alpha)^2-8\alpha}}.$$

Obviously if $\alpha \le 3 - 2\sqrt{2}$, then t_n tends to t^* monotonously. For Newton's sequence $z_n = N_f^n(z)$ and a zero ζ of f, we can prove that $\|z_{n+1} - z_n\|$ and $\frac{\|\zeta - z_n\|}{\|\zeta - z_0\|}$ are dominated by $t_{n+1} - t_n$ and $\frac{t^*-t_n}{t^*-t_0}$ respectively. Thus we obtain

THEOREM 3. 1° *If and only if $\alpha(z,f) \le 3 - 2\sqrt{2}$, $z_n = N_f^n(z_0)$ are well defined for $z_0 = z \in \mathbf{E}$ and any analytic map $f : \mathbf{E} \to \mathbf{F}$ and satisfy*

$$\|z_{n+1} - z_n\| \le (\frac{1}{2})^n \|z_1 - z_0\|, \qquad n = 0, 1, \cdots.$$

More precisely, for given $q \in (0,1)$, we have

2° *If and only if $\alpha(z,f) \le \frac{1+4q+q^2-(1+q)\sqrt{1+6q+q^2}}{2q}$, $z_n = N_f^n(z_0)$ are well defined for $z_0 = z \in \mathbf{E}$ and any analytic map $f : \mathbf{E} \to \mathbf{F}$ and satisfy*

$$\|z_{n+1} - z_n\| \le q^{2^n-1}\|z_1 - z_0\|, \qquad n = 0, 1, \cdots,$$

and

$$\|\zeta - z_n\| \le q^{2^n-1}\|\zeta - z\|, \qquad n = 0, 1, \cdots.$$

Setting $q = \frac{1}{2}$ in theorem 3, we obtain a precise bound of criterion $\alpha(z,f)$, $\frac{13-3\sqrt{17}}{4} \approx 0.157671$. That means that z is approximate zero of f for Newton's iteration and also one of the second kind of f.

1.4. Application. In this section we give an application of point estimate to find all zeros of a polynomial simultaneously. For a complex polynomial of degree d

$$f(z) = z^d + a_1 z^{d-1} + \cdots + a_{d-1}z + a_d,$$

all zeros $(z_1, \cdots, z_d) = (\zeta_1, \cdots, \zeta_d)$ satisfy the following equations

$$\begin{cases} z_1 + \cdots + z_d + a_1 = 0, \\ z_1 z_2 + \cdots + z_{d-1}z_d - a_2 = 0, \\ \cdots\cdots\cdots\cdots \\ z_1 z_2 \cdots z_d - (-1)^d a_d = 0. \end{cases}$$

Write above equations as $F(z)$ for $z = (z_1, \cdots, z_d) \in \mathbf{C}^d$. Applying Newton's iteration to F, its computational formulae are

$$z_i^{(n+1)} = z_i^{(n)} - \frac{f(z_i^{(n)})}{\prod_{j\neq i}(z_i^{(n)} - z_j^{(n)})}, \qquad i = 1, \cdots, d.$$

This iteration was first presented by Weierstrass. Zhao Fengguang et al. proved that

$$\alpha(z, F) \leq \frac{1}{\delta}\sum_{i=1}^{d} \frac{|f(z_i)|}{\prod_{j\neq i}|(z_i - z_j)|}$$

with l_1-module, where $\delta = \min_{1\leq i<j\leq d}|z_i - z_j|$. Hence we get

THEOREM 4. *Let $z = (z_1, \cdots, z_d) \in \mathbf{C}^d$ satisfy*

$$\sum_{i=1}^{d} \frac{|f(z_i)|}{\prod_{j\neq i}|(z_i - z_j)|} \leq (3 - 2\sqrt{2}) \min_{1\leq i<j\leq d} |(z_i - z_j)|,$$

then Weierstrass iteration starting from $z^{(0)} = z$ converges.

Corresponding results of error estimates and approximate zeros can also be obtained.

1.5. Deformed Newton's methods. Properly choosing a map $A_n : \mathbf{F} \to \mathbf{E}$ to substitute for $Df(z_n)^{-1}$ in Newton's iteration, the computational efficiency of the deformed Newton's iteration $z_{n+1} = z_n - A_n f(z_n)$ is often higher than Newton's. [**24**] has studied point estimates of the following deformations:

I. $A_n = Df(z_{m[n/m]})^{-1}$ for given $m \in \mathbf{N}$;

II. $A_{n+1} = 2A_n - A_n Df(z_n) A_n$;

III. $A_n = Df(\zeta_n)^{-1}, \quad \zeta_{n+1} = \zeta_n - \frac{1}{2}A_n f(z_{n+1})$

where the square brackets denotes the integer part. We state a particular case for $A_0 = Df(z_0)^{-1}$. If $\alpha \leq 3 - 2\sqrt{2}$, then three deformations above are convergent.

2. Higher-order generalization of Newton's method

2.1. Halley's family. Halley, an astronomer, after whom the Halley Comet is named, once used an iterative function for $f: H_f(z) = z - \frac{2f(z)f'(z)}{2f'(z)^2 - f(z)f''(z)}$ with convergent speed of order three. In fact, $z' = H_f(z)$ is a zero of Pade's approximation to f at z with order $[1/1]$. For $k \in \mathbf{N}$, Let $z' = H_{n,f}(z)$ be a zero of Pade's approximation to f with order $[1/n-1]$. Then we get an iterative family. The elements of $H_{1,f} = N_f, H_{2,f} = H_f, \cdots$ have increasing convergent orders. In [**27**] an analytic representation of $H_{k,f}(z)$ is given by

$$H_{k,f}(z) = z + \frac{kD^{k-1}[f(z)^{-1}]}{D^k[f(z)^{-1}]}.$$

Analogously to the form of Bernoulli iteration, there is no problem in Banach space to give the following expression

$$H_{k,f}(z) = z - \left(I - \sum_{i=2}^{k} Df(z)^{-1}\frac{D^i f(z)}{i!}\prod_{j=1}^{i-1}(H_{k-1,f}(z) - z)\right)^{-1} \cdot Df(z)^{-1}f(z).$$

Using the same majorant function h in section 1.3, we can prove

THEOREM 5. *Let $f: \mathbf{E} \to \mathbf{F}$ be analytic, $z \in \mathbf{E}$. If $\alpha = \alpha(z, f) \leq 3 - 2\sqrt{2}$, then for arbitrary $k \in \mathbf{N}$, Halley's iterative family is well defined and satisfies*

$$\|\zeta - z_{k,n}\| \leq \frac{1 + \alpha - \sqrt{(1+\alpha)^2 - 8\alpha}}{4\alpha}\varepsilon_{k,n}\|z_{1,1} - z\|,$$

where $z_{k,n} = H^n_{k,f}(z)$, $f(\zeta) = 0$ and

$$\varepsilon_{k,n} = \begin{cases} \left[\frac{1+\alpha-\sqrt{(1+\alpha)^2-8\alpha}}{1+\alpha+\sqrt{(1+\alpha)^2-8\alpha}}\left(\frac{3-\alpha-\sqrt{(1+\alpha)^2-8\alpha}}{3-\alpha+\sqrt{(1+\alpha)^2-8\alpha}}\right)^{\frac{1}{k}}\right]^{(k+1)^n-1} & \text{if } \alpha < 3 - 2\sqrt{2}. \\ \left(\frac{1}{k+1}\right)^n & \text{if } \alpha = 3 - 2\sqrt{2}. \end{cases}$$

2.2. Euler's series. Let $f: \mathbf{C} \to \mathbf{C}$ be analytic and $f'(z) \neq 0$, for a $z \in \mathbf{C}$, f_z^{-1} represents a local inverse map of f at z which maps $w = f(z)$ into z. Hence in some neighbourhood of w, $z = f_z^{-1}(w')$ can be expanded by the powers of $w' - w$

$$z' = z + \sum_{n=1}^{\infty} \frac{(f_z^{-1})^{(n)}(w)}{n!}(w' - w)^n.$$

This series is called Euler's series. Let $R(z, f)$ denote the convergent radius of this series.

In order to obtain a lower bound we take a replacement $\lambda = (z' - z)/u$ and $\omega = (w' - w)/w$ where $u = -f(z)/f'(z)$. Through this replacement Taylor series of $w' = f(z')$ at z and $z' = f_z^{-1}(w')$ at w can be reduce to $\omega = \lambda - \sum_{n=2}^{\infty} a_n\lambda^n$ and

$\lambda = \omega + \sum_{n=2}^{\infty} b_n \omega^n$, where $a_n = \frac{f^{(n)}(z)}{-f'(z)\cdot n!} u^{n-1}$. By means of Bell polynomial, we can obtain

$$b_n = \sum_{m=2}^{n} \nu_{n,m} a_m,$$

where

$$\nu_{n,m} = \sum_{j_1+2j_2+\cdots+(n-2)j_{n-2}=n-m} C_m^{j_1,\cdots,j_{n-2}} b_2^{j_1} \cdots b_{n-1}^{j_{n-2}},$$

and

$$C_m^{j_1,\cdots,j_{n-2}} = \frac{m!}{j_1! \cdots j_{n-2}!(m - j_1 - \cdots - j_{n-2})!}.$$

Hence the series $\sum_{n=1}^{\infty} B_n \alpha^{n-1} |\omega|^n$ with positive terms is a majorant one of reduced Euler's series, where $\lambda = \sum_{n=1}^{\infty} B_n \omega^n$ is the inverse of series $\omega = \lambda - \sum_{n=2}^{\infty} \lambda^n$. Thus we get

THEOREM 6. *It holds*

$$\frac{\alpha(z,f)R(z,f)}{|f(z)|} \geq 3 - 2\sqrt{2},$$

where the constant at the right is the best possible. Furthermore it holds

$$\left| \frac{(f_z^{-1})^{(n)}(w)}{n!} \right| \leq \frac{B_n}{|f'(z)|} \left(\frac{\alpha(z,f)}{|f(z)|} \right)^{n-1}, \quad \forall n \in \mathbf{N},$$

where $B_1 = 1$, $B_n = \sum_{\frac{n}{2} \leq i \leq n} (-1)^{n-1} \frac{(2i-3)!!}{(n-i)!(2i-n)!} 3^{2i-n} 2^{i-n-2}$ *are also the best possible in the inequality for all* n.

It is necessary to point out, M.H. Kim also obtains the same lower bound using complex analysis method. But in [**28**] we use the majorant series method and the result is easy to be extended to Banach space. Moreover, there is an interesting fact that coefficient B_n of the majorant series is just the number of different forms to add brackets to the term $x_1 x_2 \cdots x_n$ of n letters. For example, for $x_1x_2x_3x_4$, there are eleven forms through adding brackets: $x_1x_2x_3x_4$, $x_1(x_2x_3x_4)$, $(x_1x_2x_3)x_4$, $x_1x_2(x_3x_4)$, $x_1(x_2x_3)x_4$, $(x_1x_2)x_3x_4$, $x_1(x_2(x_3x_4))$, $x_1((x_2x_3)x_4)$, $(x_1(x_2x_3))x_4$, $((x_1x_2)x_3)x_4$, $(x_1x_2)(x_3x_4)$, thus for $n = 4$, $B_4 = 11$.

2.3. Euler's family. With the notions above, the partial sum of series of f_z^{-1} at $w' = 0$ with k terms is $z + \sum_{n=1}^{k} b_n$ which consists of a family of iterative functions. Write $E_{k,f}(z)$. This family is called Euler's family. [**10, 11**] study its convergence and complexity systematically and get a lower estimate of constructive cost of the algebraic elementary theorem. This improves results in [**12**]. Some important works about computational complexity in numerical analysis also refer to [**8, 9**] and [**16**]. Here we study the point estimate for analytic maps in Banach space. For convenience, we still discuss the convergence in complex space. But the conclusion can be easy to extend to Banach space. The majorant function in section 1.3 is also one of Euler's iterative family. The reason for this is that for $z \in \mathbf{C}$ and $f'(z) \neq 0$ we can prove

$$f(z') = \sum_{n=k+1}^{\infty} \sum_{m=2}^{n} \sum_{j_1+2j_2+\cdots+(n-1)j_{k-1}=n-m} C_m^{j_1,\cdots,j_{k-1}} b_2^{j_1} \cdots b_k^{j_{k-1}} a_m f(z).$$

Thus we have

THEOREM 7. *Let $f : \mathbf{C} \to \mathbf{C}$ be analytic and $z \in \mathbf{C}$. If $\alpha(z,f) \leq 3 - 2\sqrt{2}$, then for arbitrarily given $k \in \mathbf{N}$, iterations of Euler's family $z_{k,n} = E_{k,f}^n(z)$ are well defined and iterative sequence is convergent to a zero of f.*

Error estimate of Euler's iterative family can not be given like Halley's. One form is

$$t^* - t_{k,n+1} = (k+1) \int_{t_{k,n}}^{t^*} b_{k+1,h}(t)dt,$$

where $t_{k,n+1} = E_{k,h}^n(0)$.

3. Complexity theory for solving equation

3.1. Computational complexity for solving polynomials. Let f be a polynomial with the form $f(z) = \sum_{i=0}^d a_i z^i, a_i \in \mathbf{C}$ and $a_d = 1$. Define

$$P(1) = \{f | f(z) = \sum_{i=0}^{d} a_i z^i, a_d = 1, |a_i| < 1, i = 0, 1, \cdots, d-1\}.$$

A modified Newton's iteration is inductively defined by $z_n = T_h(z_{n-1})$, $z_0 \in \mathbf{C}$, where $T_h(z) = z - h\frac{f(z)}{f'(z)}$ for some $0 < h \leq 1$

For the polynomial set $P(1)$, [**12**] proves that

THEOREM 8. *There is a universal polynomial $S(d, \frac{1}{\mu})$ and a function $h = h(d,\mu)$ such that for degree d and $\mu, 0 < \mu < 1$, the following is true with probability $1-\mu$. Let $x_0 = 0$. Then $x_n = T_h(x_{n-1})$ is well defined and x_s satisfies $|\frac{f(x_{s+i})}{f(x_{s+i-1})}| \leq \frac{1}{2}$ and $x_{s+i} = N_1(x_s)$ for $i = 1, 2, \cdots$, where $s = S(d, \frac{1}{\mu})$.*

Remark *The polynomial $S(d, \frac{1}{\mu})$ can be taken as $(100(d+2))^9/\mu^7$.*

3.2. Complexity of Kuhn algorithm. Let **C** be complex plane. In the space $\mathbf{C} \times [-1, \infty)$, write $C_d = C \times \{d\}, d = -1, 0, 1, 2, \cdots$. Assume $\bar{z} \in \mathbf{C}$ and h given. Then we can give a triangulation to C_d with the center $\bar{z}$ and mesh diameter $\sqrt{2}2^{-d}h$ for $d \geq 0$ (see [**29**]). Note that for $d = -1$, the mesh diameter remains $\sqrt{2}h$. Therefore a simplicial triangulation T of half space $\mathbf{C} \times [-1, \infty)$ is given through relating the vertices of $\mathbf{C}_d$ and $\mathbf{C}_{d+1}$ properly. Write $V(T)$ as the set of vertices of T.

Let $f_{-1}(z) = (z - \bar{z})^n$, $f_d(z) = f(z)$, $d = 0, 1, 2, \cdots$. Define a mapping $l : V(T) \to \{1, 2, 3\}$ as follows

$$l(z, d) = \begin{cases} 1, & \text{if } -\frac{\pi}{3} \leq \arg f_d(z) \leq \frac{\pi}{3} \quad \text{or } f_d(z) = 0, \\ 2, & \text{if } \frac{\pi}{3} < \arg f_d(z) \leq \pi, \\ 3, & \text{if } -\pi < \arg f_d(z) < \frac{\pi}{3}\,. \end{cases}$$

So in $\mathbf{C} \times [-1, \infty)$, all vertices has been labeled by 1, 2 or 3 respectively. If a three-point set $\{z_1, z_2, z_3\} \subset V(T)$ satisfies $\{l(z_1), l(z_2), l(z_3)\} = \{1, 2, 3\}$, we call it a completely labeled three-point set.

Kuhn algorithm is realized by searching all three-point set with complete label on each larger or successive layers (for details see [**3**]). About the complexity of Kuhn algorithm for solving algebraic equation, [**29**] gives that

THEOREM 9. *For given $\varepsilon > 0$, to find all zeros of an algebraic equation $f(z) = 0$ of degree n within the precision ε, the number of evaluation of f are at most $O(n^3 \log n/\varepsilon)$.*

3.3. Lehmer's and Weierstrass' parallel algorithm. For $c \in \mathbf{C}$, $r \in \mathbf{R}_+$, the set of complex $\{z| \;\|z - c\| < r\}$ is called an open disk with the center c and the radius r and denoted by $(c; r)$, $[c; r]$ is the enclosure of $(c; r)$. Now we introduce Lehmer's algorithm for a polynomial.

Assume that every zero for a polynomial is in one of the disks $Z_j^{(k)} = (z_j^{(k)}; 2^{-k}r)$, $j = 1, 2, \cdots, d_k$ and every open disk has at least one zero of f for $j = 1, 2, \cdots, d_k$, we determine whether or not the zeros of f are in the following eight open disks

$$Z_{j,0}^{(k)} := (z_j^{(k)}, 2^{-k-1}r),$$

$$Z_{j,\nu}^{(k)} := (z_j^{(k)} + \frac{1 + \cos\frac{2\pi}{7}}{1 + 2\cos\frac{2\pi}{7}} \cos\frac{2(\nu - 1)\pi}{7};\ 2^{-k-1}r), \quad \nu = 1, \cdots, 7,$$

when a zero of f is in some $Z_{j,\nu}^{(k)}$, we write it as

$$Z_{j'}^{(k)} = (z_j^{(k+1)}; 2^{-1-k}r), \qquad j' := 1 + \sum_{i=1}^{j-1}\sum_{\mu=1}^{\nu-1} \chi(Z_{i,\mu}^{(k)}),$$

where

$$\chi(Z) = \begin{cases} 1, & \text{if there is at least one zero in } Z \\ 0, & \text{otherwise.} \end{cases}$$

If we can not determine whether or not the zero of f is in $Z_{j,\nu}^{(k)}$, then we replace $Z_{j,\nu}^{(k)}$ by some open disk and continue to judge (for details see [**4**]).

THEOREM 10. *The cost of computation for finding the ε-approximation of all zeros of $f \in \mathbf{P}_d(1)$ is at most $16d^2 \log \frac{2}{\varepsilon}$ with the probability one by using Lehmer's algorithm.*

If we have found an ε-approximation of a zero, we can use it to an initial value of the high-order convergent iteration.

Weierstrass' parallel algorithm is defined by

$$W_i^{(n+1)} = \mathrm{mid}W_i^{(n)} - f(\mathrm{mid}W_i^{(n)}) \prod_{j=1, j\neq i}^{d} \frac{1}{\mathrm{mid}W_i^{(n)} - W_j^{(n)}}$$

for $i \in \{1, \cdots, d\}$ and $n \in \mathbf{N}_0$.

Assume that the simple zeros $\zeta_1, \zeta_2, \cdots, \zeta_d$ of f are known in the disks $W_1^{(0)}$, $\cdots$, $W_d^{(0)}$, respectively.

Let $r^{(n)} = \max_{1\le i\le d} \mathrm{rad}W_i^{(n)}$. If the disk vector $W^{(0)} = (W_1^{(0)}, \cdots, W_d^{(0)})$ satisfies

$$r^{(0)} \le \frac{1}{3d} \min\{r := |\zeta_1 - \zeta_2| > 0 \ \ |f(\zeta_1) = f(\zeta_2) = 0\}, \tag{$*$}$$

then

$$r^{(n)} < 2(\frac{1}{2})^{2^{n-1}} r^{(0)}$$

(see [**27**]).

This condition ensures that the Weierstrass parallel algorithm has good convergence property.

THEOREM 11 ([**26**]). *For $0 < \mu < 1$ and $d \ge 2$, the computational cost for finding the disk vector satisfying $(*)$ by Lehmer's algorithm is at most $l = O(d^3 + d^2 \log \frac{1}{\mu})$ times of f-evaluation with probability $1 - \mu$.*

4. Numerical integration

4.1. Efficiency of integration approximation. Let H^k be a Sobolev space on $[0,1]$ and $H^k \to R$ be a quadrature functional with algebraic precision of at least $k-1$ degrees. Suppose $J : f \longmapsto \int_0^1 f(t)dt, h = \frac{1}{n}$. Define a complexificated quadrature functional QM_h of Q through $M_h : C[0,1] \to C[0,1]$ determined by

$$M_h f(t) = h \sum_{i=0}^{n-1} f(ih + th), \qquad \forall f \in C[0,1], \qquad \forall t \in [0,1].$$

Under the assumption that the inner product between unit element e and any element f in H^k obeys the standard normal distribution (see [**15**]), the study of integral approximate efficiency can be reduced to find the average

$$\mathcal{E}_{Q,h}^k = \mathrm{Av}_{f\in H^k} \ |(J - QM_h)f|.$$

For this problem, we have a general result [**23**].

THEOREM 12. *Let*

$$\mathcal{E}_{u,h}^k = \{(-1)^k \frac{2}{\pi}(J-Q)_t(J-Q)_s \frac{(s-t)_t^{2k-1}}{(2k-1)!}\}^{\frac{1}{2}} h^k,$$

where $L_t f(t) = Lf$, $t_+ = \max\{t, 0\}$. *If* Q *is give by*

$$Qf = \sum_{i=0}^{n-1} p_j f(t_j), \quad \sum_{j=0}^{m-1} p_j t_j^{r-1} = \frac{1}{r!}, \quad 1 \le r \le k,$$

then we have

$$\mathcal{E}_{Q,h}^k = \Big\{(-1)^k \frac{2}{\pi}\Big[\frac{1}{(2k+1)!} \sum_{i-0}^{m-1} p_i \frac{t_i^{2k} + (1-t_i)^{2k}}{(2k)!} - \sum_{l=i+1}^{m-1} p_l \frac{(t_l - t_i)^{2k-1}}{(2k-1)!}\Big]\Big\}^{\frac{1}{2}} h^k.$$

4.2. Information-based complexity for integration problem. Traub, Wasilkowski, Wozniakonski and other mathematicians establish a new framework for integration approximations by modern developments of function approximation theory.

Given an information $N(f)$ about f, the problem is which is the best approximation to a solution among all algorithms using this information.

In detail, we give a result of [**30**].

Let

$$W_p^r[a,b] = \{f | f \in AC^{r-1}[a,b], \quad \|f^r\| \le 1\}.$$

Give

$$N(f) = [f(x_1), \cdots, f^{(\nu_1 - 1)}(x_1), \cdots, f(x_n), \cdots, f^{(\nu_n - 1)}(x_n)]$$
$$\text{for} \quad a \le x_1 \le \cdots \le x_n \le b,$$

where $1 \le \nu_i \le r$ $(i = \overline{1,n})$, and $\sum_{i=1}^n \nu_i \le N$.

THEOREM 13. *For* $f \in W_p^r[a,b]$, *then there exist points* $\tau_1, \cdots, \tau_N \in (0,1)$ *such that for the information* $N^* = [f(\tau_1), \cdots, f(\tau_N)]$ *we have*

$$r(N^*) \le r(N),$$

where $r(N)$ *represents the radius of the information* N *(see* [**16**]).

This theorem says among all quadrature formulae in $W_p^r[a,b]$ with the form

$$\int_a^b f(t)dt = \sum_{i=1}^{n}\sum_{j=0}^{v_i-1} a_{i,j} f^{(j)}(x_i) + \sum_{i=1}^{l} a_i f^{(\alpha_i)}(a) + \sum_{j=1}^{m} b_i f^{(\beta_i)}(b) + R(f),$$

there exists an optimal quadrature formula with the form

$$\int_a^b f(t)dt = \sum_{i=1}^{N} d_i^* f(\tau_i) + \sum_{i=1}^{l} a_i^* f^{(\alpha_i)}(a) + \sum_{j=1}^{m} b_j^* f^{(\beta_j)}(b) + R(f),$$

where $0 \le \alpha_1 < \cdots < \alpha_l \le r-1$, $0 \le \beta_1 < \cdots < \beta_m \le r-1$, $d_i^* > 0$ $(i = \overline{1,N})$ and $a < \tau_1 < \cdots < \tau_N < b$.

4.3. N-approximation number of the integral operator. Let $L(X,Y)$ be the set of all continuous linear operator T mapping X to Y, both X and Y are linear normed spaces. For an operator $T \in L(X,Y)$, define

$$a_n(T) = \inf\{\|T - S\| \mid S \in L(X,Y) \text{ and } \operatorname{rank} S < n\},$$

where "rank$S < n$" means the image $S(X)$ is contained in a subspace of Y with the dimension less than n. [**6**] first proposed the concept $a_n(T)$, which is called n-approximation number of the operator T. [**7**] called it $(n-1)$-dimensional linear width. The study of $a_n(T)$ is becoming a very important part in the modern approximation theory as well as Kolmogorov-width.

For a special case $X = Y = L^q[0,1] (q \ge 1)$, consider the integral operator $T_K : X \to Y$,

$$T_K(x)(s) = \int_0^1 K(s,t)x(t)dt, \quad s \in [0,1].$$

When $K(s,t)$ is a nondegenerate totally positive kernel in $L^q([0,1]^2)$, [**7**] give an expression of $a_n(T_K)$. When $K(s,t)$ belongs to the Sobolev space $W_p^r([0,1]^2)$, T_K becomes a compact operator. We give an upper bound of $a_n(T_K)$:

THEOREM 14. *For $K(s,t) \in W_p^r([0,1]^2), 1 \le p \le \infty$ and $r \in N_0$, there exists a positive constant $c_1(r,p)$ such that*

$$a_n(T_K) \le c_1(r,p)\|K\|_{r,p} n^{-r},$$

where $\|\cdot\|_{r,p}$ denotes any norm in Sobolev space.

It is difficult to give a lower bound of $a_n(T_K)$ for all $K \in W_p^r([0,1]^2)$. But we have

THEOREM 15. *For all $n > 1$, there exists a kernel sequence $K_n(s,t) \in W_p^r([0,1]^2)$ with the property $\sup_n \|K_n\|_{r,p} < \infty$ and a constant $c_2(r,p) > 0$ such that*

$$a_n(T_{K_n}) \ge c_2(r,p) n^{-r}.$$

The proofs of Theorems 14 and 15 can be found in [**2**]. Now we give an application to solve Fredholm equations of the second kind using degenerated kernel methods.

Let $X = L^q[0,1](q \geq 1)$. Then Fredholm equation of the second kind is defined by

$$(I - T_K)x = y, \qquad x, y \in X,$$

where T_K is a linear integral operator with W_p^r-kernel.

Choose $K(s,t)$ such that $I - T_K$ is invertible. Then Fredholm equation is solvable for all $y \in X$. A common-used method of solving Fredholm equation is degenerated kernel method (see [**1**]). Suppose $M(T_K)$ is an operator of the dimension less than n obtained by a degenerated kernel method. Then the approximation solution of the equation satisfies

$$(I - M(T_K))\overline{x} = y$$

and the error estimate to the set of T satisfying $I - T$ invertible, denoted by H_p^r, can be denoted by

$$\nu(M) = \sup_{T \in H_p^r} \{\|(I - T_K)^{-1} - (I - M(T_K))^{-1}\|\}.$$

Write D_n the set of all degenerated method M of rank less than n. Then the optimal error ε^* is

$$\varepsilon^* = \inf\{\nu(M) \ : \ M \in D_n\}.$$

Using Theorems 14 and 15, we can obtain

THEOREM 16. *Under the condition of Theorem 10, it holds*

$$\varepsilon^* = O(n^{-r}) \qquad (n > 1).$$

Remark *As* $p = \infty$, *the estimation improves the conclusion in* [**1**].

4.4. The optimal recovery of some smooth function classes. The optimal recovery theory is a main subject of study in approximation theory. Sun and others make an effort on this field and get many important results. In this section we introduce the work in [**5**].

Given $p, q, 1 \leq p, q \leq \infty$, a set of measurable functions defined on the real line R is said to be the amalgam of L^p and l^q if each function locally forms an l^q-sequence in L^p and L^p-norm over the intervals $[n, n+1]$. Denote it by $L_{p,q}(R)$ and introduce $L_{p,q}$-norm as follows:

$$\|f\|_{p,q} = \{\sum_{n \in Z}(\int_n^{n+1} |f(x)|^p dx)^{\frac{q}{p}}\}^{\frac{1}{p}},$$

$$\|f\|_{p,\infty} = \sup_n(\int_n^{n+1} |f(x)|^q dx)^{\frac{1}{p}}, \quad 1 \leq p \leq \infty,$$

and

$$\|f\|_{\infty,q} = \Big\{ \sum_{n\in Z} (\text{ess} \sup_{n\le x\le n+1} |f(x)|)^q \Big\}^{\frac{1}{q}}, \quad 1 \le q < \infty.$$

DEFINITION 3. *Let $M \subset L_{s,s}(R)$ be a linear subspace and let $n > 0$ (not necessarily an integer) we say that the dimensional index of $M \le n$ if*

$$\underline{\lim}_{N\to\infty} \frac{\dim(M|_{[-N,N]})}{2N} \le n,$$

where

$$M|_{[-N,N]} = \{f(t)\chi_N(t) : f \in M, \chi_N(t) \text{ is the characteristic function of } [-N,N]\}.$$

We write it as $\operatorname{cdim}(M) \le n$ for convenience. Define infinite dimensional Kolmogorov n-width d_n and linear n-width δ_n (see [**5**]) as follows:

$$d_n(W^r_{p,q}, L_q) = \inf_{M, \operatorname{cdim}(M)\le n} \sup_{f\in W^r_{p,q}} \inf_{g\in M} \|f-g\|_{L_q(R)}$$

and

$$\delta_n(W^r_{p,q}, L_q) = \inf_S \sup_{f\in W^r_{p,q}} \|f - S(f)\|_{L_q},$$

where S is the set of all linear operator with $\operatorname{cdim}(S(W^r_{p,q}(R))) \le n$.

As $p=\infty$, [**5**] obtains

THEOREM 17. $\forall q, 1\le q<\infty$

(i) *For any $n \in Z_+$*

$$\begin{aligned} d_n(W^r_{\infty,q}, L_q) &= \delta_n(W^r_{\infty,q}, L_q) = E(W^r_{\infty,q}, S_{n,r-1}, L_q) \\ &= \sup_{f\in W^r_{\infty,q}} \|f - s_{n,r}(f)\|_{L_q(R)} = n^{-r}\|\phi_{1r}\|_{L_q[0,1]}, \end{aligned}$$

where $s_{n,r}(f)$ is the cardinal interpolating operator from $W^r_{\infty,q}$ to $S_{n,r-1}$

(ii) *For any $n > 0 (n \in \mathbf{R}_+)$*

$$\begin{aligned} n^{-r}\|\phi_{1r}\|_{L_q[0,1]} &\le d_n(W^r_{\infty,q}, L_q) \le \delta(W^r_{\infty,q}, L_q) \\ &\le n^{-r}\|\phi_{1r}\|_{L_q[0,1]} + O(n^{-r-1}), \quad n\to\infty, \end{aligned}$$

where ϕ_{1r} is the Euler spline of degree r with period π and

$$S_{n,r-1} = \Big\{ s(t) \in C^{r-2}(R) : s^{(r)}(t) = 0, t \in \Big(\frac{j}{n}, \frac{j+1}{n}\Big), j \in \mathbf{Z} \Big\}.$$

References

1. S. Heinrich, *On the optimal error of degenerate kernel methods*, J. Integral Equations, **9**(1985), 251–266.
2. Zhengda Huang, *n-approximation number of integral operator*(in Chinese), Acta Mathematica Sinica, (1993)(to appear).
3. H. Kuhn, *A new proof of the fundamental theorem of algebra*, in Mathematical programming study 1, M.L. Balinski(ed.), North-Holland, Amsterdam, 1974.
4. D.H. Lehmer, *A machine method for solving polynomial equations*, J. Assoc. Comp. Math., **8**(1961), 151–162.
5. Y.P. Liu, Y.S. Sun, *Infinite dimensional Kolmogorov width and optimal interpolation on Sobolev-Wiener space*, Science in China (Ser. A), **35**(1992), 1162–1172.
6. A. Pietsch, *Operator Ideals*, Verlag der Wissenschaften Berlin, 1976.
7. A. Pinkus, *n-widths in Approximation Theory*, Springer-Verlag, Berlin, Heidelberg, New York, Tokyo, 1985.
8. M. Shub, *Some remarks on dynamical systems and numerical analysis*, in "Dynamical Systems and Partial Differential Equations: Proceedings of the VII BLAM", Universidad Simon Boliver, Caracas, 1986, pp. 69–92.
9. ———, *The geometry and topology of dynamical systems and algorithms for numerical problems*, in Proceedings of the 1983 Beijing Symposium on Differential Geometry and Differential Equations, Editor Liao Shantao, Science Press, Beijing, China, 1986, pp. 231–260.
10. M. Shub, S. Smale, *Computational complexity: On the geometry of polynomials and a theory of cost: Part I*, Ann. Scient. Ec. Norm. Sup., 4 serie, t. 18, 1985, 107 142.
11. ———, *Computational Complexity: On the geometry of polynomials and a theory of cost: II*, SIAM J. Comput., **15**:1(1986), 145–161.
12. S. Smale, *The fundamental theorem of algebra and complexity*, Bull Amer. Math. Sco., **4**:1(1981), 1–36.
13. ———, *Algorithms for solving equations*, written for the International Congress of Mathematicians, 1986.
14. ———, *Newton's method estimates from data at one point*, in Proceedings of a Conference in Honor of Gail Young, Laramie, Springer, N.Y., 1986.
15. ———, *On the efficiency of algorithms of analysis*, Bull (New Ser.) Amer. Math. Soc., **13**:2(1985), 87–121.
16. J.F. Traub, G.W. Wasilkowski, H. Wozniakowski, *Information-Based Complexity*, Academic Press, INC, 1988.
17. X.H. Wang, *Convergence on an iterative procedure* (in Chinese), Kexue Tong Bao, **20**:12(1975), 55–59.
18. ———, *Error estimates on some numerical methods of equation finding* (in Chinese), Acta Mathematica Sinica, **22**:5(1979), 638–642.
19. ———, *Convergent neighbourhood on Newton's method* (in Chinese), KeXue TongBao, Special Issue of Math., Phy. and Chemistry, 1980, 36–37.
20. X.H. Wang, D.F. Han, *On dominating sequence method in the point estimate and Smale theorem*, Science in China (Ser. A), **33**:2(1990), 135–144.
21. ———, *Domain and point estimates on Newton's iteration*, Chinese J. Numer. Math. & Appl., **12**:3(1990), 1–8.
22. ———, *Precise judgment of approximate zero of the second kind for Newton's iteration*, A Friendly Collection of Mathematics Papers I, Jilin Univ. Press, Changchun, China, 1990, pp. 22–24.
23. ———, *Computational complexity in numerical integrals*, Science in China (Ser. A), **34**:1(1991), 42–47.
24. X.H. Wang, D.F. Han, F.Y. Sun, *Point estimates on some deformed Newton's methods* (in Chinese), Math. Num. Sin., **12**:2(1990), 145–156.
25. X.H. Wang, G.X. Shen, D.F. Han, *Some remarks on the algorithm of solving equations*, Acta Mathematica Sinica (New Ser.) (Chinese J. of Mathematics), 1991, to appear.
26. X.H. Wang, X.H. Xuan, *Random polynomial space and computational complexity theory*,

Science in China(Ser. A), **30**:7(1987), 637–684.

27. X.H. Wang, S.M. Zheng, *A family of parallel and interval iterations for finding simultaneously all roots of a polynomial with rapid convergences*, J. Comput. Math., **2**:1(1984), 70–76.
28. X.H. Wang, S.M. Zheng, D.F. Han, *The convergence of Euler's series, Euler's and Halley's families* (in Chinese), Acta Math. Sinica, **33**:6(1990), 721–738.
29. Z.K. Wang, S.L. Xu, *The approximate zeros and computational complexity theory*, Science in China (Ser. A), **27**:1(1984), 8–15.
30. A.A. Zensykbaev, *Extremality of monosplines of minimal deficiency* (in Russian), Izv. Akad. Nauk SSSR Ser. Mat., **46**(1982), 1175–1198.

DEPARTMENT OF MATHEMATICS, HANGZHOU UNIVERSITY, HANGZHOU 310028, CHINA

Contemporary Mathematics
Volume **163**, 1994

Viscosity Splitting Schemes

YING LONG-AN

1. Introduction

The numerical simulation of viscous incompressible flow with high Reynold's number is a difficult problem because there is a thin boundary layer, vortices are created near the boundary, and turbulence may be caused. This is one of the most important singular perturbation problems requiring special algorithms.

Chorin developed an algorithm for this problem [4]. In his scheme three different features of flow, convection, diffusion, and generation of vorticity from the boundary, were split into three steps. This algorithm was formulated by Chorin, Hughes, McCracken and Marsden as a product formula [5]

$$u(nk) = (H(k) \circ \phi \circ E(k))^n u_0. \tag{1.1}$$

Here u_0 is the initial data, $E(t)$ is an Euler solver which produces the solutions of inviscid Euler equation, ϕ is a "Vortices creation operator" which simulates the creation of vortices along the boundary, $H(t)$ is a Stokes solver which produces the solution of linear evolution Stokes equation, and k is the length of time steps. The advantage of this algorithm lies on that only the third operator depends on the Reynold's number, but it is a simple linear operator which can be easily solved.

Beale and Majda proved the convergence of the above scheme for the Cauchy problems [2], evidently there was no operator ϕ in their scheme.

The difficulty of the mathematical study for initial-boundary value problems is that the boundary condition for the Euler equation is different from that of the Navier-Stokes equation and the changing of boundary conditions in each time step creates singularities of the approximate solutions. In [5] explicit formulas for the operators ϕ and $H(t)$ were given, namely $u \rightarrow \phi(u) = \tilde{u}$ is the odd extension of u, and

$$H(t) = P(\exp\{t\triangle\}\tilde{u} \mid \Omega), \tag{1.2}$$

1991 *Mathematics Subject Classification.* **76D05, 76M25, 65M12.**

where Ω is the domain, $P : (L^2(\Omega))^2 \to X$ is an orthogonal projection, $X =$ closure in$(L^2(\Omega))^2$ of $\{u \in (C_0^\infty(\Omega))^2;\ \nabla \cdot u = 0\}$. No convergence result was given. And it is worth to notice that after extension the velocity field $\tilde{u}$ is no longer solenoidal. At this case a heat equation for velocity is not equivalent to a Stokes equation. Therefore formula (1.2) is a little unnatural.

Benfatto and Pulvirenti proved the convergence of (1.1) in the half plane [3], where the operator ϕ is different from that of [5]. According to [3], the tangent component of velocity on the whole plane is an odd extension, but the normal component is an even extension, which imposes a distribution vortex sheet with line support and a Neŭmann boundary condition for vorticity.

For general bounded domains, Alessandrini, Douglis and Fabes considered a simplified scheme [1]

$$u(nk) = (H(k) \circ E(k))^n u_0, \tag{1.3}$$

where the creation of vortices and the diffusion are combined in one step $H(k)$ which produces the solutions of the Stokes equation in conjunction with exact zero boundary value. To avoid some technical difficulties, the solutions of the Euler equation were replaced by polynomials, then convergence was proved. Zheng and Huang [22] proved the convergence of (1.3) without any modification of the operator $E(k)$, the rate of convergence in $L^\infty(0,T;L^2(\Omega))$ is $O(k^{\frac{3}{4}-\epsilon})$, $\epsilon > 0$. This result was improved by Zhang [19], the rate of convergence in the same space is $O(k)$. Finally the author obtained the optimal result [16], the rate in $L^\infty(0,T;H^r(\Omega))$ is $O(k)$ for $0 \le r < \frac{1}{2}$ and $O(k^{\frac{s-r}{s-\theta}})$ for $\frac{1}{2} \le r \le s$, where $0 < \theta < \frac{1}{2}$, $2 \le s < \frac{5}{2}$.

In a series of papers [9]-[15] the author considered formula (1.1) in some generalized sense, where ϕ is a bounded operator in some Sobolev spaces and $H(k)$ solves homogeneous or nonhomogeneous Stokes equation with homogeneous or nonhomogeneous boundary conditions. It was observed that if ϕ is a bounded operator in H^s, $s < \frac{5}{2}$, and in $H(k)$ both equation and boundary condition are homogeneous, then the limit of the approximate solution exists but it is another function rather than the true solution [9]. However some schemes with nice convergence property, strongly in $L^\infty(0,T;H^s(\Omega))$ were given. Error bounds were obtained for the following cases: two and three dimensional problems, simply and multi connected domains, interior and exterior problems.

In this paper we will give a unifying theory for different schemes. First we consider the necessary conditions for convergence. Then we consider sufficient conditions for linear and nonlinear problems. To fix our attention, we consider three dimensional bounded domains here, our proof is also valid for two dimensional case. The error bounds for nonlinear problems improve the results of the author's earlier papers.

2. Necessary conditions for convergence

Let $\Omega \subset \mathbb{R}^3$ be a bounded, simply connected domain with sufficiently smooth boundary $\partial\Omega$. The initial boundary value problems of the Navier-Stokes equations for viscous incompressible flow are

$$\frac{\partial u}{\partial t} + (u \cdot \nabla)u + \frac{1}{\rho}\nabla p = \nu \triangle u + f, \quad \text{in}\Omega, \tag{2.1}$$

$$\nabla \cdot u = 0, \quad \text{in}\Omega, \tag{2.2}$$

$$u \mid_{x \in \partial\Omega} = 0, \tag{2.3}$$

$$u \mid_{t=0} = u_0, \tag{2.4}$$

where $u = (u_1, u_2, u_3)$ is the velocity, p is the pressure, ν, ρ are positive constants, and f is the body force.

Let us consider the following scheme: we solve $\tilde{u}_k$, $\tilde{p}_k$, u_k, p_k on interval $[ik, (i+1)k)$, $i = 0, 1, \cdots$, which satisfy

$$\frac{\partial \tilde{u}_k}{\partial t} + (\tilde{u}_k \cdot \nabla)\tilde{u}_k + \frac{1}{\rho}\nabla \tilde{p}_k = f_1^{(k)}, \tag{2.5}$$

$$\nabla \cdot \tilde{u}_k = 0, \tag{2.6}$$

$$\tilde{u}_k \cdot n \mid_{x \in \partial\Omega} = 0, \quad \tilde{u}_k(ik) = u_k(ik - 0), \tag{2.7}$$

$$\frac{\partial u_k}{\partial t} + \frac{1}{\rho}\nabla p_k = \nu \triangle u_k + f_2^{(k)}, \tag{2.8}$$

$$\nabla \cdot u_k = 0, \tag{2.9}$$

$$u_k \mid_{x \subset \partial\Omega} = f_3^{(k)}, \quad u_k(ik) = \Theta \tilde{u}_k((i+1)k - 0), \tag{2.10}$$

where $u_k(-0) = u_0$, n is the unit normal vector, Θ is the vorticity creation operator. To meet the needs for some algorithms, the boundary condition for u_k is not necessary homogeneous, but we should assume that

$$f_3^{(k)} \cdot n \mid_{x \in \partial\Omega} = 0. \tag{2.11}$$

To state the necessary conditions, denote by v_k the solution of

$$\frac{\partial v_k}{\partial t} + \frac{1}{\rho}\nabla \pi_k = \nu \triangle v_k,$$

$$\nabla \cdot v_k = 0,$$

$$v_k \mid_{x \in \partial\Omega} = f_3^{(k)}, \quad v_k(0) = 0,$$

and define Stokes operator $A = -P\triangle$ with domain $D(A) = X \cap (H_0^1(\Omega))^3 \cap (H^2(\Omega))^3$. Then we have

THEOREM 1. *If $\Theta : X \to X$ is a bounded operator, e^{-kA} and Θ is exchangeable in $X \cap (H_0^1(\Omega))^3$, $\Theta^2 = \Theta$ and $\Theta v_k(ik) = v_k(ik)$, moreover, if $(f_1^{(k)}, f_2^{(k)})$ tends to (f_1, f_2) in $(L^\infty(0,T; (L^2(\Omega))^3))^2$ as $k \to +0$, then if $(\tilde{u}_k, u_k)$ tends to*

(u, u) in $L^\infty(0,T;(H^2(\Omega))^3) \times L^\infty(0,T;(L^2(\Omega))^3)$ for any $u_0 \in X \cap (C_0^\infty(\Omega))^3$, where u is the solution of problem (2.1)–(2.4), then

$$\lim_{k\to+0} v_k = 0, \quad \text{in} \;\; L^\infty(0,T;(L^2(\Omega))^3), \tag{2.12}$$

$$\Theta = I, \qquad \text{in} \;\; X \cap (H_0^1(\Omega))^3, \tag{2.13}$$

$$Pf - Pf_2 - \Theta P f_1 = (I - \Theta) P(u \cdot \nabla)u, \tag{2.14}$$

where I is the operator of identity.

PROOF. We define $\tilde{u}'(t) = \tilde{u}_k(t) - v_k(ik)$ and $u'(t) = u_k(t) - v_k(t)$ on $[ik,(i+1)k)$, then

$$\begin{aligned}
&\frac{\partial \tilde{u}'}{\partial t} + P(\tilde{u}_k \cdot \nabla)\tilde{u}_k = Pf_1^{(k)},\\
&\tilde{u}'(ik) = u'(ik-0),\\
&\frac{\partial u'}{\partial t} = -\nu A u' + Pf_2^{(k)},\\
&u' \,|_{x\in\partial\Omega} = 0,
\end{aligned}$$

$$u'(ik) = \Theta \tilde{u}_k((i+1)k-0) - v_k(ik) = \Theta \tilde{u}'((i+1)k-0).$$

The expressions for the solutions are

$$\tilde{u}'(t) = u'(ik-0) + \int_{ik}^t P(f_1^{(k)} - (\tilde{u}_k \cdot \nabla)\tilde{u}_k) d\tau, \tag{2.15}$$

$$u'(t) = e^{-\nu(t-ik)A}\Theta\tilde{u}'((i+1)k-0) + \int_{ik}^t e^{-\nu(t-\tau)A} P f_2^{(k)} d\tau,$$

thus

$$\begin{aligned}
u'(t) =& e^{-\nu(t-ik)A}\Theta(u'(ik-0) + \int_{ik}^{(i+1)k} P(f_1^{(k)} - (\tilde{u}_k\cdot\nabla)\tilde{u}_k)d\tau)\\
&+ \int_{ik}^t e^{-\nu(t-\tau)A} P f_2^{(k)} d\tau.
\end{aligned}$$

By induction

$$\begin{aligned}
u'(t) =& \Theta e^{-\nu t A} u_0 + \sum_{i=0}^{j-1} \Theta e^{-\nu(t-ik)A} \int_{ik}^{(i+1)k} \Theta P(f_1^{(k)} - (\tilde{u}_k\cdot\nabla)\tilde{u}_k)d\tau\\
&+ \sum_{i=0}^{j-1} \Theta \int_{ik}^{(i+1)k} e^{-\nu(t-\tau)A} P f_2^{(k)} d\tau\\
&+ e^{-\nu(t-ik)A} \int_{jk}^{(j+1)k} \Theta P(f_1^{(k)} - (\tilde{u}_k\cdot\nabla)\tilde{u}_k)d\tau\\
&+ \int_{jk}^t e^{-\nu(t-\tau)A} P f_2^{(k)} d\tau,
\end{aligned}$$

where $j = [\frac{t}{k}]$. Letting $k \to +0$, we evaluate the limit and obtain

$$\lim_{k\to+0} u'(t) = \Theta(e^{-\nu tA}u_0 + \int_0^t e^{-\nu(t-\tau)A}(\Theta P(f_1 - (u\cdot\nabla)u) + Pf_2)d\tau).$$

But

$$\lim_{k\to+0} v_k(t) = \lim_{k\to+0} u_k(t) - \lim_{k\to+0} u'(t).$$

The existence of the right hand side implies the existence of the left, which is denoted by $v(t)$. Then

$$u(t) - v(t) = \Theta(e^{-\nu tA}u_0 + \int_0^t e^{-\nu(t-\tau)A}(\Theta P(f_1 - (u\cdot\nabla)u) + Pf_2)d\tau).$$

Let $t \to +0$, then the limit of the right hand side exists in $L^2(\Omega)$. Since $v(0) = 0$, we have

$$u_0 = \Theta u_0.$$

But $u_0 \in X\cap(C_0^\infty(\Omega))^3$ is arbitrary and $C_0^\infty(\Omega)$ is dense in $H_0^1(\Omega)$, we get (2.13). Let $x \to \partial\Omega$, by the property of the operator e^{-tA}, we have

$$\lim_{x\to\partial\Omega}(u(x,t) - v(x,t)) = 0.$$

Then the boundary condition for u implies $v\mid_{x\in\partial\Omega}= 0$. Hence v is the solution of homogeneous Stokes equation with zero boundary and initial data, by uniqueness $v \equiv 0$, which proves (2.12). Finally, we have

$$u(t) - e^{-\nu tA}u_0 + \int_0^t e^{-\nu(t-\tau)A}P(f - (u\cdot\nabla)u)d\tau,$$

which gives (2.14). □

Let us give some examples:

a. If $f_1^{(k)} = f$, $f_2^{(k)} = 0$, $f_3^{(k)} = 0$, then (2.14) implies $\Theta = I$. All assumptions and conclusions of Theorem 1 are satisfied. This scheme was studied in [16]. On the contrary, let Θ be such a bounded projection which is I on $X\cap(H_0^1(\Omega))^3$ but not I on the space $X\cap(H^1(\Omega))^3$, then the assumptions of Theorem 1 are satisfied but (2.14) is violated. So the scheme (2.5)–(2.10) can not converge. This is what we have indicated in [9].

b. More generally, if $f_1^{(k)} = f$, $f_2^{(k)} = 0$, $f_3^{(k)} = g(t)\tilde{u}_k((i+1)k-0)$ with g bounded, then we also get $\Theta = I$, and this scheme was given in [15].

c. If $f_3^{(k)} = 0$, $f_2^{(k)} = f_4^{(k)} + \Phi\tilde{u}_k((i+1)k-0)$, $f_4^{(k)} \to f_4$ and $f_1 + f_4 = f$, where Φ is an operator, then by (2.15)

$$\tilde{u}_k((i+1)k-0) = u_k(ik-0) + \int_{ik}^{(i+1)k} P(f_1^{(k)} - (\tilde{u}_k\cdot\nabla)\tilde{u}_k)d\tau,$$

then

$$f_2^{(k)} = f_4^{(k)} + \Phi u_k(ik-0) + \int_{ik}^{(i+1)k} \Phi P(f_1^{(k)} - (\tilde{u}_k\cdot\nabla)\tilde{u}_k)d\tau.$$

(2.14) implies

$$(I-\Theta)P((u\cdot\nabla)u-f_1)+P\lim_{k\to+0}\Big(\Phi u_k(ik-0)+\int_{ik}^{(i+1)k}\Phi P(f_1^{(k)}-(\tilde{u}_k\cdot\nabla)\tilde{u}_k)d\tau\Big)=0. \tag{2.16}$$

We take Θ a bounded projection from $X\cap(H^1(\Omega))^3$ to $X\cap(H_0^1(\Omega))^3$ and $\Phi=\frac{1}{k}(I-\Theta)$, then $\Phi u_k(ik-0)=0$, and (2.16) (2.13) and all assumptions for Θ are satisfied, this is what we have considered in [10]–[14].

3. Sufficient conditions — linear problems

We notice that the norm $\|A^\alpha u\|_0$ is equivalent to $\|u\|_{2\alpha}$ in $D(A^\alpha)$ for all nonnegative α, and $D(A^\alpha)=X\cap(H^{2\alpha}(\Omega))^3$ for $0\le\alpha<\frac{1}{4}$, [11] then we can estimate the error of scheme (2.5)–(2.10) for linear problems.

In this section we consider the Stokes equation

$$\frac{\partial u}{\partial t}+\frac{1}{\rho}\nabla p=\nu\triangle u+f,\quad \text{in } \Omega, \tag{3.1}$$

in conjunction with (2.2)–(2.4). The viscosity splitting scheme for (3.1) is

$$\frac{\partial\tilde{u}_k}{\partial t}+\frac{1}{\rho}\nabla\tilde{p}_k=f_1^{(k)} \tag{3.2}$$

with (2.6)–(2.10). Let us consider a couple of special cases. We assume that the solution u of (3.1) (2.2)–(2.4) belongs to $L^\infty(0,T;(H^s(\Omega))^3)$, $s\in[0,\frac{5}{2})$. For the case c of Section 2 [10]

$$\begin{aligned}u_k(t)=&e^{-\nu tA}u_0+\sum_{i=0}^{[\frac{t}{k}]}e^{-\nu(t-ik)A}\int_{ik}^{(i+1)k}\Theta Pf_1^{(k)}(\tau)d\tau\\&+\sum_{i=0}^{[\frac{t}{k}]-1}\int_{ik}^{(i+1)k}e^{-\nu(t-\tau)A}\Big\{Pf_4^{(k)}(\tau)+\frac{1}{k}\int_{ik}^{(i+1)k}(I-\Theta)Pf_1^{(k)}(\zeta)d\zeta\Big\}d\tau\\&+\int_{[\frac{t}{k}]k}^{t}e^{-\nu(t-\tau)A}\Big\{Pf_4^{(k)}(\tau)+\frac{1}{k}\int_{[\frac{t}{k}]k}^{([\frac{t}{k}]+1)k}(I-\Theta)Pf_1^{(k)}(\zeta)d\zeta\Big\}d\tau.\end{aligned} \tag{3.3}$$

THEOREM 2. *If Θ is a bounded operator in the norm $\|\cdot\|_s$, $s\ge1$, $u_0\in(H^{\frac{5}{2}}(\Omega))^3\cap D(A)$, the solution u of (3.1) (2.2)–(2.4) is sufficiently smooth, and*

$$\|f(t)-f_1^{(k)}(t)-f_4^{(k)}(t)\|_{\frac{1}{2}}\le Ck, \tag{3.4}$$

$$\|f_1^{(k)}(t)\|_{\frac{5}{2}}+\|\frac{d}{dt}f_1^{(k)}(t)\|_1\le C, \tag{3.5}$$

then

$$\|u(t)-u_k(t)\|_s+\|u(t)-\tilde{u}_k(t)\|_s\le Ck,\quad s<\frac{5}{2}. \tag{3.6}$$

PROOF. We have

$$u(t) = e^{-\nu t A} u_0 + \int_0^t e^{-\nu(t-\tau)A} Pf(\tau) d\tau,$$

then by (3.3)

$$\begin{aligned} & u(jk) - u_k(jk-0) \\ &= \int_0^{jk} e^{-\nu(jk-\tau)A} P(f(\tau) - f_1^{(k)}(\tau) - f_4^{(k)}(\tau)) d\tau \\ &+ \sum_{i=0}^{j-1} \int_{ik}^{(i+1)k} \left(e^{-\nu(jk-\tau)A} - e^{-\nu(jk-ik)A} \right) \Theta P f_1^{(k)}(\tau) d\tau \\ &+ \sum_{i=0}^{j-1} \int_{ik}^{(i+1)k} e^{-\nu(jk-\tau)A} \frac{1}{k} \int_{ik}^{(i+1)k} (I - \Theta) P(f_1^{(k)}(\tau) - f_1^{(k)}(\zeta)) d\zeta d\tau. \end{aligned} \tag{3.7}$$

Suffice it to prove (3.6) for $s \in [2, \frac{5}{2})$, thus

$$\begin{aligned} & \| \int_0^{jk} e^{-\nu(jk-\tau)A} P(f(\tau) - f_1^{(k)}(\tau) - f_4^{(k)}(\tau)) d\tau \|_s \\ & \leq C \| A^{\frac{s}{2}} \int_0^{jk} e^{-\nu(jk-\tau)A} P(f(\tau) - f_1^{(k)}(\tau) - f_4^{(k)}(\tau)) d\tau \|_0 \\ &= C \| \int_0^{jk} A^{1-\frac{s_1-s}{2}} e^{-\nu(jk-\tau)A} A^{\frac{s_1}{2}-1} P(f(\tau) - f_1^{(k)}(\tau) - f_4^{(k)}(\tau)) d\tau \|_0 \\ & \leq C \int_0^{jk} (jk-\tau)^{-1+\frac{s_1-s}{2}} \| f(\tau) - f_1^{(k)}(\tau) - f_2^{(k)}(\tau) \|_{s_1-2} d\tau \leq Ck, \end{aligned}$$

where $s < s_1 < \frac{5}{2}$, and

$$\begin{aligned} & \| \sum_{i=0}^{j-1} \int_{ik}^{(i+1)k} \left(e^{-\nu(jk-\tau)A} - e^{-\nu(jk-ik)A} \right) \Theta P f_1^{(k)}(\tau) d\tau \|_s \\ & \leq C \| \sum_{i=0}^{j-1} \int_{ik}^{(i+1)k} A^{\frac{s}{2}} e^{-\nu(jk-\tau)A} \left(I - e^{-\nu(\tau-ik)A} \right) \Theta P f_1^{(k)}(\tau) d\tau \|_0 \\ &= C \| \sum_{i=0}^{j-1} \int_{ik}^{(i+1)k} A^{1-\frac{s_1-s}{2}} e^{-\nu(jk-\tau)A} \\ & \quad \cdot \left(I - e^{-\nu(\tau-ik)A} \right) A^{\frac{s_1}{2}-1} \Theta P f_1^{(k)}(\tau) d\tau \|_0 \\ & \leq C \sum_{i=0}^{j-1} \int_{ik}^{(i+1)k} (jk-\tau)^{-1+\frac{s_1-s}{2}} \\ & \quad \cdot \| \left(I - e^{-\nu(\tau-ik)A} \right) A^{\frac{s_1}{2}-1} \Theta P f_1^{(k)}(\tau) \|_0 d\tau \end{aligned}$$

$$\leq C \sum_{i=0}^{j-1} \int_{ik}^{(i+1)k} (jk-\tau)^{-1+\frac{s_1-s}{2}} ((\tau - ik)\nu)$$

$$\cdot \|AA^{\frac{s_1}{2}-1} \Theta P f_1^{(k)}(\tau)\|_0 d\tau \leq Ck,$$

and similarly

$$\|\sum_{i=0}^{j-1} \int_{ik}^{(i+1)k} e^{-\nu(jk-\tau)A} \frac{1}{k} \int_{ik}^{(i+1)k} (I-\Theta)P \left(f_1^{(k)}(\tau) - f_1^{(k)}(\zeta) \right) d\zeta d\tau \|_s$$

$$= \|\sum_{i=0}^{j-1} \int_{ik}^{(i+1)k} e^{-\nu(jk-\tau)A} \frac{1}{k} \int_{ik}^{(i+1)k} \int_{\zeta}^{\tau} (I-\Theta)P \frac{d}{d\xi} f_1^{(k)}(\xi) d\xi d\zeta d\tau \|_s$$

$$\leq Ck,$$

therefore

$$\|u(jk) - u_k(jk-0)\|_s \leq Ck.$$

We set $j = [\frac{t}{k}]$, then

$$u(t) - \tilde{u}_k(t) = (u(t) - u(jk)) + (u(jk) - u_k(jk-0)) - \int_{jk}^{t} P f_1^{(k)}(\tau) d\tau,$$

$$u(t) - u_k(t)$$

$$= (u(jk) - u_k(jk-0)) + \int_{jk}^{t} e^{-\nu(t-\tau)A} P \left(f(\tau) - f_1^{(k)}(\tau) - f_4^{(k)}(\tau) \right) d\tau$$

$$+ \int_{jk}^{t} \left(e^{-\nu(t-\tau)A} - e^{-\nu(t-jk)A} \right) \Theta P f_1^{(k)}(\tau) d\tau$$

$$+ \int_{jk}^{t} e^{-\nu(t-\tau)A} \frac{1}{k} \int_{jk}^{(j+1)k} (I-\Theta)P \left(f_1^{(k)}(\tau) - f_1^{(k)}(\zeta) \right) d\zeta d\tau$$

$$- \int_{t}^{(j+1)k} e^{-\nu(t-ik)A} \Theta P f_1^{(k)}(\tau) d\tau.$$

The estimate for them are similar. □

For the case a of Section 2, we have [22]

$$u_k(t) = e^{-\nu t A} u_0 + \sum_{i=0}^{[\frac{t}{k}]} \int_{ik}^{(i+1)k} e^{-\nu(t-ik)A} f(\tau) d\tau.$$

And the following result was proved in [16]:

THEOREM 3. *If* $u_0 \in (H^{\frac{5}{2}}(\Omega))^3 \cap D(A)$, $f \in L^\infty(0,T;(H^{\frac{5}{2}}(\Omega))^3)$, *then*

$$\|\tilde{u}_k(t)\|_s \leq C,$$

$$\|u(t) - \tilde{u}_k(t)\|_r \leq \begin{cases} Ck, & 0 \leq r < \frac{1}{2}, \\ Ck^{\frac{s-r}{s-\theta}}, & \frac{1}{2} \leq r \leq s, \end{cases}$$

where $s < \frac{5}{2}$, $0 < \theta < \frac{1}{2}$.

Likewise, we may get the estimates for other schemes.

4. Sufficient conditions — nonlinear problems

We state some technical lemmas, the proofs of some of which can be found in the references, thus omitted.

LEMMA 1 ([14]). *If* $s > 2$, $v \in (H^s(\Omega))^3$, $0 \le r \le 1$, *then*

$$\|(v \cdot \nabla)v\|_r \le C\|v\|_1^{2-\frac{1+2r}{2s-2}}\|v\|_s^{\frac{1+2r}{2s-2}}.$$

LEMMA 2. *If* $2 \le s \le 3$, $u \in (H^2(\Omega))^3$, $v \in (H^s(\Omega))^3$, *then*

$$\|(u \cdot \nabla)v\|_{s-1} \le C\|u\|_2\|v\|_s. \tag{4.1}$$

PROOF. We define a linear operator $Lv = (u \cdot \nabla)v$ for fixed u, then

$$\|(u \cdot \nabla)v\|_1 \le \|u\|_{0,\infty}\|v\|_2 + \|u\|_{1,4}\|v\|_{1,4} \le C\|u\|_2\|v\|_2,$$

$$\|(u \cdot \nabla)v\|_2 \le \|u\|_{0,\infty}\|v\|_3 + \|u\|_2\|v\|_{1,\infty} \le C\|u\|_2\|v\|_3,$$

which proves (4.1) for $s = 2$ and 3. The interpolation theorem [7] gives (4.1) for all s. $\square$

LEMMA 3 ([16]). *If* $0 < r < \frac{1}{2}$,

$$B = e^{tA}P\sum_{\ell=1}^{3}\frac{\partial}{\partial x_\ell}(v_\ell w),$$

then

$$\|B\|_1 \le Ct^{\frac{r}{2}-1}\|v\|_2\|w\|_r, \qquad \forall v \in (H^2(\Omega))^3, \quad w \in (H^1(\Omega))^3.$$

LEMMA 4 ([10]). *If* $\|u_0\|_3 \le M_1$, $u_0 \in X$, *then there exists a constant* $C_1 > 0$ *such that if*

$$k_0 = \frac{1}{C_1(M_1 + \sup_{0\le t\le T}\|f(t)\|_3 + 1)}, \tag{4.2}$$

and $0 \le t \le k_0$, *then the solution* u *of problem*

$$\begin{aligned}&\frac{\partial u}{\partial t} + (u \cdot \nabla)u + \frac{1}{\rho}\nabla p = f,\\&\nabla \cdot u = 0,\\&u \cdot n|_{x\in\partial\Omega} = 0, \qquad u|_{t=0} = u_0\end{aligned}$$

satisfies

$$\|u\|_s \le C_2\left(\|u_0\|_s + 1\right), \qquad \forall s \in [0, 3], \tag{4.3}$$

where the constant C_2 *depends only on the domain* Ω, $\sup_{0\le t\le T}\|f(t)\|_3$ *and constants* s, T.

Let us consider the case c of Section 2 as an example. We assume that the operator Θ is bounded in the norm $\|\cdot\|_s$ for $s > \frac{1}{2}$, an example of which was given in [14]. Besides, we assume that the solution u is sufficiently smooth. Lemma 4 is applied to the Euler part of the scheme (2.5)–(2.10). Thus we should estimate the H^3 norm of u_k.

LEMMA 5 ([10]). *If* $2 \le s < \frac{5}{2}$, $k \le 1$, $\|\tilde{u}_k(t)\|_s \le M_2$, *and*

$$\|f_4^{(k)}(t)\|_1 + \|\frac{d}{dt} f_4^{(k)}(t)\|_1 \le C_3, \tag{4.4}$$

$$\|f_1^{(k)}(t)\|_1 \le C_4, \tag{4.5}$$

for $t \in [ik, (i+1)k)$, *then*

$$\|u_k(t)\|_3 \le C_5 (t - ik)^{\frac{s-3}{2}}$$

on the same interval, where the constant C_5 *depends only on the domain* Ω, *operator* Θ, *and constants* M_2, C_3, C_4, ν, s.

The estimates of the H^s norm of $\tilde{u}_k$ appears in Lemma 5, thus we need the following:

LEMMA 6 ([16]). *If* $2 < s < \frac{5}{2}$, (4.4) (4.5) *holds for all* i, *and there exists a constant* $M_0 > 0$ *such that*

$$\|\tilde{u}_k(t)\|_1 \le M_0,$$

and

$$\|\tilde{u}_k(t)\|_s \le C_2 \left(\|u_k(ik - 0)\|_s + 1 \right), \qquad t \in [ik, (i+1)k), \tag{4.6}$$

then

$$\|\tilde{u}_k(t)\|_s \le M_1, \tag{4.7}$$

where the constant M_1 *depends only on the domain* Ω, *and constants* M_0, C_2, C_3, C_4, s, T, ν.

PROOF. Let us estimate $\|u_k(jk - 0)\|_s$, where u_k is expressed by (3.3). Let $r \in \left(\frac{1}{2}, s - \frac{3}{2}\right)$, then the second term is dominated by

$$\begin{aligned}
&\|\sum_{i=0}^{j-1} e^{-\nu(jk-ik)A} \int_{ik}^{(i+1)k} \Theta P f_1^{(k)}(\tau) d\tau \|_s \\
&\le C \|\sum_{i=0}^{j-1} A^{\frac{s}{2}} e^{-\nu(jk-ik)A} \int_{ik}^{(i+1)k} \Theta P f_1^{(k)}(\tau) d\tau \|_0 \\
&= C \|\sum_{i=0}^{j-1} A^{\frac{s-r}{2}} e^{-\nu(jk-ik)A} \int_{ik}^{(i+1)k} A^{\frac{r}{2}} \Theta P f_1^{(k)}(\tau) d\tau \|_0 \\
&\le C \int_0^{jk} (jk - \tau)^{\frac{r-s}{2}} d\tau \cdot \sup_{0 \le \tau < jk} \|f_1^{(k)}(\tau)\|_r.
\end{aligned}$$

The other terms can be estimated in a similar way. Then we have

$$\|u_k(jk-0)\|_s \leq C\Big(\|u_0\|_s + \sup_{0\leq\tau<jk}\|f_1^{(k)}(\tau)\|_r + \sup_{0\leq\tau<jk}\|f_4^{(k)}(\tau)\|_r\Big).$$

The estimates for scheme (2.5)–(2.10) are obtained if $f_1^{(k)}(\tau)$ is replaced by $f_1^{(k)}(\tau)-(\tilde{u}_k\cdot\nabla)\tilde{u}_k$, hence

$$\|u_k(jk-0)\|_s \leq C + C\sup_{0\leq\tau<jk}\|(\tilde{u}_k\cdot\nabla)\tilde{u}_k\|_r.$$

By Lemma 1 and (4.6)

$$\|\tilde{u}_k(t)\|_s \leq C + C\sup_{0\leq\tau<jk}\|\tilde{u}_k\|_1^{2-\frac{1+2r}{2s-2}}\|\tilde{u}_k\|_s^{\frac{1+2r}{2s-2}}.$$

Noting that $\frac{1+2r}{2s-2}<1$, (4.7) is easily obtained. □

The estimate of the H^1 norm of $\tilde{u}_k$ appears in Lemma 6. To obtain this estimate and get the desired error estimate, we consider an auxiliary problem: replacing $(\tilde{u}_k\cdot\nabla)\tilde{u}_k$ in (2.5) by $(u\cdot\nabla)u$, where u is the exact solution of (2.1) -(2.4), we get a linear problem. Denote by $\tilde{u}^*, u^*$ the corresponding solutions. Then the estimate of $u-\tilde{u}^*$ and $u-u^*$ is given by Theorem 2.

LEMMA 7. *If* $2<s<\frac{5}{2}$,

$$\|f(t)-f_1^{(k)}(t)-f_4^{(k)}(t)\|_{\frac{1}{2}} \leq C_6 k, \tag{4.8}$$

$$\|f_1^{(k)}(t)\|_{\frac{5}{2}} + \|\frac{d}{dt}f_1^{(k)}(t)\|_1 \leq C_7, \tag{4.9}$$

and $\|\tilde{u}_k\|_s\leq M_2$, *then*

$$\|u^*(t)-u_k(t)\|_1 + \|\tilde{u}^*(t)-\tilde{u}_k(t)\|_1 \leq C_8 k,$$

where the constant C_8 *depends only on the domain* Ω, *constants* ν, s, T, C_6, C_7, M_2, *and the exact solution* u.

PROOF. For the sake of notational convenience, we set $\tilde{f}=(\tilde{u}_k\cdot\nabla)\tilde{u}_k-(u\cdot\nabla)u$, and

$$s_i = \begin{cases}(ik,(i+1)k), & i=0,1,\cdots,[\frac{t}{k}]-1,\\ ([\frac{t}{k}]k,t), & i=[\frac{t}{k}],\end{cases}$$

then

$$u^*(t)-u_k(t) = I_1+I_2+I_3,$$

where

$$I_1 = \sum_{i=0}^{[\frac{t}{k}]}\int_{s_i}e^{-\nu(t-\tau)A}\frac{1}{k}\int_{ik}^{(i+1)k}P\tilde{f}(\zeta)d\zeta d\tau,$$

$$I_2 = \sum_{i=0}^{[\frac{t}{k}]}\int_{s_i}\Big(e^{-\nu(t-ik)A}-e^{-\nu(t-\tau)A}\Big)\frac{1}{k}\int_{ik}^{(i+1)k}\Theta\tilde{f}(\zeta)d\zeta d\tau,$$

$$I_3 = e^{-\nu(t-[\frac{t}{k}]k)A}([\frac{t}{k}] + 1 - \frac{t}{k}) \int_{[\frac{t}{k}]k}^{([\frac{t}{k}]+1)k} \Theta \tilde{f}(\tau) d\tau.$$

Now

$$\tilde{f} = (u \cdot \nabla)(\tilde{u}_k - \tilde{u}^*) + (u \cdot \nabla)(\tilde{u}^* - u) - ((u - \tilde{u}^*) \cdot \nabla)\tilde{u}_k - ((\tilde{u}^* - \tilde{u}_k) \cdot \nabla)\tilde{u}_k,$$

and by Theorem 2 and (4.8) (4.9) we have

$$\|u(t) - u^*(t)\|_s + \|u(t) - \tilde{u}^*(t)\|_s \leq Ck.$$

Lemma 3 implies

$$\|I_1\|_1 \leq C \sum_{i=0}^{[\frac{t}{k}]} \int_{s_i} (t-\tau)^{\frac{r}{2}-1} (\sup_{ik \leq \zeta < (i+1)k} \|\tilde{u}^*(\zeta) - \tilde{u}_k(\zeta)\|_r + k) d\tau. \tag{4.10}$$

Let us derive an estimate for $\tilde{u}^* - \tilde{u}_k$. Set $\tilde{\omega}_k = \text{curl } \tilde{u}_k$, $\tilde{\omega}^* = \text{curl } \tilde{u}^*$ and $\omega = \text{curl } u$, then by (2.5) we have

$$\frac{\partial(\tilde{\omega}^* - \tilde{\omega}_k)}{\partial t} + (u \cdot \nabla)\omega - (\omega \cdot \nabla)u - (\tilde{u}_k \cdot \nabla)\tilde{\omega}_k + (\tilde{\omega}_k \cdot \nabla)\tilde{u}_k = 0.$$

It is easy to see that all nonlinear terms of the above equation are bounded in $L^2(\Omega)$, therefore

$$\left\| \frac{\partial(\tilde{\omega}^* - \tilde{\omega}_k)}{\partial t} \right\|_0 \leq C.$$

By integration we obtain

$$\|(\tilde{\omega}^* - \tilde{\omega}_k)(t)\|_0 \leq \|(\tilde{\omega}^* - \tilde{\omega}_k)(ik)\|_0 + Ck, \qquad t \in [ik, (i+1)k).$$

The norms $\|\text{curl } v\|_0$ and $\|v\|_1$ are equivalent if $v \in (H^1(\Omega))^3 \cap X$, [8] hence

$$\begin{aligned} \|(\tilde{u}^* - \tilde{u}_k)(t)\|_1 &\leq C(\|(\tilde{u}^* - \tilde{u}_k)(ik)\|_1 + k) \\ &= C(\|(u^* - u_k)(ik-0)\|_1 + k). \end{aligned} \tag{4.11}$$

Substituting (4.11) into (4.10) gives

$$\|I_1\|_1 \leq C \sum_{i=0}^{[\frac{t}{k}]} \int_{s_i} (t-\tau)^{\frac{r}{2}-1} (\|(u^* - u_k)(ik-0)\|_1 + k) d\tau.$$

By Lemma 2

$$\begin{aligned} \|I_2\|_1 &= \| \sum_{i=0}^{[\frac{t}{k}]} \int_{s_i} \int_{ik}^{\tau} \nu A e^{-\nu(t-\xi)A} d\xi \cdot \frac{1}{k} \int_{ik}^{(i+1)k} \Theta \tilde{f}(\zeta) d\zeta d\tau \|_1 \\ &\leq C \| \sum_{i=0}^{[\frac{t}{k}]} \int_{s_i} \int_{ik}^{\tau} \nu A^{\frac{3}{2}} e^{-\nu(t-\xi)A} d\xi \cdot \frac{1}{k} \int_{ik}^{(i+1)k} \Theta \tilde{f}(\zeta) d\zeta d\tau \|_0 \\ &= C \| \sum_{i=0}^{[\frac{t}{k}]} \int_{s_i} \int_{ik}^{\tau} \nu A^{2-\frac{s}{2}} e^{-\nu(t-\xi)A} d\xi \cdot \frac{1}{k} \int_{ik}^{(i+1)k} A^{\frac{s-1}{2}} \Theta \tilde{f}(\zeta) d\zeta d\tau \|_0 \end{aligned}$$

$$
\begin{aligned}
&\leq C\sum_{i=0}^{[\frac{t}{k}]}\int_{s_i}\int_{ik}^{\tau}(t-\xi)^{\frac{s}{2}-2}d\xi\cdot\frac{1}{k}\int_{ik}^{(i+1)k}\|\tilde{f}(\zeta)\|_{s-1}d\zeta d\tau \\
&\leq Ck\int_0^t(t-\xi)^{\frac{s}{2}-2}d\xi\leq Ck.
\end{aligned}
$$

Finally

$$\|I_3\|_1\leq Ck\sup_{[\frac{t}{k}]k\leq\tau<([\frac{t}{k}]+1)k}\|\Theta\tilde{f}(\tau)\|_1\leq Ck.$$

Therefore

$$\|u^*(t)-u_k(t)\|_1\leq C\sum_{i=0}^{[\frac{t}{k}]}\int_{s_i}(t-\tau)^{\frac{r}{2}-1}\|(u^*-u_k)(ik-0)\|_1d\tau+Ck.$$

Set $\psi(t)=\sup_{0\leq\tau<t}\|(u^*-u_k)(\tau)\|_1$, then

$$\|u^*(t)-u_k(t)\|_1\leq C\int_0^t(t-\tau)^{\frac{r}{2}-1}\psi(\tau)d\tau+Ck.$$

Taking the supremum of the left hand side of the above inequality with respect to t, we obtain

$$\psi(t)\leq C\int_0^t(t-\tau)^{\frac{r}{2}-1}\psi(\tau)d\tau+Ck,$$

which yields the estimate for u^*-u_k. The estimate for $\tilde{u}^*-\tilde{u}_k$ follows from (4.11). □

Using a standard induction procedure [12], we apply Theorem 2 and Lemmas 4-7 to obtain the following result.

THEOREM 4. *If the solution u of (2.1)–(2.4) is sufficiently smooth, Θ is a bounded operator in the norm $\|\cdot\|_s$, $s>\frac{1}{2}$, and*

$$
\begin{aligned}
&\|f(t)-f_1^{(k)}(t)-f_4^{(k)}(t)\|_{\frac{1}{2}}\leq Ck, \\
&\|f_1^{(k)}(t)\|_3+\|\frac{d}{dt}f_1^{(k)}(t)\|_1+\|f_4^{(k)}(t)\|_1+\|\frac{d}{dt}f_4^{(k)}(t)\|_1\leq C,
\end{aligned}
$$

then there is a constant $k_0>0$, such that

$$
\begin{aligned}
&\|\tilde{u}_k(t)\|_s\leq M, \qquad 0\leq t\leq T, \\
&\max(\|u(t)-u_k(t)\|_1,\ \|u(t)-\tilde{u}_k(t)\|_1)\leq M'k, \quad 0\leq t\leq T,
\end{aligned}
$$

where $0<k\leq k_0$, $0\leq s<\frac{5}{2}$, and the constants M, M' are independent of k.

The results for case a are given in [16] and the results for case b are given in [15] [20] [21].

REFERENCES

1. R. Alessandrini, A. Douglis & E. Fabes, *An approximate layering method for the Navier-Stokes equations in bounded cylinders*, Ann. Mat. Pura Appl., **135**(1983), 329–347.
2. J.T. Beale & A. Majda, *Rates of Convergence for viscous splitting of the Navier-Stokes equations*, Math. Comp., **37**(1981), 243–259.
3. G. Benfatto & M. Pulvirenti, *Convergence of Chorin-Marsden product formula in the half-plane*, Comm. Math. Phys., **106**(1986), 427–458.
4. A.J. Chorin, *Numerical study of slightly viscous flow*, J. Fluid Mech., **57**(1973), 785–796.
5. A.J. Chorin, T.J.R. Hughes, M.F. McCracken & J.E. Marsden, *Product formulas and numerical algorithms*, Comm. Pure Appl. Math., **31**(1978), 205–256.
6. D. Liang, *A viscosity splitting method for the Navier-Stokes equations* (in Chinese), Chin. Ann. Math., 13A, **6**(1992), 708–717.
7. J.L. Lions & E. Magenes, *Nonhomogeneous Boundary Value Problems and Applications*, Springer-Verlag, 1972
8. R. Temam, *Navier-Stokes Equations, Theory and Numerical Analysis*, North Holland, 1984
9. L.-a. Ying, *Convergence study for viscous splitting in bounded domains*, Lecture Notes in Math., vol. 1297(1987), Springer-Verlag, 184–202.
10. L.-a. Ying, *Viscosity splitting method for three dimensional Navier-Stokes equations*, Acta Math. Sinica, New Series, **4**:3(1988), 210–226.
11. L.-a. Ying, *Viscosity splitting method in bounded domains*, Science in China, Series A, **32**:8(1989), 908–921.
12. L.-a. Ying, *On the viscosity splitting method for initial boundary value problems of the Navier-Stokes equations*, Chin. Ann. Math., **10B**:4(1989), 487–512.
13. L.-a. Ying, *Viscous splitting for the unbounded problem of the Navier-Stokes equations*, Math. Comp., **55**:191(1990), 89–113.
14. L.-a. Ying, *The viscosity splitting solutions of the Navier-Stokes equations*, J. Partial Differential Equations, **3**:4(1990), 31–48.
15. L.-a. Ying, *Viscosity splitting scheme for the Navier-Stokes equations*, Numer. Math. for PDE, **7**(1991), 317–338.
16. L.-a. Ying, *Optimal error estimates for a viscosity splitting formula*, in "Proceedings of the 2nd Conference on Numerical Methods for Partial Differential Equations" (Eds. L. Ying, B. Guo), World Scientific, 1991, 139–147.
17. P. Zhang, *Viscous splitting for the exterior problem of the Navier-Stokes equations*, (in Chinese), Acta. Scientiarum Naturalium Universitatis Pekinensis, **27**:3(1991), 264–280.
18. P. Zhang, *Viscous splitting for nonzero tangent boundary value*, (in Chinese), Numer. Math., a Journal of Chinese Universities, **14**:2(1992), 99–110.
19. P. Zhang, *A sharp estimate of simplified viscosity splitting scheme*, J. Comput. Math., **11**:3(1993), 205–210.
20. P. Zhang, *A family of viscosity splitting scheme for the Navier-Stokes equations*, J. Comput. Math., **11**:1(1993), 20–36.
21. P. Zhang, *A symmetrical viscosity splitting scheme for the Navier-Stokes equation*, Numer. Math., J. Chin. Univ., **1**:1(1992), 97–113.
22. Q. Zheng & M. Huang, *A simplified viscosity splitting method for solving the initial boundary value problems of Navier-Stokes equation*, J. Comput. Math., **10**:1(1992), 39–56.

DEPARTMENT OF MATHEMATICS, PEKING UNIVERSITY, BEIJING 100871, CHINA

Contemporary Mathematics
Volume **163**, 1994

Natural Boundary Element Method and Adaptive Boundary Element Method

—Some New Developments In Boundary Element Methods In China

YU DE-HAO

Since the end of 1970's the boundary element methods have been rapidly developed and widely applied in China. In recent ten years many Chinese mathematicians published their creative results in the research on boundary element methods. In this paper we shall give a brief introduction on some of them, especially on the natural boundary element method and the adaptive boundary element method.

1. Natural boundary element method

As we well know, many boundary value problems can be reduced into integral equations on the boundary, which has the advantages of reducing the number of dimensions by 1 as well as of the capability to treat problems involving infinite domain. There are many different ways for the reduction of the boundary value problems, which lead to various boundary element methods. Since the end of 1970's, Feng Kang and Yu De-hao have developed a new boundary element method, i.e., canonical or natural boundary element method (see [7–11], [24–33], [19]). It is based on the natural reduction via the Green's formulae and the Green's functions. The natural integral equations, which are the relations between the Dirichlet boundary values and their corresponding Neumann boundary values, are hypersingular. By the methods of Green's functions, Fourier transform or Fourier series, and complex analysis, the natural integral equations and Poisson integral formulae for harmonic, biharmonic, plane elasticity and Stokes problems are obtained. In order to overcome the difficulty produced by the hypersingularity of integral kernel, a new numerical method for solving these

1991 *Mathematics Subject Classification.* 65N38.

hypersingular integral equations is proposed. The coupling of natural boundary element method with finite element method is also suggested. The corresponding convergence and error estimates are given. Moreover, an approximation of boundary conditions at infinity for problems involving infinite domain is obtained by the natural boundary reduction. Natural boundary element method is quite different from other boundary element methods. It preserves faithfully the essential characteristics of the original problem and is fully compatible with the finite element method. It can be coupled with finite element method naturally and directly. It also have some advantages in numerical calculation.

1.1. Natural boundary reduction and hypersingular integral equations. Consider an elliptic differential operator of order $2m$

$$A = \sum_{|p|,|q|\leq m} (-1)^{|p|}\partial^p(a_{pq}(x))\partial^q = A^* \tag{1.1}$$

with its associated bilinear form

$$D(u,v) = \sum_{|p|,|q|\leq m} \int_\Omega a_{pq}(x)\partial^p u \partial^q v\, dx \tag{1.2}$$

on a bounded domain Ω with smooth boundary Γ. Let $\gamma = (\gamma_0, \gamma_1, \cdots, \gamma_{m-1})$ be the Dirichlet differential boundary operator and $\beta = (\beta_0, \beta_1, \cdots, \beta_{m-1})$ be the Neumann differential boundary operator, which is complementary to the operator γ. Then

$$D(u,v) = \int_\Omega vAu\, dx + \sum_{i=0}^{m-1} \int_\Gamma \beta_i u \cdot \gamma_i v\, ds. \tag{1.3}$$

Let $V = H^m(\Omega), V_0 = H_0^m(\Omega), V(A) = \{u \in V \,|_{Au=0}\}, Tr = \prod_{i=0}^{m-1} H^{m-i-\frac{1}{2}}(\Gamma)$. Because of the existence and uniqueness of the solution of the Dirichlet problem

$$\begin{cases} Au = 0, & \text{in } \Omega, \\ \gamma u = u_0, & \text{on } \Gamma, \end{cases} \tag{1.4}$$

where $u_0 \in Tr$, the inverse operator $P : Tr \to V(A)$ of the trace operator γ exists, and the product operator $\mathcal{K} = \beta P : Tr \to Tr'$ is a pseudodifferential operator. Thus we have

$$\beta u = \mathcal{K}\gamma u, \qquad \forall u \in V(A) \tag{1.5}$$

and

$$u = P\gamma u, \qquad \forall u \in V(A). \tag{1.6}$$

They are called the natural integral equation and the Poisson integral formula respectively. Let

$$\hat{D}(u_0, v_0) = \langle \mathcal{K}u_0, v_0 \rangle, \qquad u_0, v_0 \in Tr, \tag{1.7}$$

where $\langle \cdot, \cdot \rangle$ denotes the dual pairing between Tr and Tr'. Then

$$D(u,v) = \hat{D}(\gamma u, \gamma v), \qquad \forall u \in V(A), v \in V \tag{1.8}$$

and

$$D(Pu_0, Pv_0) = \hat{D}(u_0, v_0), \qquad \forall u_0, v_0 \in Tr. \tag{1.9}$$

Consider the Neumann problem

$$\begin{cases} Au = 0, & \text{in } \Omega, \\ \beta u = f, & \text{on } \Gamma, \end{cases} \tag{1.10}$$

where f satisfies the compatibility condition

$$\langle f, \gamma v \rangle = 0, \quad \forall v \in V(A, \beta) = \{v \in V \mid_{Av=0, \beta v=0}\}. \tag{1.11}$$

Then (1.10) is equivalent to the natural integral equation on Γ

$$\mathcal{K} u_0 = f, \tag{1.12}$$

or to its corresponding variational form

$$\begin{cases} \text{Find} \quad u_0 \in Tr, \quad \text{such that} \\ \hat{D}(u_0, v_0) = \langle f, v_0 \rangle, \quad \forall v_0 \in Tr. \end{cases} \tag{1.13}$$

Now if $D(u,v)$ is a V-elliptic continuous symmetric bilinear form on $V/V(A,\beta)$, from (1.8) and the Lax-Milgram Theorem the existence and uniqueness of the solution of (1.13) in $Tr/Tr(\mathcal{K})$ are obtained, where

$$Tr(\mathcal{K}) = \{v_0 \in Tr \mid \mathcal{K} v_0 = 0\}.$$

The natural integral operators $\mathcal{K}$ are hypersingular. We need some special method for solving these integral equations. When $\Omega \subset \mathbf{R}^2$ is an interior or exterior circular domain, the kernels of integral equations reduced from many elliptic boundary value problems, e.g., harmonic, biharmonic, plane elasticity and Stokes problems, are only contain the hypersingular terms of type $-\frac{1}{4\pi \sin^2 \frac{\theta - \theta'}{2}}$. We proposed a 'series expansion method' as follows. Let $\{L_i(\theta)\}_{i=1,\cdots,N}$ be the basis functions of the boundary element space, we only need to calculate following integral

$$\begin{aligned} q_{ij} &= \int_\Gamma \int_\Gamma \left(-\frac{1}{4\pi \sin^2 \frac{\theta - \theta'}{2}}\right) L_j(\theta') L_i(\theta) \, d\theta' d\theta \\ &= \langle -\frac{1}{4\pi \sin^2 \frac{\theta}{2}} * L_j(\theta), L_i(\theta) \rangle, \end{aligned} \tag{1.14}$$

where $*$ denotes the convolution. Using an important formula for generalized functions

$$-\frac{1}{4\pi \sin^2 \frac{\theta}{2}} = \frac{1}{2\pi} \sum_{-\infty}^{\infty} |n| e^{in\theta} = \frac{1}{\pi} \sum_{n=1}^{\infty} n \cos n\theta, \tag{1.15}$$

we can get

$$q_{ij} = \frac{1}{\pi}\sum_{n=1}^{\infty} n \int_0^{2\pi}\int_0^{2\pi} \cos n(\theta-\theta')L_i(\theta)L_j(\theta')\,d\theta' d\theta,$$

which can be calculated exactly. Particularly, when $\{L_i(\theta)\}_{i=1,\cdots,N}$ be the piecewise linear basis functions on uniform mesh, we have

$$q_{ij} = a_{|i-j|}, \qquad i,j = 1,\cdots,N,$$

where

$$a_k = \frac{4N^2}{\pi^3}\sum_{n=1}^{\infty}\frac{1}{n^3}\sin^4\frac{n\pi}{N}\cos\frac{nk}{N}2\pi, \quad k = 0,1,\cdots,N-1, \tag{1.16}$$

which is a convergent series, $Q = [q_{ij}]_{N\times N}$ is a symmetric circulant and semi-positive definite matrix with rank $N-1$.

1.2. Natural boundary element method for some typical equations. We list here some representations of natural integral equations and Poisson integral formulae for 4 typical elliptic boundary value problems, i.e., harmonic, biharmonic, plane elasticity and Stokes problems, over some typical 2-D domains, and give some results for their boundary element solutions. For details we can see [9], [24–33] and [49].

1.2.1. *Harmonic equation.*

$$\Delta u = 0, \qquad \gamma u = u\mid_\Gamma, \quad \beta u = \frac{\partial u}{\partial n}\mid_\Gamma .$$

a. $\Omega =$ upper half-plane

$$u_n(x) = -\frac{1}{\pi x^2} * u_0(x), \tag{1.17}$$

$$u(x,y) = \frac{y}{\pi(x^2+y^2)} * u_0(x), \quad y > 0. \tag{1.18}$$

b. $\Omega =$ interior unit circle

$$u_n(\theta) = -\frac{1}{4\pi\sin^2\frac{\theta}{2}} * u_0(\theta), \tag{1.19}$$

$$u(r,\theta) = \frac{1-r^2}{2\pi(1+r^2-2r\cos\theta)} * u_0(\theta), \quad 0 \le r < 1. \tag{1.20}$$

We also have the results for interior or exterior circle domain with radius R.

When Ω is an arbitrary simply-connected domain, if the holomorphic function $w = F(z)$ conformally maps $z \in \Omega$ onto the interior unit circle, then for Ω, the natural integral kernel and the Poisson integral kernel are given by

$$K(z,z') = -\frac{|F'(z)F'(z')|}{\pi|F(z)-F(z')|^2}, \qquad z,z' \in \Gamma \tag{1.21}$$

$$P(z,z') = \frac{|F'(z')|(1-|F(z)|^2)}{2\pi|F(z)-F(z')|^2}, \qquad z \in \Omega, z' \in \Gamma. \tag{1.22}$$

THEOREM 1.1. *The natural integral operator $\mathcal{K}$ given by (1.19) is a pseudo-differential operator of order 1, which satisfies*

$$\mathcal{K}^2 = -\frac{\partial^2}{\partial\theta^2},$$

and

$$\frac{\sqrt{2}}{2}\|f\|_s \leq \|\mathcal{K}f\|_{s-1} \leq \|f\|_s.$$

THEOREM 1.2. *If $u_n \in H^{-\frac{1}{2}}(\Gamma)$ satisfies the compatibility condition, then the variational problem corresponding to (1.19) has one and only one solution u_0 in $H^{\frac{1}{2}}(\Gamma)/P_0$, and*

$$\|u_0\|_{H^{\frac{1}{2}}(\Gamma)/P_0} \leq \sqrt{2}\|u_n\|_{-\frac{1}{2},\Gamma}.$$

The natural integral equation (1.19) can be solved numerically by series expansion method. We have got the stiffness matrices for piecewise linear, piecewise quadratic and piecewise cubic Hermite elements. We also have convergence and error estimates in energy norm, L_2-norm and C-norm (see [25], [29]), e.g., we have

THEOREM 1.3. *If $u_0 \in H^{k+1}(\Gamma), k \geq 1$, and the interpolation operator Π : $H^{\frac{1}{2}}(\Gamma) \to S_N$ satisfies*

$$\|v_0 - \Pi v_0\|_{s,\Gamma} < Ch^{k+1-s}|v_0|_{k+1,\Gamma}, \quad \forall v_0 \in H^{k+1}(\Gamma), 0 \leq s \leq k+1,$$

then

$$\|u_0 - u_0^h\|_{\hat{D}} \leq Ch^{k+\frac{1}{2}}\|u_0\|_{k+1,\Gamma}.$$

1.2.2. *Biharmonic equation (plate-bending problem).*

$$\Delta^2 u = 0, \qquad \gamma u = (u|_\Gamma, \frac{\partial u}{\partial n}|_\Gamma), \quad \beta u = (Tu, Mu),$$

where

$$\begin{cases} Tu = \{-\dfrac{\partial\Delta u}{\partial n} + (1-\nu)\dfrac{\partial}{\partial s}[(\dfrac{\partial^2 u}{\partial x^2} - \dfrac{\partial^2 u}{\partial y^2})n_x n_y + \dfrac{\partial^2 u}{\partial x\partial y}(n_y^2 - n_x^2)]\}_\Gamma, \\ Mu = [\nu\Delta u + (1-\nu)(\dfrac{\partial^2 u}{\partial x^2}n_x^2 + \dfrac{\partial^2 u}{\partial y^2}n_y^2 + 2\dfrac{\partial^2 u}{\partial x\partial y}n_x n_y)]_\Gamma. \end{cases}$$

a. $\Omega =$ upper half-plane

$$\begin{cases} Tu(x) = \dfrac{2}{\pi x^2} * u_0''(x) + (1+\nu)u_n''(x) \\ Mu(x) = (1+\nu)u_0''(x) - \dfrac{2}{\pi x^2} * u_n(x), \end{cases} \tag{1.23}$$

$$u(x,y) = \frac{2y^3}{\pi(x^2+y^2)^2} * u_0(x) - \frac{y^2}{\pi(x^2+y^2)} * u_n(x), \qquad y > 0. \tag{1.24}$$

b. $\Omega =$ interior unit circle

$$\begin{cases} Tu(\theta) = -(1+\nu)u_0''(\theta) + \dfrac{1}{2\pi\sin^2\frac{\theta}{2}} * u_0''(\theta) \\ \qquad + (1+\nu)u_n''(\theta) + \dfrac{1}{2\pi\sin^2\frac{\theta}{2}} * u_n(\theta) \\ Mu(\theta) = (1+\nu)u_0''(\theta) + \dfrac{1}{2\pi\sin^2\frac{\theta}{2}} * u_0(\theta) \\ \qquad + (1+\nu)u_n(\theta) - \dfrac{1}{2\pi\sin^2\frac{\theta}{2}} * u_n(\theta). \end{cases} \tag{1.25}$$

$$\begin{aligned} u(r,\theta) = & \frac{(1-r^2)^2(1-r\cos\theta)}{2\pi(1+r^2-2r\cos\theta)^2} * u_0(\theta) \\ & - \frac{(1-r^2)^2}{4\pi(1+r^2-2r\cos\theta)} * u_n(\theta), \quad 0 \le r < 1. \end{aligned} \tag{1.26}$$

We also have the results for interior or exterior circle domain with radius R (see [24]).

THEOREM 1.4. *The natural integral operator $\mathcal{K}$ given by* (1.25) *maps $H^{\frac{3}{2}}(\Gamma)\times H^{\frac{1}{2}}(\Gamma)$ to $H^{-\frac{3}{2}}(\Gamma)\times H^{-\frac{1}{2}}(\Gamma)$, and its corresponding bilinear form $\hat{D}(u_0,u_n;v_0,v_n)$ is symmetric, continuous and positive definite on $[H^{\frac{3}{2}}(\Gamma)\times H^{\frac{1}{2}}(\Gamma)]/P_1$.*

THEOREM 1.5. *If the boundary load $(Tu, Mu) = (t,m) \in H^{-\frac{3}{2}}(\Gamma)\times H^{-\frac{1}{2}}(\Gamma)$ satisfies the compatibility condition, then the variational problem corresponding to* (1.25) *has one and only one solution (u_0,u_n) in $[H^{\frac{3}{2}}(\Gamma)\times H^{\frac{1}{2}}(\Gamma)]/P_1$, and*

$$\|(u_0,u_n)\|_{[H^{\frac{3}{2}}(\Gamma)\times H^{\frac{1}{2}}(\Gamma)]/P_1} \le \frac{2\sqrt{2}}{1-\nu}\|(t,m)\|_{H^{-\frac{3}{2}}(\Gamma)\times H^{-\frac{1}{2}}(\Gamma)}.$$

The natural integral equation (1.25) can be solved numerically by series expansion method. We have got the stiffness matrix for piecewise cubic Hermite elements. We also have convergence and error estimates in energy norm, L_2-norm and C-norm (see [25, 29]), e.g., we have

THEOREM 1.6. *If $(u_0,u_n)\in H^{k_0+1}(\Gamma)\times H^{k_1+1}(\Gamma)$, and the interpolation operators Π_0 and Π_1 satisfy*

$$\|v-\Pi_i v\|_{s,\Gamma} \le Ch^{k_i+1-s}|v|_{k_i+1,\Gamma}, \quad \forall v\in H^{k_i+1}(\Gamma), 0\le s\le k_i+1, i=0,1,$$

then

$$\|(u_0,u_n)-(u_0^h,u_n^h)\|_{\hat{D}} \le C(h^{k_0-\frac{1}{2}}|u_0|_{k_0+1,\Gamma} + h^{k_1+\frac{1}{2}}|u_n|_{k_1+1,\Gamma}).$$

1.2.3. *Plane elasticity equations.*

$$\mu\Delta\vec{u} + (\lambda+\mu)\text{grad div}\,\vec{u} = 0, \qquad \gamma\vec{u} = \vec{u}\,|_\Gamma, \quad \beta\vec{u} = \vec{g},$$

where $g_i = \sum_{j=1}^2 \sigma_{ij}(\vec{u})n_j\,|_\Gamma$, $i = 1, 2$. Let $a = \lambda + 2\mu$, $b = \mu$.

a. Ω = upper half-plane

$$\begin{cases} g_1(x) = -\dfrac{2ab}{\pi(a+b)x^2} * u_1(x,0) - \dfrac{2b^2}{a+b}\delta'(x) * u_2(x,0) \\ g_2(x) = \dfrac{2b^2}{a+b}\delta'(x) * u_1(x,0) - \dfrac{2ab}{\pi(a+b)x^2} * u_2(x,0), \end{cases} \tag{1.27}$$

$$\begin{cases} u_1(x,y) = [\dfrac{y}{\pi(x^2+y^2)} + \dfrac{(a-b)y(x^2-y^2)}{\pi(a+b)(x^2+y^2)^2}] * u_1(x,0) \\ \qquad + \dfrac{2(a-b)xy^2}{\pi(a+b)(x^2+y^2)^2} * u_2(x,0) \\ u_2(x,y) = \dfrac{2(a-b)xy^2}{\pi(a+b)(x^2+y^2)^2} * u_1(x,0) \\ \qquad + [\dfrac{y}{\pi(x^2+y^2)} - \dfrac{(a-b)y(x^2-y^2)}{\pi(a+b)(x^2+y^2)^2}] * u_2(x,0), \end{cases} \quad y > 0. \tag{1.28}$$

b. Ω = exterior unit circle

$$\begin{cases} g_1(\theta) = -\dfrac{2ab}{a+b}\dfrac{1}{4\pi\sin^2\frac{\theta}{2}} * u_1(1,\theta) + \dfrac{2b^2}{a+b}u_2'(1,\theta) \\ g_2(\theta) = -\dfrac{2b^2}{a+b}u_1'(1,\theta) - \dfrac{2ab}{a+b}\dfrac{1}{4\pi\sin^2\frac{\theta}{2}} * u_2(1,\theta), \end{cases} \tag{1.29}$$

or in polar decomposition,

$$\begin{cases} g_r(\theta) = K_{rr} * u_r(1,\theta) + K_{r\theta} * u_\theta(1,\theta) \\ g_\theta(\theta) = K_{\theta r} * u_r(1,\theta) + K_{\theta\theta} * u_\theta(1,\theta) \end{cases} \tag{1.30}$$

where

$$K_{rr} = K_{\theta\theta} = -\frac{ab}{2\pi(a+b)\sin^2\frac{\theta}{2}} + \frac{2b^2}{a+b}\delta(\theta) + \frac{ab}{\pi(a+b)}$$

$$K_{r\theta} = -K_{\theta r} = -\frac{ab}{\pi(a+b)}\text{ctg}\frac{\theta}{2} + \frac{2b^2}{a+b}\delta'(\theta).$$

We also have the Poisson integral formula and the results for interior or exterior circle domain with radius R (see [28, 29, 33]).

THEOREM 1.7. *The natural integral operator $\mathcal{K}$ given by (1.29) or (1.30) maps $H^{\frac{1}{2}}(\Gamma)^2$ to $H^{-\frac{1}{2}}(\Gamma)^2$, and its corresponding bilinear form $\hat{D}(\vec{u}_0, \vec{v}_0)$ is symmetric, continuous and positive definite on $H^{\frac{1}{2}}(\Gamma)^2/\mathcal{R}(\Gamma)$, where $\mathcal{R}(\Gamma) = \gamma\mathcal{R}$ is the set of boundary value functions of all functions in $\mathcal{R}$,*

$$\mathcal{R} = \{(C_1 - C_3y, C_2 + C_3x) \mid C_1, C_2, C_3 \in \mathbf{R}\}.$$

THEOREM 1.8. *If the boundary load* $\vec{g} \in H^{-\frac{1}{2}}(\Gamma)^2$ *satisfies the compatibility condition, then the variational problem corresponding to* (1.29) *or* (1.30) *has one and only one solution* $\vec{u}_0$ *in* $H^{\frac{1}{2}}(\Gamma)^2/\mathcal{R}(\Gamma)$, *and*

$$\|\vec{u}_0\|_{H^{\frac{1}{2}}(\Gamma)^2/\mathcal{R}(\Gamma)} \leq \frac{a+b}{\sqrt{2}b(a-b)}\|\vec{g}\|_{H^{-\frac{1}{2}}(\Gamma)^2}.$$

The natural integral equation (1.29) or (1.30) can be solved numerically. We have got the stiffness matrix for piecewise linear elements. We also have convergence and error estimates in energy norm, L_2-norm and C-norm (see [28,29]), e.g., we have

THEOREM 1.9. *If* $\vec{u}_0 \in H^{k+1}(\Gamma)^2, k \geq 1$, *the interpolation operator* Π *satisfies*

$$\|v_0 - \Pi v_0\|_{s,\Gamma} \leq Ch^{k+1-s}\|v_0\|_{k+1,\Gamma} \quad \forall v_0 \in H^{k+1}(\Gamma), s = 0, 1,$$

then

$$\|\vec{u}_0 - \vec{u}_0^h\|_{\hat{D}} \leq Ch^{k+\frac{1}{2}}\|\vec{u}_0\|_{k+1,\Gamma}.$$

1.2.4. *Stokes equations.*

$$\begin{cases} -\nu\Delta\vec{u} + \operatorname{grad} p = 0 \\ \operatorname{div}\vec{u} = 0, \end{cases} \tag{1.31}$$

$\gamma\vec{u} = \vec{u}\mid_\Gamma$, $\beta\vec{u} = \vec{g}$, where $g_i = \sum_{j=1}^2 \sigma_{ij}(\vec{u}, p)n_j \mid_\Gamma$, $i = 1, 2$.

In [32] we have given the representation of solution of equations (1.31) via two complex variable functions:

$$\begin{cases} u_1(x,y) = Re[-\varphi'(z)\bar{z} + \varphi(z) - \psi(z)] \\ u_2(x,y) = Im[\varphi'(z)\bar{z} + \varphi(z) + \psi(z)] \\ p(x,y) = -4\nu Re\varphi'(z), \end{cases} \tag{1.32}$$

where $\varphi(z)$ and $\psi(z)$ are analytic functions in Ω.

a. $\Omega =$ upper half-plane

$$\begin{cases} g_1(x) = -\dfrac{2\nu}{\pi x^2} * u_1(x,0) \\ g_2(x) = -\dfrac{2\nu}{\pi x^2} * u_2(x,0), \end{cases} \tag{1.33}$$

$$\begin{cases} u_1(x,y) = \dfrac{2x^2y}{\pi(x^2+y^2)^2} * u_1(x,0) + \dfrac{2xy^2}{\pi(x^2+y^2)^2} * u_2(x,0) \\ u_2(x,y) = \dfrac{2xy^2}{\pi(x^2+y^2)^2} * u_1(x,0) + \dfrac{2y^3}{\pi(x^2+y^2)^2} * u_2(x,0) \qquad y > 0. \\ p(x,y) = 2\nu[\dfrac{2xy}{\pi(x^2+y^2)^2} * u_1(x,0) + \dfrac{y^2-x^2}{\pi(x^2+y^2)^2} * u_2(x,0)] \end{cases} \tag{1.34}$$

b. Ω =exterior unit circle

$$\begin{cases} g_1(\theta) = -\dfrac{2\nu}{4\pi\sin^2\frac{\theta}{2}} * u_1(1,\theta) \\ g_2(\theta) = -\dfrac{2\nu}{4\pi\sin^2\frac{\theta}{2}} * u_2(1,\theta), \end{cases} \tag{1.35}$$

$$\begin{cases} u_1(r,\theta) = P(r,\theta) * u_1(1,\theta) + \dfrac{r^2-1}{2r^2} \\ \qquad \times \{\cos 2\theta[(-r\dfrac{\partial}{\partial r}P(r,\theta)) * u_1(1,\theta) - \dfrac{\partial}{\partial\theta}P(r,\theta) * u_2(1,\theta)] \\ \qquad + \sin 2\theta[\dfrac{\partial}{\partial\theta}P(r,\theta) * u_1(1,\theta) + (-r\dfrac{\partial}{\partial r}P(r,\theta)) * u_2(1,\theta)]\} \\ u_2(r,\theta) = P(r,\theta) * u_2(1,\theta) + \dfrac{r^2-1}{2r^2} \\ \qquad \times \{\sin 2\theta[(-r\dfrac{\partial}{\partial r}P(r,\theta)) * u_1(1,\theta) - \dfrac{\partial}{\partial\theta}P(r,\theta) * u_2(1,\theta)] \\ \qquad - \cos 2\theta[\dfrac{\partial}{\partial\theta}P(r,\theta) * u_1(1,\theta) + (-r\dfrac{\partial}{\partial r}P(r,\theta)) * u_2(1,\theta)]\} \\ p(r,\theta) = \dfrac{2\nu}{r}\{\cos\theta[(-r\dfrac{\partial}{\partial r}P(r,\theta)) * u_1(1,\theta) - \dfrac{\partial}{\partial\theta}P(r,\theta) * u_2(1,\theta)] \\ \qquad + \sin\theta[\dfrac{\partial}{\partial\theta}P(r,\theta) * u_1(1,\theta) + (-r\dfrac{\partial}{\partial r}P(r,\theta)) * u_2(1,\theta)]\}, \end{cases} \tag{1.36}$$

where

$$P(r,\theta) = \frac{r^2-1}{2\pi(1+r^2-2r\cos\theta)}, \qquad r > 1.$$

We also have got the natural integral equations and Poisson integral formulae for interior or exterior circle domain with radius R, and also have some results for natural integral operator $\mathcal{K}$, for its corresponding variational problem, and for the convergence and error estimates of its boundary element approximation (see [32, 46, 49]).

1.3. The coupling of natural BEM with FEM and the approximation of boundary conditions at infinity. For general domains we can use the coupling of natural boundary element method with classical finite element method (see [27, 29, 42, 46, 49]). For example, let Ω be a domain bounded by two sides Γ_1 and Γ_2 of a concave angle $\alpha(\pi < \alpha \leq 2\pi)$ and a smooth curve Γ. When $\alpha = 2\pi$, the domain contains a crack. Consider the boundary value problem

$$\begin{cases} \Delta u = 0, & \text{in } \Omega, \\ \partial_n u = 0, & \text{on } \Gamma_1 \cup \Gamma_2, \\ \partial_n u = f, & \text{on } \Gamma, \end{cases} \tag{1.37}$$

where $f \in H^{-\frac{1}{2}}(\Gamma)$ satisfies the compatibility condition $\int_\Gamma f\,ds = 0$. Let

$$D(u,v) = \iint_\Omega \nabla u \cdot \nabla v\,dx,$$

then (1.37) is equivalent to the variational problem

$$\begin{cases} \text{Find} \quad u \in H^1(\Omega) \qquad \text{such that} \\ D(u,v) = \int_\Gamma fv\,ds \qquad \forall v \in H^1(\Omega). \end{cases} \tag{1.38}$$

Draw in Ω an arc Γ' dividing Ω into Ω_1 and a sector Ω_2. Then (1.38) is equivalent to following problem

$$\begin{cases} \text{Find} \quad u \in H^1(\Omega) \qquad \text{such that} \\ D_1(u,v) + \hat{D}_2(\gamma' u, \gamma' v) = \int_\Gamma fv\,ds \qquad \forall v \in H^1(\Omega_1) \end{cases} \tag{1.39}$$

where γ' is the trace operator mapping $H^1(\Omega)$ onto $H^{\frac{1}{2}}(\Gamma')$,

$$D_1(u,v) = \iint_{\Omega_1} \nabla u \cdot \nabla v\,dx,$$

$$\hat{D}_2(u_0, v_0) = -\frac{\pi}{4\alpha^2} \int_0^\alpha \left(\frac{1}{\sin^2 \frac{\theta-\theta'}{2\alpha}\pi} + \frac{1}{\sin^2 \frac{\theta+\theta'}{2\alpha}\pi}\right) u_0(\theta') v_0(\theta)\,d\theta'\,d\theta.$$

From the approximate problem of (1.39) we can get a system of linear algebraic equations $QU = b$, where $Q = Q_1 + Q_2$, the first part can be obtained by the finite element method, and the second part is given by some formulae in [27].

We also have the convergence and error estimates in energy norm, L_2-norm and C-norm for this coupling method (see [27, 29]), e.g., we have

THEOREM 1.10. *If* $u \in H^{k+1}(\Omega_1), k \geq 1$, Π *satisfies*

$$\|v - \Pi v\|_{1,\Omega_1} \leq Ch^j \|v\|_{j+1,\Omega_1}, \qquad \forall v \in H^{j+1}(\Omega_1),\ j = 1, \cdots, k,$$

then

$$(\|u - u_h\|_{D_1}^2 + \|\gamma' u - \gamma' u_h\|_{\hat{D}_2}^2)^{\frac{1}{2}} \leq Ch^k \|u\|_{k+1,\Omega_1}.$$

Now we consider a problem over an unbounded domain. For example, consider the problem over an exterior domain Ω with smooth boundary Γ:

$$\begin{cases} -\Delta u = 0, \qquad \text{in } \Omega, \\ \dfrac{\partial u}{\partial n} = f, \qquad \text{on } \Gamma, \\ u \text{ is bounded at infinity.} \end{cases} \tag{1.40}$$

The treatment of such problem by the classical finite element method is often a difficulty, because a simple replacement of the infinite domain by a bounded domain can hardly produce the demanded accuracy. The coupling of the natural boundary element method with the finite element method has provided an approach to solve this problem.

Draw a circle Γ_R with radius R enclosing Γ. Ω is divided into Ω_i and Ω_e. Then the natural integral equation on Γ_R for the exterior domain Ω_e is just the exact boundary condition on the artificial boundary Γ_R, i.e., the problem (1.40) is equivalent to

$$\begin{cases} -\Delta u = 0, & \text{in } \Omega_i, \\ \frac{\partial u}{\partial n} = f, & \text{on } \Gamma \end{cases} \tag{1.41}$$

and

$$\frac{\partial u}{\partial r}(R,\theta) = \frac{1}{4\pi R \sin^2 \frac{\theta}{2}} * u(R,\theta), \qquad \text{on } \Gamma_R. \tag{1.42}$$

We can solve it by the coupling of natural BEM with FEM as above. From the integral boundary condition (1.42) on Γ_R we can also get a series of approximate differential boundary conditions or approximate integral boundary conditions as follows:

$$\begin{cases} N=1: & \frac{\partial u}{\partial r} = -\frac{1}{R}u \quad \text{or} \quad \frac{\partial u}{\partial r} = \frac{1}{R}\frac{\partial^2 u}{\partial \theta^2} \\ N=2: & \frac{\partial u}{\partial r} = \frac{1}{R}(-\frac{2}{3}u + \frac{1}{3}\frac{\partial^2 u}{\partial \theta^2}) \\ N=3: & \frac{\partial u}{\partial r} = \frac{1}{R}(\frac{74}{60}\frac{\partial^2 u}{\partial \theta^2} + \frac{15}{60}\frac{\partial^4 u}{\partial \theta^4} + \frac{1}{60}\frac{\partial^6 u}{\partial \theta^6}) \\ N=4: & \frac{\partial u}{\partial r} = \frac{1}{R}(-\frac{4}{7}u + \frac{41}{90}\frac{\partial^2 u}{\partial \theta^2} + \frac{1}{36}\frac{\partial^4 u}{\partial \theta^4} + \frac{1}{1260}\frac{\partial^6 u}{\partial \theta^6}) \\ \cdots\cdots \end{cases} \tag{1.43}$$

or

$$\frac{\partial u}{\partial r}(R,\theta) = -\frac{1}{\pi R}\sum_{n=1}^{N} n \int_0^{2\pi} u(R,\theta')\cos n(\theta-\theta')\, d\theta'. \tag{1.44}$$

Let u be the solution of (1.40) and u_N be the solution of (1.41) and (1.44), we have

THEOREM 1.11. *If* $u \in H^1(\Omega_i) \cap H^{k-\frac{1}{2}}(\Gamma_a)$, $k \geq 1$, $R \geq \sigma a$, $\sigma > 1$, *then*

$$\|u-u_N\|_{H^1(\Omega_i)/P_0} \leq C N^{1-k} (\frac{a}{R})^N \|u\|_{k-\frac{1}{2},\Gamma_a},$$

where C *is a constant independent of* N *and* R, a *is the radius of* Γ_a *which is the smallest circle enclosing* Γ.

This estimate reveals the relationship between the error and the approximate grade of boundary condition and the radius of the artificial boundary.

2. Adaptive boundary element method

The adaptive procedures, which utilize the currently available computed information for steering the computational process, play an increasingly decisive role in scientific calculations. According to a-posteriori error estimates of the approximate solution, the mesh or approximation structure is automatically changed so as to improve the quality of the numerical solutions. In recent ten years, the adaptive procedures have been applied in finite element methods efficiently (see [1, 40, 41]). But the adaptive boundary element methods have been investigated only very recently (see [34-39], [21, 22]). Because of essential differences between boundary integral equations and differential equations, there are some additional difficulties for developing adaptive boundary element methods due to the nonlocality of integral operators. A first rigorous mathematical analysis of the corresponding h-versions was given by D. Yu based on strong coercivity (see [34]). Then W. Wendland and D. Yu extended these investigations to general strongly elliptic boundary integral equations for Galerkin boundary element method as well as for nodal collocation with odd degree splines (see [21]), and obtained for strongly elliptic boundary integral equations and for the Galerkin boundary element method that the residual can serve as a local error indicator, where only a local property, i.e., K-meshes, is assumed (see [22]). These meshes can be used for adaptive refinement and therefore are desirable for a feedback method based on these local a-posteriori estimates.

2.1. The pseudo-locality and its discrete analogue. Consider an integral equation on the sufficiently smooth boundary Γ of some given two-dimensional bounded domain,

$$Au = \int_\Gamma a(s,t)u(t)\,dt = f(s), \tag{2.1}$$

where $A : H^\gamma(\Gamma) \to H^{-\gamma}(\Gamma)$ is a strongly elliptic pseudodifferential operator on Γ of the order $\alpha = 2\gamma$. We assume that (2.1) has a unique solution. The corresponding variational problem is

$$\begin{cases} \text{Find} \quad u \in H^\gamma(\Gamma) \quad \text{such that} \\ A(u,v) = \int_\Gamma fv\,ds \qquad \forall v \in H^\gamma(\Gamma), \end{cases} \tag{2.2}$$

where

$$A(u,v) = (Au, v) = \int_\Gamma vAu\,ds.$$

Let u and u_h be the solution of (2.2) and its boundary element solution respectively, and let

$$e = u - u_h \tag{2.3}$$

be the error. Then the residual

$$R = Ae = f - Au_h \tag{2.4}$$

is computable when u_h is known. We hope to estimate the distribution of e on Γ by using the residual R. As we know, it is the reasons for the remarkable efficiency of adaptive finite element methods that the differential operators are local operators and the finite element approximation process also has a local character. But boundary integral operators are nonlocal, even with piecewise polynomial basis functions having small supports, the corresponding boundary element stiffness matrices are fully populated. Then we investigate the pseudo-locality of boundary integral operators, which are pseudodifferential operators on the boundary, and its discrete analogue in terms of the so called influence index. It is the pseudo-locality of boundary integral operators and the locality of boundary element basis functions that allow the successful application of adaptive procedures to boundary element methods. If the influence index is much smaller than the number of basis functions, the adaptive procedure for boundary element method will be very efficient (see [34, 21]).

For a pseudodifferential operator $P(x, D)$, the pseudolocal property is usually defined by

$$\text{sing supp}\, P(x,D)w \subset \text{sing supp}\, w, \quad \forall w \in \mathcal{E}'(\Omega)$$

where the singular support of a distribution w is the complement of the open set on which w is smooth. We have the following lemma.

LEMMA 2.1. *If the kernel of the boundary integral operator is given by*

$$A(s,t) = \log|x-y| \qquad or \qquad A(s,t) = |x-y|^{-\alpha}$$

with some $\alpha > 0$, where the points x and y on Γ correspond to s and t respectively, then the operator A has the pseudolocal property.

Let $A = D + K$, where D maps $H^\gamma(\Gamma)$ onto $H^{-\gamma}(\Gamma)$ and K maps $H^\gamma(\Gamma)$ into $H^{-\gamma+\delta}(\Gamma)$ with some $\delta > 0$. For D we require strong coercivity, i.e.,

$$(Dv, v) \geq C\|v\|_\gamma^2, \qquad \forall v \in H^\gamma(\Gamma), \tag{2.5}$$

where C is a positive constant. Then we define the *influence index* of D with respect to the set of basis functions $\{L_Q(s)\}$ as follows. Here for the boundary element method we divide Γ into N elements Γ_j corresponding to the intervals $[s_{j-1}, s_j)$ of the arc length s. By $h_j = s_j - s_{j-1}$ we denote the local mesh width. Let $\{L_Q(s)\}_{Q\in J}$ be a basis of the function space $H_h^\gamma(\Gamma)$ of piecewise polynomials of degree p associated with the partition $\{\Gamma_j\}$ satisfying $H_h^\gamma(\Gamma) \subset H^\gamma(\Gamma)$, and J denote the set of given interpolation nodes on Γ associated with $H_h^\gamma(\Gamma)$. The number of points in J is $M = N$ for $p = 0$ and $M = pN$ for $p \geq 1$.

ASSUMPTION 2.2. *The set J can be partitioned into subsets $J_1, J_2, \cdots, J_k$, such that for any $l = 1, 2, \cdots, k$, and for all $P \neq Q \in J_l$,*

$$D(v_P, v_Q) \leq \frac{1}{|J_l|} D(v_P, v_P)^{\frac{1}{2}} D(v_Q, v_Q)^{\frac{1}{2}}, \quad \forall v_P \in H_P,\, v_Q \in H_Q \tag{2.6}$$

where $|J_l|$ *denotes the number of nodes in* J_l,

$$H_Q = \{v \in H^\gamma(\Gamma) \mid v = 0 \, on \, \Gamma \setminus \Gamma_Q\}, \quad \Gamma_Q = \, supp \, L_Q(s).$$

Such partitions always exist, e.g., take $k = M$ and $|J_l| = 1$. We also have $1 \leq k \leq M$. Hence there exists the minimum ρ of these k. We call ρ the influence index of D with respect to $\{L_Q\}$. ρ depends on D and $\{\Gamma_Q\}$.

The following proposition was shown in [34].

PROPOSITION 2.3. *For* $p = 0$ *or* 1, *and in the cases* (ii)-(iv) *under the additional restriction of uniform partitions there holds:*

(i) *if* $D(s,t) = constant > 0$, *then* $\rho = M$;
(ii) *if* $D(s,t)$ *is of type* $log|x-y|$, *then* $\rho = \alpha M$, *where* $0 < \alpha < 1$;
(iii) *if* $D(s,t)$ *is of type* $|x-y|^{-1}$, *then* $\rho \sim M^{\frac{1}{2}}$;
(iv) *if* $D(s,t)$ *is of type* $|x-y|^{-2}$, *then* $\rho \sim M^{\frac{1}{3}}$;
(v) *if* $D(s,t) = C\delta^{(k)}(s-t)$, $\quad k = 0, 1, \cdots$, *where* $\delta^{(k)}(\cdot)$ *is the* k*-th derivative of the Dirac* δ*-function, then* $\rho = 1$ *for* $p = 0$ *and* $\rho = 2$ *for* $p = 1$.

From this proposition we can see that the stronger the singularity of D, the smaller will be the influence index ρ.

2.2. A-posteriori local error estimates for h-version on K-meshes. In order to use adaptive methods, we need some a-posteriori error estimates of approximate solutions. Obviously, the continuity of A and A^{-1} implies directly

$$C_1\|R\|_{-\gamma} \leq \|e\|_\gamma \leq C_2\|R\|_{-\gamma}. \tag{2.7}$$

Many works on adaptive boundary element methods are based on it. But (2.7) is only the global estimate of the error on Γ. It does not give us any information of the distribution of the error. The a-posteriori error estimates used in adaptive methods should be local and computable. We can use them to get error indicators which are used for guiding the mesh refinement or the improvement of approximation structure. Different adaptive methods maybe have same or different local error indicators. Moreover, for the h-version, p-version or hp-version of adaptive methods, the constants appeared in the local error indicators should be independent of h, p, or h and p respectively.

Because in the h-version of adaptive boundary element methods the local refinement of meshes are allowed, the local error estimates obtained on uniform or quasiuniform mesh families are not desirable as a-posteriori error bounds or indicators in adaptive methods. Here we suppose that the family of meshes consists of K-meshes, where a mesh is called a K-mesh if for any two subintervals $\Delta = \bar{\Delta}$ and $\Delta' = \bar{\Delta}'$ of the partition with $\Delta \cap \Delta' \neq \emptyset$ there holds

$$K^{-1} \leq |\Delta'|/|\Delta| \leq K \tag{2.8}$$

with fixed $K \geq 1$ for the whole family and $|\Delta|, |\Delta'|$ the length of Δ and Δ', respectively. Mostly, adaptive meshes can be constructed as K-meshes. Obviously, uniform or quasiuniform meshes are K-meshes, but conversely, it is not so.

For approximating spaces $S_h = S_h^{k,t}$, $0 \le k < t$, we require the following properties:

$$S_h \subseteq H^k(\Gamma), \tag{2.9}$$

$$\inf_{v_h \in S_h} \|v - v_h\|_i \le C h^{m-i} \|v\|_m, \quad 0 \le i \le m \le t, i \le k, \quad \forall v \in H^m(\Gamma), \tag{2.10}$$

where the constant C is independent of v and h. As usual, a typical finite element function space will satisfy (2.9), (2.10) and some 'inverse relation'. Here, however, we only assume S_h to satisfy (2.9) and (2.10), since we want to apply our results to adaptive meshes with (2.8), on which S_h usually can not satisfy the inverse relation. We have following result (see [22]).

THEOREM 2.4. *Let A be a strongly elliptic pseudodifferential operator of order α with $-2 \le \alpha \le 2$, let $t-1 \ge k \ge |\alpha| + \min(0, \frac{\alpha}{2})$, and let the boundary elements be defined on a family of K-meshes. Let $u \in H^k(I_2) \cap H^\alpha(\Gamma) \cap L_2(\Gamma)$, $I_1 \subset\subset I_2 \subset \Gamma$. Then for $0 < h \le h_0$, the local error of the Galerkin boundary element method satisfies the estimate*

$$\|u - u_h\|_{L_2(I_1)} \le C(h^{k+1}\|u\|_0 + h^k\|R\|_{-\alpha} + \|R\|_{H^{-\alpha}(I_2)}) \tag{2.11}$$

for $-2 \le \alpha < 0$ and

$$\begin{aligned}\|u - u_h\|_{L_2(I_1)} \le & C(h^{\min(k+\alpha,t)-\gamma+1}\|u\|_\gamma \\ & + h^{\min(k+\alpha,t)}\|R\|_0 + h^\alpha\|R\|_{L_2(I_2)})\end{aligned} \tag{2.12}$$

for $0 \le \alpha \le 2$.

Remark. The boundary integral operators in most applications have operators of orders $\alpha = -1, \alpha = 0$ and $\alpha = 1$. For these cases, (2.11) and (2.12), respectively, specialize to the following estimates:

$$\begin{aligned}\alpha = -1: &\quad \|u - u_h\|_{0,I_1} \le C(h^{k+1}\|u\|_{0,\Gamma} + h^k\|R\|_{1,\Gamma} + \|R\|_{1,I_2}),\\ \alpha = 0: &\quad \|u - u_h\|_{0,I_1} \le C(h^{k+1}\|u\|_{0,\Gamma} + h^k\|R\|_{0,\Gamma} + \|R\|_{0,I_2}),\\ \alpha = 1: &\quad \|u - u_h\|_{0,I_1} \le C(h^{k+\frac{3}{2}}\|u\|_{\frac{1}{2},\Gamma} + h^{k+1}\|R\|_{0,\Gamma} + h\|R\|_{0,I_2}).\end{aligned}$$

Here, the right hand side of (2.11) and (2.12) consists of the local norm of the residual which is computable by using the approximate solution, and additional terms of higher order which are neglegible for the local error indicator.

2.3. A-posteriori local error estimates for p-version. Let $I_1 \subset\subset I_2 \subseteq \Gamma$ and $\omega \in C_0^\infty(\Gamma)$ be a cutfunction with

(i) $\omega = 1$ in I_1,
(ii) $\operatorname{supp} \omega \subset\subset I_2$,
(iii) $0 \le \omega \le 1$.

Then from

$$A\omega e = \omega A e + (A\omega - \omega A)e = \omega R + (A\omega - \omega A)e,$$

we get

$$\|A\omega e\|_m \leq \|\omega R\|_m + \|(A\omega - \omega A)e\|_m,$$
$$\|A\omega e\|_m \geq \|\omega R\|_m - \|(A\omega - \omega A)e\|_m.$$

Since A and A^{-1} are continuous pseudodifferential operators of order α and $-\alpha$ respectively, and the commutator $A\omega - \omega A$ is a continuous operator of order $\alpha - 1$, we have

$$\|\omega e\|_{\alpha+m} \leq C\|A\omega e\|_m \leq C_1\|\omega R\|_m + C_2\|e\|_{\alpha-1+m},$$
$$\|\omega e\|_{\alpha+m} \geq C\|A\omega e\|_m \geq C_3\|\omega R\|_m - C_4\|e\|_{\alpha-1+m}.$$

Then

$$\|e\|_{\alpha+m,I_1} \leq C_1\|R\|_{m,I_2} + C_2\|e\|_{\alpha-1+m,\Gamma}, \tag{2.13}$$
$$\|e\|_{\alpha+m,I_2} \geq C_3\|R\|_{m,I_1} - C_4\|e\|_{\alpha-1+m,\Gamma}. \tag{2.14}$$

If $\|e\|_{\alpha-1+m}$ is an additional term of higher order with respect to h (for h-version), or p^{-1} (for p-version), or h and p^{-1} (for hp-version), we can asymptotically use $C_1\|R\|_{m,I_2}$ as an upper estimate of $\|e\|_{\alpha+m,I_1}$, and $C_3\|R\|_{m,I_1}$ as a lower estimate of $\|e\|_{\alpha+m,I_2}$. These local error estimates are directly obtained by the property of pseudodifferential operator A.

For h-version, by

$$\|e\|_{\alpha-1} \leq Ch\|e\|_\alpha, \tag{2.15}$$

we can get

$$\|e\|_{\alpha,I_1} \leq C_1\|R\|_{0,I_2} + C_2 h\|e\|_{\alpha,\Gamma}, \tag{2.16}$$
$$\|e\|_{\alpha,I_2} \geq C_3\|R\|_{0,I_1} - C_4 h\|e\|_{\alpha,\Gamma} \tag{2.17}$$

from (2.13) and (2.14).

For p-version, we need some results about the interpolation error estimates as follows.

LEMMA 2.5. *If $w \in H^s(\Gamma)$, $V_p(\Gamma) \subset H^\gamma(\Gamma)$ is the approximation space for the p-version, then for $p = 1, 2, \cdots$ there exists $w_p \in V_p(\Gamma)$ such that*

$$\|w - w_p\|_{H^{\frac{1}{2}}(\Gamma)} \leq Cp^{-s+\frac{1}{2}}\|w\|_{H^s(\Gamma)}, \quad s > \frac{1}{2} \tag{2.18}$$

for $\gamma = \frac{1}{2}$, and

$$\|w - w_p\|_0 \leq Cp^{-s}\|w\|_s, \quad s > 0 \tag{2.19}$$

for $\gamma = 0$, where the constant C is independent of u and p, but depends on s and the partition on Γ.

Using (2.13), (2.14) and lemma 2.5, we can get following local error estimates for p-version.

THEOREM 2.6. *Let A be a strongly elliptic pseudodifferential operator of order $\alpha = 1$ or 0, and the interpolate estimate for p-version* (2.18) *or* (2.19) *hold. Let $I_1 \subset\subset I_2 \subset \Gamma$ and p be sufficiently large. Then the Galerkin boundary element method satisfies the estimates*

$$\|e\|_{1,I_1} \le C_1\|R\|_{0,I_2} + C_2 p^{-1}\|e\|_{1,\Gamma}$$
$$\|e\|_{1,I_2} \ge C_3\|R\|_{0,I_1} - C_4 p^{-1}\|e\|_{1,\Gamma}$$

for $\gamma = \frac{1}{2}$, or

$$\|e\|_{0,I_1} \le C_1\|R\|_{0,I_2} + C_2 p^{-1}\|e\|_{0,\Gamma}$$
$$\|e\|_{0,I_2} \ge C_3\|R\|_{0,I_1} - C_4 p^{-1}\|e\|_{0,\Gamma}$$

for $\gamma = 0$.

PROOF. Let $t \le \gamma$ and $A^*w = \psi$, we have

$$\begin{aligned}\|e\|_t &\le \sup_{\|\psi\|_{-t}\le 1} |(e,\psi)| = \sup|(e,A^*w)| \\ &= \sup|(e,A^*(w-\chi)) + (e,A^*\chi)| = \sup|(e,A^*(w-\chi))| \\ &\le \sup\|e\|_\gamma\|A^*(w-\chi)\|_{-\gamma} \le C\|e\|_\gamma \sup\|w-\chi\|_\gamma, \quad \forall\chi\in V_p(\Gamma).\end{aligned} \tag{2.20}$$

Using (2.18) and (2.20) with $\gamma = \frac{1}{2}, s = 1-t$, and

$$\|w\|_{1-t} \le C\|\psi\|_{-t},$$

we get

$$\|e\|_t \le C\|e\|_{\frac{1}{2}} \sup p^{-\frac{1}{2}+t}\|w\|_{1-t} \le Cp^{-\frac{1}{2}+t}\|e\|_{\frac{1}{2}}.$$

Then

$$\|e\|_0 \le Cp^{-\frac{1}{2}}\|e\|_{\frac{1}{2}} \le Cp^{-\frac{1}{2}}\|e\|_0^{\frac{1}{2}}\|e\|_1^{\frac{1}{2}},$$

i.e.,

$$\|e\|_0 \le Cp^{-1}\|e\|_1. \tag{2.21}$$

By (2.13), (2.14) with $\alpha = 1$, $m = 0$ and (2.21), we obtain

$$\|e\|_{1,I_1} \le C_1\|R\|_{0,I_2} + C_2 p^{-1}\|e\|_{1,\Gamma}, \tag{2.22}$$
$$\|e\|_{1,I_2} \ge C_3\|R\|_{0,I_1} - C_4 p^{-1}\|e\|_{1,\Gamma}. \tag{2.23}$$

For $t \le \gamma = 0$, using (2.19) and (2.20), we have

$$\|e\|_t \le Cp^t\|e\|_0.$$

Then

$$\|e\|_{-1} \le Cp^{-1}\|e\|_0. \tag{2.24}$$

Hence, by (2.13) and (2.14) with $\alpha = m = 0$, we get

$$\|e\|_{0,I_1} \le C_1\|R\|_{0,I_2} + C_2 p^{-1}\|e\|_{0,\Gamma}, \tag{2.25}$$
$$\|e\|_{0,I_2} \ge C_3\|R\|_{0,I_1} - C_4 p^{-1}\|e\|_{0,\Gamma}. \tag{2.26}$$

□

Using some interpolate estimates for hp-version, we can also get the corresponding local error estimates.

Remark. There are some developments in other directions in the boundary element methods in China. Zhu Jialin worked on an indirect BEM for the solutions of harmonic, biharmonic, Stokes and diffusion equations, and gave corresponding convergence and error estimates (see [50–55]). Han Houde discussed BEM for Signorini problem, contact problem and some integro-differential equations, and gave a new variational formulation for the coupling of FEM-BEM (see [12–18]). Lin Qun and his students studied the extrapolation method for the approximations to the solution of some boundary integral equations of the second kind (see [19, 20, 23]). There are also many engineering applications of BEM in China, for which we can see [2–6].

References

1. Babuška, I., Yu, De-hao, *Asymptotically exact a-posteriori error estimator for biquadratic elements*, Finite Elements in Analysis and Design, **3**(1987), 341–354.
2. Du, Qing-hua (ed.), Proceedings of First Conference of Boundary Element Methods in Engineering, (in Chinese), Chongqing, 1985.
3. Du, Qing-hua (ed.), *Boundary Element*, Pergamon Press, 1986.
4. Du, Qing-hua, Tanaka, M. (eds.), *Theory and Applications of Boundary Element Methods*, Tsinghua University Press, Beijing, 1988.
5. Du, Qing-hua (ed.), Proceedings of Second Conference of Boundary Element Methods in Engineering, (in Chinese), Nanning, 1988.
6. Du, Qing-hua, Yao, Zhen-han, *One decade of Engineering Research on BIE-BEM in China*, Boundary Elements X, vol.1, C.A. Brebbia ed., Springer-Verlag, Berlin, 621–630, 1988.
7. Feng, Kang, *Differential versus integral equations and finite versus infinite elements*, Mathematica Numerica Sinica, **2**:1(1980), 100–105.
8. Feng, Kang, *Canonical boundary reduction and finite element method*, Proceedings of International Invitational Symposium on the Finite Element Method (1981, Hefei), Science Press, Beijing, 1982.
9. Feng, Kang, Yu, De-hao, *Canonical integral equations of elliptic boundary value problems and their numerical solutions*, Proceedings of the China-France Symposium on Finite Element Methods, Feng Kang and J.L. Lions eds., Science press, Beijing, 211–252, 1983.
10. Feng, Kang, *Finite element method and natural boundary reduction*, Proceedings of the International Congress of Mathematicians, Warszawa, 1439–1453, 1983.
11. Feng, Kang, *Asymptotic radiation conditions for reduced wave equation*, Journal of Computational Mathematics, **2**:2(1984), 130–138.
12. Han, Hou-de, Wu, Xiao-nan, *Approximation of infinite boundary condition and its application to finite element methods*, Journal of Computational Mathematics, **3**:2(1985), 179–192.
13. Han, Hou-de, Wu, Xiao-nan, *The mixed finite element method for Stokes equations on unbounded domains*, J. Sys. Sci. & Math. Scis., **5**:2(1985), 121–132.
14. Han, Hou-de, *The boundary finite element methods for Signorini problems*, Numerical Methods for Partial Differential Equations, Lecture Notes in Mathematics, No.1297, Springer-Verlag, 38–49, 1987.
15. Han, Hou-de, Guan, Zhi, Yu, Chongqing, *The canonical boundary element analysis of a model for an electropaint process—the canonical boundary element approximation of a Signorini problem*, Applied Mathematics, A Journal of Chinese Universities, **3**:1(1988), 101–111.
16. Han, Hou-de, *Boundary integro-differential equations of elliptic boundary value problems and their numerical solutions*, Scientia Sinica, Series A, **31**:10(1988), 1153–1165.

17. Han, Hou-de, Hsiao, G.C., *The boundary element method for a contact problem*, Theory and Applications of Boundary Element Methods, Du, Q. and Tanaka, M. eds., Tsinghua University Press, Beijing, 33–38, 1988.
18. Han, Hou-de, *A new class of variational formulations for the coupling of finite and boundary element methods*, Journal of Computational Mathematics, **8**:3(1990), 223–232.
19. Lin, Qun, Liu, Jia-quan, *Extrapolation method for Fredholm integral equations with non-smooth kernels*, Numerishe Mathematik, **35**(1980), 459–464.
20. Shi, Jun, *A high accuracy method for second kind integral equations and its application to boundary integral equations.* (in Chinese), Doctor Thesis, Institute for System Science, Academia Sinica, Beijing, 1990.
21. Wendland, W.L., Yu, De-hao, *Adaptive boundary element methods for strongly elliptic integral equations*, Numerische Mathematik, **53**(1988), 539–558.
22. Wendland, W.L., Yu, De-hao, *A-posteriori local error estimates of boundary element methods with some pseudo-differential equations on closed curves*, Bericht Nr.20, Uni. Stuttgart, Math. Institut A, Juni 1989; J. Comp. Math., **10**:3(1992), 273–289.
23. Xie, Rui-feng, *A high accuracy method for second kind boundary integral equations* (in Chinese), Doctor Thesis, Institute for System Science, Academia Sinica, Beijing, 1988.
24. Yu, De-hao, *Canonical integral equations of biharmonic elliptic boundary value Problems*, Mathematica Numerica Sinica, **4**:3(1982), 330–336.
25. Yu, De-hao, *Numerical solutions of harmonic and biharmonic canonical integral equations in interior or exterior circular domains*, Journal of Computational Mathematics, **1**:1(1983), 52–62.
26. Yu, De-hao, *Numcrical solution of harmonic canonical integral equation over sector with crack and concave angle* (in Chinese), Journal on Numerical Methods and Computer Applications, **4**:3(1983), 183–188.
27. Yu, De-hao, *Coupling canonical boundary element method with FEM to solve harmonic problem over cracked domain*, Journal of Computational Mathematics, **1**:3(1983), 195–202.
28. Yu, De-hao, *Canonical boundary element method for plane elasticity problems*, Journal of Computational Mathematics, **2**:2(1984), 180–189.
29. Yu, De-hao, *Canonical boundary reduction and natural boundary element method* (in Chinese), Doctor Thesis, Computing Center, Academia Sinica, 1984.
30. Yu, De-hao, *Error estimates for the canonical boundary element method*, Proceedings of the 1984 Beijing Symposium on Differential Geometry and Differential Equations, Feng Kang ed., Science Press, Beijing, 343–348, 1985.
31. Yu, De-hao, *Approximation of boundary conditions at infinity for a harmonic equation*, Journal of Computational Mathematics, **3**:3(1985), 219–227.
32. Yu, De-hao, *Canonical integral equations of Stokes problem*, Journal of Computational Mathematics, **4**:1(1986), 62–73.
33. Yu, De-hao, *A system of plane elasticity canonical integral equations and its application*, Journal of Computational Mathematics, **4**:3(1986), 200–211.
34. Yu, De-hao, *A-posteriori error estimates and adaptive approaches for some boundary element methods*, Boundary Elements IX, Vol.1, C.A.Brebbia, W.L.Wendland and G.Kuhn eds., Springer-Verlag, Berlin, 241–256, 1987.
35. Yu, De-hao, *Self-adaptive boundary element methods*, ZAMM **68**:5(1988), 435–437.
36. Yu, De-hao, *Some mathematical aspects of adaptive boundary element methods*, Theory and Applications of Boundary Element Methods, Q. Du and M. Tanaka eds., Tsinghua University Press, Beijing, 297–304, 1988.
37. Yu, De-hao, *Adaptive boundary element methods and local error estimates* (in Chinese), Proceedings of Second Conference of BEM in Engineering, Vol. 1, Nanning, 13–23, 1988.
38. Yu, De-hao, *Some new developments in mathematical theory of boundary element methods* (in Chinese), The Theory of the Combination Method of Analytic and Numerical Solutions and Applications in Engineering, Li Jia-bao ed., Hunan University Press, Changsha, 228–232, 1990.
39. Yu, De-hao, *Mathematical foundation of adaptive boundary element methods*, Lectures on the Second World Congress on Computational Mechanics, Stuttgart, 308–311, 1990;

Comp. Meth. Appl. Mech. Engrg., **91**(1991), 1237–1243.

40. Yu, De-hao, *Asymptotically exact a-posteriori error estimator for elements of bi-even degree*, Math. Numer. Sinica, **13**:1(1991), 89–101; Chinese J. of Numer. Math. and Appl., **13**:2(1991), 64–78.
41. Yu, De-hao, *Asymptotically exact a-posteriori error estimator for elements of bi-odd degree*, Math. Numer. Sinica, **13**:3(1991), 307–314; Chinese J. of Numer. Math. and Appl., **13**:4(1991), 82–90.
42. Yu, De-hao, *A direct and natural coupling of BEM and FEM*, Boundary Elements XIII, C.A. Brebbia, G.S. Gipson eds., Comp. Mech. Publ., Southampton, 995–1004, 1991.
43. Yu, De-hao, *The p-adaptive boundary element method and its a-posteriori error estimator*, Proc. of the 4th China-Japan Symposium on BEM, Q. Du, M. Tanaka eds., Inter. Acad. Publ., Beijing, 3–8, 1991.
44. Yu, De-hao, *Extraction of higher derivatives in FEM for Poisson equation and plane elasticity problem* (in Chinese), Math. Numer. Sinica, **14**:1(1992), 107–117.
45. Yu, De-hao, *Extraction methods for derivatives in finite element approximations of Stokes problem* (in Chinese), Math. Numer. Sinica, **14**:2(1992), 184–193.
46. Yu, De-hao, *The coupling of natural BEM and FEM for Stokes problem on unbounded domain*, Math. Numer. Sinica, **14**:3(1992), 371–378; Chinese J. of Numer. Math. and Appl., **14**:4(1992), 111–120.
47. Yu, De-hao, *Adaptive finite and boundary element methods and the numerical evaluation of hypersingular integrals on K-meshes*, Inter. Conf. on Comp. of Diff. Eqs. and Dyn. Sys., Beijing, 1992.
48. Yu, De-hao, *The numerical evaluation of hypersingular integrals in boundary element method* (in Chinese), Proc. of 3rd Conf. on FEM in China, 7–10, 1992.
49. Yu, De-hao, *Mathematical Theory of Natural Boundary Element Method* (in Chinese), Science Press, Beijing, 1993.
50. Zhu, Jia-lin, *A boundary integral equation method for the stationary Stokes problem in 3D*, Boundary Elements V, C.A. Brebbia ed., Springer-Verlag, Berlin, 1983.
51. Zhu, Jia-lin, *The boundary integral equation method for solving the Dirichlet problem of a biharmonic equation*, Mathematica Numerica Sinica, 6:3(1984), 278–288.
52. Zhu, Jia-lin, *The boundary integral equation method for the 3-D stationary Stokes problem*, Mathematica Numerica Sinica, **7**:1(1985), 40–49.
53. Zhu, Jia-lin, *A boundary integral equation method for solving stationary Stokes problem*, Mathematica Numerica Sinica, **8**:3(1986), 281–289.
54. Zhu, Jia-lin, *An indirect boundary element method in the solution of the diffusion equation*, Boundary Elements VIII, C.A. Brebbia ed., Springer-Verlag, 1986.
55. Zhu, Jia-lin, *Asymptotic error estimates of BEM for two-dimensional flow problems*, Theory and Applications of Boundary Element Methods, M. Tanaka and Q. Du eds., Pergamon Press, 1987.

COMPUTING CENTER, ACADEMIA SINICA, P.O. BOX 2719, BEIJING 100080, CHINA.

Contemporary Mathematics
Volume 163, 1994

Trust Region Algorithms for Nonlinear Programming*

YUAN YA-XIANG

ABSTRACT. Nonlinear programming, or nonlinear optimization, is to minimize or maximize a nonlinear function, possibly subject to finitely many algebraic equations and inequalities. Trust region algorithms are a class of numerical algorithms for optimization. In this paper we review some main results of trust region algorithms for nonlinear optimization.

1. Introduction

Nonlinear programming, or nonlinear optimization, is to minimize or maximize a nonlinear function, possibly subject to finitely many algebraic equations and inequalities. It normally has the following form:

$$\min_{x \in \Re^n} \quad f(x) \tag{1.1}$$

$$\text{subject to} \quad c_i(x) = 0 \qquad i = 1, 2, \ldots, m_e; \tag{1.2}$$

$$c_i(x) \geq 0 \qquad i = m_e + 1, \ldots, m \tag{1.3}$$

where $f(x)$ and $c_i(x)$ ($i = 1, \ldots, m$) are real functions defined in $\Re^n$, and $m \geq m_e$ are two non-negative integers.

Numerical algorithms for nonlinear programming problem (1.1)-(1.3) are iterative. There are mainly two types of algorithms, one is line search algorithms and the other is trust region algorithms.

Line search algorithms are commonly used and they have been extensively studied. In a line search algorithm, at the beginning of each iteration, the current iterate x_k is available and a search direction d_k is computed. Then a line search is carried out to obtain a step size $\alpha_k > 0$, and the next iterate point is set

1991 *Mathematics Subject Classification.* 65k10.
Key words and phrases. trust region, nonlinear optimization, convergence.
* This work was supported by a grant from the National Science Foundation of China.

by $x_{k+1} = x_k + \alpha_k d_k$. The algorithm repeat this process until some optimal conditions are satisfied within given tolerance.

Trust region algorithms are relatively new algorithms. The main idea of trust region algorithms are as follows. At the beginning of each iteration, a trust region is available, which is normally a neighborhood centered at the current iterate x_k. A subproblem is solved in the trust region, giving the solution s_k which is called the trial step. Then some tests are carried out to decide whether this trial step should be accepted. If the trial step is acceptable, we let $x_{k+1} = x_k + s_k$ otherwise $x_{k+1} = x_k$. The size of the new trust region also depends on the trial step. Roughly speaking, if the trial step is reasonably good the size of trust region would be increased or remain unchanged, otherwise it is reduced. In order to make the trust region subproblem easy to solve, the trust region is normally $\{x \mid ||x_k|| \leq \Delta_k\}$ which is a general ball centered at x_k with radius $\Delta_k > 0$, where $||\cdot||$ is some norm in $\Re^n$. In this case, updating the trust region radius is equivalent to adjust the size of the trust region. In a practical algorithm, $||\cdot||$ is usually $||\cdot||_2$ or $||\cdot||_\infty$.

Most researches on trust region algorithms are mainly started in the 80s. Hence trust region algorithms are less mature than line search algorithms, and by now the applications of trust region algorithms are not as widely as that of line search algorithms. However, trust region methods have two advantages. One is that they are reliable and robust, another is that they have very strong convergence properties. In the recent ICIAM 91 conference and the 14th International Symposium on Mathematical Programming, invited talks are given on trust region algorithms (Conn, 1991; Dennis, 1991). More and more attentions will be drawn to trust region algorithms.

2. Levenberg-Marquardt method and trust region

Trust region algorithms are developed mainly in the last twenty years, and most of the researches on trust region are done in the last ten years. But the technique of trust region is, in some sense, equivalent to that of the classical Levenberg-Marquardt method which is a method for nonlinear least squares problems and which was first given by Levenberg (1944) and re-derived by Marquardt (1963).

For the nonlinear least squares problem

$$\min_{x \in \Re^n} ||F(x)||_2^2 \tag{2.1}$$

where $F(x) = (f_1(x), \ldots, f_m(x))^T$ and $f_i(x)(i = 1, \ldots, m)$ are continuous differentiable functions in $\Re^n$. The Gauss-Newton method for problem (2.1) is iterative, and at the current iterate x_k, the Gauss-Newton step is

$$d_k = -A(x_k)^+ F(x_k) \tag{2.2}$$

where $A(x) = \nabla F(x)^T$ is the Jacobi matrix, and A^+ is the generalized inverse of A. It is easy to see that the Gauss-Newton step is a solution of the subproblem

$$\min_{d \in \Re^n} ||F(x_k) + A(x_k)d||_2^2 \tag{2.3}$$

which is an approximation to the original problem (2.1) near the current iterate x_k. One difficulty of using the Gauss-Newton step is that the Jacobi matrix $A(x_k)$ may be ill conditioned. The Levenberg-Marquardt method, a very successful nonlinear least squares solver, chooses the step as follows

$$d_k = -(A(x_k)^T A(x_k) + \lambda_k I)^{-1} A(x_k)^T F(x_k) \tag{2.4}$$

where $\lambda_k \geq 0$ is a parameter which is updated from iteration to iteration (see, Moré, 1978). The original idea of Levenberg-Marquardt method is introducing the parameter λ_k to overcome the ill condition of $A(x_k)$, or in other words, to prevent $||d_k||_2$ being too large. As it is easily seen that (2.4) is the solution of

$$\min_{d \in \Re^n} ||F(x_k) + A(x_k)d||_2^2 + \lambda_k ||d||_2^2 \ . \tag{2.5}$$

Subproblem (2.5) is a modification of (2.3). The additional term $\lambda_k||d||_2^2$ can be viewed as a penalty term which prevents $||d_k||$ being too large.

Define

$$\Delta_k = ||(A(x_k)^T A(x_k) + \lambda_k I)^{-1} A(x_k)^T F(x_k)||_2 \ , \tag{2.6}$$

then it is not difficult to prove that the Levenberg-Marquardt step (2.4) is also a solution to the following problem

$$\min_{d \in \Re^n} ||F(x_k) + A(x_k)d||_2^2 \tag{2.7}$$

$$\text{subject to} \qquad ||d||_2 \leq \Delta_k \ . \tag{2.8}$$

Now it is obvious that problem (2.7)-(2.8) is a trust region subproblem. It is in this sense that we can view the classical Levenberg-Marquardt method as a trust region algorithm.

However, a trust region algorithm usually modifies the trust region radius Δ_k from iteration to iteration directly, while the Levenberg-Marquardt method updates the parameter λ_k, which in turn modifies the value Δ_k from (2.6) implicitly. Modifying Δ_k directly has the advantage of controlling and monitoring the length of d_k easily.

3. Unconstrained optimization

In this section, we consider trust region algorithms for unconstrained optimization problem:

$$\min_{x\in\Re^n} f(x) \tag{3.1}$$

where $f(x)$ is a continuous differentiable function in $\Re^n$. At each iteration, a trial step is calculated by solving the subproblem

$$\min_{d\in\Re^n} g_k^T d + \frac{1}{2} d^T B_k d = \phi_k(d) \tag{3.2}$$

$$\text{subject to} \qquad ||d||_2 \le \Delta_k, \tag{3.3}$$

where $g_k = \nabla f(x_k)$ is the gradient at the current approximate solution, B_k is an $n\times n$ symmetric matrix which approximates the Hessian of $f(x)$ and $\Delta_k > 0$ is a trust region radius. Let s_k be a solution of (3.2)-(3.3). The predicted reduction is defined by the reduction in the approximate model, that is

$$Pred_k = \phi_k(0) - \phi_k(s_k). \tag{3.4}$$

Unless the current point x_k is a stationary point and B_k is positive semi-definite, the predicted reduction $Pred_k$ is always positive. The actual reduction is the reduction in the objective function:

$$Ared_k = f(x_k) - f(x_k + s_k). \tag{3.5}$$

And we define the ratio between the actual reduction and the predicted reduction by

$$r_k = \frac{Ared_k}{Pred_k} \tag{3.6}$$

which is used to decide whether the trial step is acceptable and to adjust the new trust region radius.

A general trust region algorithm for unconstrained optimization can be given as follows.

ALGORITHM 3.1.

Step 1 *Given* $x_1 \in \Re^n$, $\Delta_1 > 0$, $\epsilon \ge 0$,
$0 < c_3 < c_4 < 1 < c_1$, $0 \le c_0 \le c_2 < 1$, $c_2 > 0$, $k := 1$.

Step 2 *If* $||g_k||_2 \le \epsilon$ *then stop;*
Solve (3.2)-(3.3) *giving* s_k.

Step 3 *Compute* r_k;

$$x_{k+1} = \begin{cases} x_k & \text{if } r_k \le c_0 \\ x_k + s_k & \text{otherwise} \end{cases} \tag{3.7}$$

Choose Δ_{k+1} *that satisfies*

$$\Delta_{k+1} \in \begin{cases} [c_3 ||s_k||_2, \; c_4 \Delta_k] & \text{if } r_k < c_2 \\ [\Delta_k, \; c_1 \Delta_k] & \text{otherwise;} \end{cases} \tag{3.8}$$

Step 4 update B_{k+1}*;*
$k := k + 1$*; go to Step 2.*

The constants c_i ($i = 0, \dots, 4$) can be chosen by users. Typical values are $c_0 = 0$, $c_1 = 2$, $c_2 = c_3 = 0.25$, $c_4 = 0.5$. For other choices of those constants, please see Fletcher (1982a, 1987), Moré (1983), Powell (1984a) etc.. The parameter c_0 is usually zero (e.g. Fletcher, 1980; Powell, 1975) or a small positive constant (e.g. Duff, Nocedal and Reid, 1987; Sorensen, 1982b). The advantage of using zero c_0 is that a trial step is accepted whenever the objective function is reduced. Hence it would not throw away a "good point", which is a desirable property especially when the function evaluations are very expensive. But, the price we pay for letting $c_0 = 0$ is that the global convergence result is only

$$\liminf_{k \to \infty} \quad ||g_k||_2 = 0 \tag{3.9}$$

instead of

$$\lim_{k \to \infty} ||g_k||_2 = 0 \tag{3.10}$$

which can be achieved if $c_0 > 0$. However, given a positive tolerance ϵ, (3.9) is sufficient to guarantee a finite termination of (3.1).

The subproblem (3.2)-(3.3) has been studied by many authors. And the following lemma is well known (for example, see Gay, 1981; Moré and Sorensen, 1983):

LEMMA 3.2. *A vector* $d^* \in \Re^n$ *is a solution of the problem*

$$\min_{d \in \Re^n} g^T d + \frac{1}{2} d^T B d = \phi(d) \tag{3.11}$$

$$\text{subject to} \qquad ||d||_2 \leq \Delta \tag{3.12}$$

where $g \in \Re^n$, $B \in \Re^{n \times n}$ *is a symmetric matrix, and* $\Delta > 0$, *if and and only if there exists* $\lambda^* \geq 0$ *such that*

$$(B + \lambda^* I) d^* = -g \tag{3.13}$$

and that $B + \lambda^* I$ *is positive semi-definite,* $||d^*||_2 \leq \Delta$ *and*

$$\lambda^* (\Delta - ||d^*||_2) = 0. \tag{3.14}$$

By considering the maximum reduction of the quadratic model $\phi(d)$ along the steepest descent direction $-g$ in the trust region we can easily prove the following result.

LEMMA 3.3 (POWELL, 1975). *If d^* is a solution of (3.11)-(3.12), it follows that*

$$\phi(0) - \phi(d^*) \geq \frac{1}{2}||g||_2 \min\{\Delta, ||g||_2/||B||_2\}. \tag{3.15}$$

In the simple case where B is positive definite and

$$||B^{-1}g||_2 \leq \Delta , \tag{3.16}$$

the solution for (3.11)-(3.12) is $d^* = -B^{-1}g$. Otherwise, solving (3.11)-(3.12) is equivalent to compute a non-negative parameter λ^* such that $(B + \lambda^* I)$ is positive semi-definite, and that there exists d^* satisfies (3.13) and

$$||d^*||_2 = \Delta. \tag{3.17}$$

If $(B + \lambda^* I)$ is positive definite, we can write $d^* = -(B + \lambda^* I)^{-1}g$ and (3.17) is equivalent to

$$||(B + \lambda^* I)^{-1}g||_2 = \Delta. \tag{3.18}$$

As (3.18) is a nonlinear equation with one unknown λ^*, Newton's method can be use to solve the equivalent nonlinear equation:

$$\frac{1}{||(B + \lambda I)^{-1}g||_2} - \frac{1}{\Delta} = 0, \tag{3.19}$$

because the left hand side of (3.19) is very close to a linear function of λ. The case where $B + \lambda^* I$ has zero eigenvalues is called the "hard case", and any solution will be in the following form:

$$d^* = -(B + \lambda^* I)^+ g + v \tag{3.20}$$

where v is a vector in the null space of $B + \lambda^* I$. For more detailed discussions of subproblem (3.11)-(3.12), see Gay (1981) and Moré and Sorensen (1983).

The global convergence analyses of trust region algorithms depend on the fact that the predicted reduction satisfies (3.15). Hence, instead of solving (3.2)-(3.3) exactly, we can compute a trial step s_k that satisfies

$$\phi_k(0) - \phi_k(s_k) \geq \tau ||g_k||_2 \min\{\Delta_k, ||g_k||_2/||B_k||_2\} , \tag{3.21}$$

where τ is some positive constant. To compute a vector s_k satisfying (3.21) is usually much easier than solving (3.2)-(3.3) exactly. The vector s_k can be calculated by dog-leg type techniques or by searching in the two dimensional space spanned by the steepest descent direction and the Newton's step. For more details, please see Dennis and Mei (1979), Powell (1970b), Shultz, Schnabel and Byrd (1985) and Thomas (1979). The subproblem (3.2)-(3.3) can also be solved approximately by a preconditioned conjugate gradient method which can be regarded as a generalized dog-leg technique (see, Steihaug, 1983).

The first convergence result for Algorithm 3.1 is given by Powell (1975):

THEOREM 3.4. *If $f(x)$ is bounded below, if $g(x) = \nabla f(x)$ is uniformly continuous, if s_k satisfies* (3.21), *if there exists a positive constant β_1 such that*

$$||B_k||_2 \leq \beta_1(1 + \sum_{i=1}^{k} ||s_k||_2) \tag{3.22}$$

holds for all k, and if $\epsilon = 0$ is chosen in Algorithm 3.1, *it follows that*

$$\liminf_{k\to\infty} \quad ||g_k||_2 = 0. \tag{3.23}$$

Powell's result is strengthen by Shultz, Schnabel and Byrd (1985) with some additional conditions:

THEOREM 3.5. *Under the conditions of Theorem* 3.4, *if $c_0 > 0$ and $||B_k||_2$ is uniformly bounded, the sequence $\{x_k\}$ generated by Algorithm* 3.1 *satisfies*

$$\lim_{k\to\infty} ||g_k||_2 = 0. \tag{3.24}$$

The condition (3.22) is weaker than the uniformly boundedness of B_k, and it can be satisfied when B_k is updated by Powell's symmetric Broyden formula (see, Powell, 1970a, 1975). Similar to that of Dennis and Moré (1974), Powell(1975) proves the following superlinear convergence result.

THEOREM 3.6. *Under the conditions of Theorem* 3.4, *if the sequence generated by Algorithm* (3.1) *converges to x^*, if $\nabla^2 f(x)$ is continuous in a neighborhood of x^* and $\nabla^2 f(x^*)$ is positive definite, if $s_k = -B_k^{-1} g_k$ whenever $||s_k||_2 < \Delta_k$, and if the condition*

$$\lim_{k\to\infty} ||g_{k+1} - g_k - B_k s_k||_2/||s_k||_2 = 0 \tag{3.25}$$

is satisfied, the sequence x_k converges to x^* Q-superlinearly in the sense that

$$\lim_{k\to\infty} ||x_{k+1} - x^*||_2/||x_k - x^*||_2 = 0. \tag{3.26}$$

It is also shown by Powell (1975) that B_k updated by the PSB formula gives the limit (3.25), consequently superlinear convergence follows. However, Powell's superlinear convergence result requires that B_k is updated at every iteration, even at a failed iteration. As updating B_k requires the evaluation of $g(x_k + s_k)$, therefore even at an unacceptable point x_k+s_k which satisfies $f(x_k+s_k) > f(x_k)$ we still have to compute $g(x_k + s_k)$. Khalfan (1989) suggests update B_k by the following formula

$$B_{k+1} = B_k + 2[f(x_k + s_k) - f(x_k) - s_k^T g_k - \frac{1}{2} s_k^T B_k s_k] \frac{s_k s_k^T}{||s_k||_2^4} \tag{3.27}$$

whenever $f(x_k+s_k) \geq f(x_k)$, thus there is no needs to compute $g(x_k+s_k)$ at such iterations. Superlinear convergence remains true after this modification. More details can be seen in Khalfan (1989). Another direct corollary of Theorem 3.6 is that Newton's method with trust region converges superlinearly.

Theorem 3.4 is also true if the condition 3.22 is replaced by

$$\sum_{k=1}^{\infty} \frac{1}{M_k} = \infty , \tag{3.28}$$

where

$$M_k = \max_{1 \le i \le k} ||B_i||_2 + 1 . \tag{3.29}$$

Condition (3.28) is satisfied if B_k is updated by the BFGS formula, $f(x)$ is convex and $\nabla^2 f(x)$ is uniformly bounded. And it is also shown by Powell (1984a) that the condition (3.28) can not be further relaxed.

The exact solution d^* of (3.11)-(3.12) also satisfies that

$$\phi(0) - \phi(d^*) \ge -\frac{1}{2}\sigma_n(B)\Delta^2, \tag{3.30}$$

where $\sigma_n(B)$ is the smallest eigenvalue of B. Inequality (3.30) can be easily verified by studying the maximum reduction of $\phi(d)$ in the direction of the eigenvector of B corresponding to eigenvalue $\sigma_n(B)$. For Newton's method with trust region where $B_k = \nabla^2 f(x_k)$, if the computed trial step s_k satisfies (3.21) and

$$\phi_k(0) - \phi_k(s_k) \ge -\bar{\tau}\sigma_n(B_k)\Delta_k^2 \tag{3.31}$$

for some positive constant $\bar{\tau}$, and if the sequence x_k converges to a point x^*, it can be proved that x^* must satisfy the second order necessary condition, that is $\nabla^2 f(x^*)$ is positive semi-definite. If $\nabla^2 f(x^*)$ is positive definite, and if $s_k = -(\nabla^2 f(x^*))^{-1} * \nabla f(x^*)$ whenever $||s_k||_2 \le \Delta_k$, Newton's method with trust regions converges quadratically. More details about Newton's method with trust region techniques can be found in Fletcher (1987), Shultz, Schnabel and Byrd (1985), Sorensen (1982a) and Steihaug (1983).

The following two lemmas are given by Nocedal and Yuan (1991):

LEMMA 3.7. *If d^* is a solution of (3.11)-(3.12), $||d^*||_2 = \Delta$, and $\lambda^* \ge 0$ satisfies (3.13)-(3.14), we have that*

$$0 \le \lambda^* \le ||g||_2/\Delta - \sigma_n(B) \tag{3.32}$$

and that

$$d^{*T} g \le -\frac{||g||_2^2}{\sigma_1(B) - \sigma_n(B) + ||g||_2/\Delta}, \tag{3.33}$$

where $\sigma_1(B)$ and $\sigma_n(B)$ are the largest and smallest eigenvalues of B.

LEMMA 3.8. *If d^* is a solution of (3.11)-(3.12), then we have*

$$d^{*T} g \le -\frac{1}{2}||g||_2 \min\{\Delta, \quad ||g||_2/2||B||_2\}. \tag{3.34}$$

Based on (3.34), Nocedal and Yuan (1991) calculate a trial step s_k in the form

$$s_k = -(B_k + \lambda_k I)^{-1} g_k \tag{3.35}$$

for some non-negative parameter λ_k such that $B_k + \lambda_k I$ is positive definite and s_k satisfies that

$$s_k^T g_k \leq -\hat{\tau} ||g_k||_2 \min\{\Delta_k, \quad ||g_k||_2/||B_k||_2\} , \tag{3.36}$$

where $\hat{\tau}$ is some positive constant. From (3.36), we can see that the trial step s_k is also a "sufficiently" descent direction in the sense that the angle between s_k and $-g_k$ will be bounded away from $\pi/2$ if $||g_k||_k$ is bounded away from zero, and if $||s_k||_2$ and $||B_k||_2$ are bounded above. Using this property of s_k, if the trial step is unacceptable, back tracking can be used to search an acceptable point. For more details of this trust region with back tracking algorithm, please see Nocedal and Yuan (1991).

4. Constrained optimization

One of the most successful line search algorithms for constrained optimization is the sequential quadratic programming method which computes the search direction d_k by solving the following subproblem:

$$\min_{d \in \Re^n} g_k^T d + \frac{1}{2} d^T B_k d = \phi_k(d) \tag{4.1}$$

$$s.t. \quad c_i(x_k) + d^T \nabla c_i(x_k) = 0 \qquad i = 1, 2, \ldots, m_e; \tag{4.2}$$

$$c_i(x_k) + d^T \nabla c_i(x_k) \geq 0 \qquad i = m_e + 1, \ldots, m \tag{4.3}$$

which is an approximation to problem (1.1)-(1.3) near the current iterate x_k. Trust region algorithms require the trial step lies in the trust region, which means we have one more constraint of the form:

$$||d||_2 \leq \Delta_k \tag{4.4}$$

where $\Delta_k > 0$ is the current trust region radius. One obvious difficulty of applying sequential quadratic programming method with trust region is that the constraints (4.2)-(4.4) may have no solutions. There are two ways to overcome this difficulty.

The first is to solve the following subproblem

$$\min_{d \in \Re^n} g_k^T d + \frac{1}{2} d^T B_k d = \phi_k(d) \tag{4.5}$$

$$s.t. \quad \theta c_i(x_k) + d^T \nabla c_i(x_k) = 0 \qquad i = 1, 2, \ldots, m_e; \tag{4.6}$$

$$\theta c_i(x_k) + d^T \nabla c_i(x_k) \geq 0 \qquad i = m_e + 1, \ldots, m \tag{4.7}$$

$$||d||_2 \leq \Delta_k \tag{4.8}$$

where $\theta \in (0, 1]$. In order to make θ close to 1, a penalty term $\sigma(\theta - 1)^2$ can be added to the objective function in (4.5). Trust region algorithms that use (4.5)-(4.8) can be found in Byrd, Schnabel and Shultz (1987) and Vardi (1985). In these two algorithms given by Byrd, Schnabel and Shultz (1987) and Vardi (1985), the L_1 exact penalty function

$$P(x) - f(x) + \sigma(\sum_{i=1}^{m_e} |c_i(x)| + \sum_{i=m_e+1}^{m} \max[0, -c_i(x)]) \tag{4.9}$$

is used as a merit function to decide whether a trial step should be acceptable. Because of the nonsmoothness of the L_1 exact penalty function (4.9), the Marotos effect may occur, which can prevent superlinear convergence of $\{x_k\}$ (see, Marotos, 1978). Therefore, if the trial step s_k is unacceptable, Byrd, Schnabel and Shultz (1987) computes a second order correction step:

$$\bar{s}_k = -A_k(A_k^T A_k)^{-1} c(x_k + s_k) \tag{4.10}$$

where $A_k = A(x_k) = \nabla c(x_k)^T$, and uses $s_k + \bar{s}_k$ as a new trial step. Such second order correction step can be traced back to Mayne and Polak (1982) and Coleman and Conn (1982), and the idea is similar to that of the second order correction step for nonsmooth optimization suggested by Fletcher (1982b). As the second order correction step is used, Byrd, Schnabel and Shultz (1987) proves the Q-superlinear convergence of their algorithm.

Another way to overcome the possible inconsistency of constraints (4.2)-(4.4), only for equality constrained optimization problems, is to solve the subproblem:

$$\min_{d \in \Re^n} g_k^T d + \frac{1}{2} d^T B_k d = \phi_k(d) \tag{4.11}$$

$$s.\ t. \quad \sum_{i=1}^{m} (c_i(x_k) + d^T \nabla c_i(x_k))^2 \leq \xi_k \tag{4.12}$$

$$||d||_2 \leq \Delta_k \tag{4.13}$$

proposed by Celis, Dennis and Tapia (1985), where $\xi_k \geq 0$ is chosen by some techniques.

Celis, Dennis and Tapia (1985) uses

$$\xi_k = \sum_{i=1}^{m} (c_i(x_k) + (s_k^{CP})^T \nabla c_i(x_k))^2 \tag{4.14}$$

where s_k^{CP} is the Cauchy point which is the minimum of the least squares of the linearized constraint violations

$$\sum_{i=1}^{m} (c_i(x_k) + d^T \nabla c_i(x_k))^2 \tag{4.15}$$

along the steepest direction in the trust region. More details can be found in Celis (1985). The convergence analyses of the Celis-Dennis-Tapia algorithm are given by El-Alem (1988).

Powell and Yuan (1991) requires that the parameter ξ_k satisfies

$$\min_{||d||_2 \le b_1\Delta_k} ||c_k + A_k^T d||_2^2 \le \xi_k \le \min_{||d||_2 \le b_2\Delta_k} ||c_k + A_k^T d||_2^2 \tag{4.16}$$

where $c_k = c(x_k) = (c_1(x), \dots, c_m(x))^T$, $A_k = A(x_k) = \nabla c(x_k)^T$ and $b_1 \ge b_2$ are two constants in $(0, 1)$. The value ξ_k can be given by

$$\xi_k = ||c_k + A_k^T d(\mu_k)||_2^2 \tag{4.17}$$

where

$$d(\mu_k) = -(A_k A_k^T + \mu_k I)^{-1} c_k \tag{4.18}$$

for any $\mu_k \ge 0$ provided that $||d(\mu_k)||_2/\Delta_k \in [b_2,\ b_1]$. The penalty function that Powell and Yuan (1991) uses as the merit function is Fletcher's differentiable penalty function:

$$P_k(x) = f(x) + \lambda(x)^T c(x) + \sigma_k ||c(x)||_2^2 \tag{4.19}$$

where $\sigma_k > 0$ is the current penalty parameter, and $\lambda(x)$ is the least squares solution of

$$\min_{\lambda \in \Re^m} ||g(x) - A(x)\lambda||_2^2 \ . \tag{4.20}$$

The advantage of using a smooth exact penalty function is that Marotos effect would not occur. But, one undesirable property of the smooth penalty function (4.19) is that the standard predicted reduction of $P_k(x)$ would depends on $\nabla\lambda(x)^T$ which requires the evaluations of the Hessian of the objective function and the constraints. In order to avoid calculations of any second order derivatives, the predicted reduction is defined by

$$\begin{aligned} Pred_k = &- (g_k - A_k\lambda_k)^T s_k - \frac{1}{2} s_k^T B_k \hat{s}_k \\ &- (\lambda(x_k + s_k) - \lambda_k)^T (c_k + \frac{1}{2} A_k^T s_k) \\ &- \sigma_k (||c_k + A_k^T s_k||_2^2 - ||c_k||_2^2) \end{aligned} \tag{4.21}$$

where s_k is the trial step and

$$\hat{s}_k = (I - A_k A_k^+) s_k \tag{4.22}$$

is the orthogonal projection of s_k into the null space of A_k^T.

The following algorithm is given by Powell and Yuan (1991):

ALGORITHM 4.1.

Step 1 Given $x_1 \in \Re^n$, $\Delta_1 > 0$, $\epsilon \ge 0$,
$0 < b_2 \le b_1 < 1$, $\sigma_1 > 0$, $k := 1$.

Step 2 *if* $||c_k||_2 + ||g_k + A_k\lambda_k||_2 \le \epsilon$ *then stop;*
calculate the value ξ_k*;*
Solve problem (4.11)-(4.13) *which gives* s_k.

Step 3 *Calculate* $Pred_k$ *from formula* (4.21);
if the inequality

$$Pred_k \ge \frac{1}{2}\sigma_k(||c_k||_2^2 - ||c_k + A_k^T s_k||_2^2) \tag{4.23}$$

fails then increase σ_k *to the value*

$$\sigma_k^{new} = 2\sigma_k^{old} + \max\{0, \frac{2Pred_k^{old}}{||c_k + A_k^T s_k||_2^2 - ||c_k||_2^2}\}, \tag{4.24}$$

which ensures that the new $Pred_k$ *satisfies* (4.23).

Step 4 *Compute the ratio*

$$r_k = \frac{P_k(x_k) - P_k(x_k + s_k)}{Pred_k} \tag{4.25}$$

and let

$$x_{k+1} = \begin{cases} x_k & \text{if } r_k \le 0 , \\ x_k + s_k & \text{otherwise} ; \end{cases} \tag{4.26}$$

$$\Delta_{k+1} = \begin{cases} \max[\Delta_k,\ 4||s_k||_2], & r_k > 0.9 , \\ \Delta_k & 0.1 \le r_k \le 0.9 , \\ \min[\Delta_k/4,\ ||s_k||_2/2], & r_k < 0.1 ; \end{cases} \tag{4.27}$$

Generate B_{k+1}, *set* $\sigma_{k+1} = \sigma_k$*;*
$k := k + 1$ *and go to Step 2.*

Convergence results of the above algorithm are given by Powell and Yuan (1991). We state the two main results as follows, for the proofs of these results, please see Powell and Yuan (1991).

THEOREM 4.2. *Assume that* $\{x_k\}$, $\{s_k\}$, *and* $\{B_k\}$ *are uniformly bounded, if* $A(x)$ *has full column rank for all* k, *the sequence* $\{x_k\}$ *will not be bounded away from Kuhn-Tucker points of problem* (1.1)-(1.3), *that is*

$$\liminf_{k\to\infty} \ [||c_k||_2 + ||g_k - A_k\lambda_k||_2] = 0 . \tag{4.28}$$

THEOREM 4.3. *Under the conditions of Theorem 4.2, if* x_k *converges to* x^* *and the second order sufficient condition holds at* x^*, *and if*

$$\lim_{k\to\infty} \max_{A_k^T d = 0, ||d||_2 \le 1} \frac{|d^T(B_k - W(x^*, \lambda^*))s_k|}{||s_k||_2} = 0 , \tag{4.29}$$

we have that

$$\lim_{k\to\infty} r_k = 1 , \tag{4.30}$$

and x_k converges to x^* Q-superlinearly, where

$$W(x^*, \lambda^*) = \nabla^2 f(x^*) - \sum_{i=1}^{m} \lambda_i^* \nabla^2 c_i(x^*) \tag{4.31}$$

and $\lambda^* = (\lambda_1^*, \ldots, \lambda_m^*)^T$ are the Lagrange multipliers such that

$$\nabla f(x^*) - \sum_{i=1}^{m} \lambda_i^* \nabla c_i(x^*) = 0 \ . \tag{4.32}$$

For trust region subproblem (4.5)-(4.8), One way to solve it is to calculate a range step and a null step separately (for example, see Omojokun, 1989; Byrd, Schnabel, and Shultz, 1987). If the trust region constraint (4.8) is replace by

$$||d||_\infty \leq \Delta_k \ , \tag{4.33}$$

the subproblem can be written as a quadratic programming problem, hence it can be solved by any methods for quadratic programming problems.

The subproblem (4.11)-(4.13) seems more difficult to solve. The following two lemmas are given by Yuan (1990):

LEMMA 4.4. *Let d^* is a solution of*

$$\min_{d \in \Re^n} g^T d + \frac{1}{2} d^T B d = \phi(d) \tag{4.34}$$

$$s.t. ||A^T d + c||_2 \leq \xi \tag{4.35}$$

$$||d||_2 \leq \Delta \tag{4.36}$$

where $g \in \Re^n$, $A \in \Re^{n \times m}$, $c \in \Re^m$, $\Delta > 0$, B is a symmetric matrix and

$$\xi > \min_{||d||_2 \leq \Delta} ||A^T d + c||_2 \ , \tag{4.37}$$

there exist Lagrange multipliers $\lambda^* \geq 0$ and $\mu^* \geq 0$ such that

$$(B + \lambda^* I + \mu^* A A^T) d^* = -(g + \mu^* A c) \tag{4.38}$$

and the complementarity conditions

$$\lambda^* [\Delta - ||d^*||_2] = 0 \tag{4.39}$$

$$\mu^* [\xi - ||A^T d^* + c||_2] = 0 \tag{4.40}$$

are satisfied. Furthermore, if the multipliers λ^*, μ^* are unique, the matrix

$$H(\lambda^*, \mu^*) = B + \lambda^* I + \mu^* A A^T \tag{4.41}$$

has at most one negative eigenvalue.

LEMMA 4.5. *Under the conditions of Lemma 4.4, let Ω be all the Lagrange multipliers $\lambda \geq 0$, $\mu \geq 0$ such that (4.38)-(4.40) hold, then there exists $(\lambda, \mu) \in \Omega$ such that the matrix $H(\lambda, \mu)$ has at most one negative eigenvalue.*

If B is positive definite, problem (4.34)-(4.36) can be solved by a dual method. The dual problem has only two variables λ and μ and the constraints for the dual are $\lambda \geq 0$ and $\mu \geq 0$. Moreover, the gradient and Hessian of the dual objective function can be easily computed. Thus the truncated Newton method can be applied to solve the dual problem (see, Yuan, 1988). But, usually B_k is not positive definite in (4.11), as one advantage of trust region algorithms over line search algorithms is that there is no positive definite requirement for B_k.

For a general symmetric matrix B, it is not clear whether there exists a cheap way to solve (4.34)-(4.36). One way of solving (4.34)-(4.36) approximately is to compute the minimum of the objective function $\phi(d)$ in the 2-dimensional subspace $span\{g, (A^T)^+c\}$ within the two constraints (4.35)-(4.36) (for example, see Dennis and Williamson, 1991).

Trust region algorithms for constrained optimization can be constructed based on exact penalty functions. Yuan (1992) gives a trust region algorithm that is based on the L_∞ penalty function, namely the trial step s_k at every iteration is obtained by minimizing an approximation of the L_∞ penalty function

$$P_\sigma(x) = f(x) + \sigma_k ||c^-(x)||_\infty \tag{4.42}$$

within a trust region, where $\sigma_k > 0$ is a penalty parameter that is updated automatically and where $c^-(x)$ is the constraint violation function that is defined by

$$c_i^-(x) = c_i(x), \qquad i = 1, \dots, m_e; \tag{4.43}$$

$$c_i^-(x) = \min[c_i(x), 0], \ i = m_e + 1, \dots, m. \tag{4.44}$$

The subproblem has the following form

$$\min_{d \in \Re^n} g_k^T d + \frac{1}{2} d^T B_k d + \sigma_k ||(c_k + A_k^T d)^-||_\infty, \tag{4.45}$$

$$s.t. \quad ||d||_\infty \leq \Delta_k. \tag{4.46}$$

Under certain conditions, it is shown that the algorithm is convergent, and for all sufficiently large k, the trial step s_k is also a solution of the following problem:

$$\min_{d \in \Re^n} \ g_k^T d + \frac{1}{2} d^T B_k d \tag{4.47}$$

subject to

$$c_i(x_k) + d^T \nabla c_i(x_k) = 0, \qquad i = 1, 2, \dots, m_e \tag{4.48}$$

$$c_i(x_k) + d^T \nabla c_i(x_k) \geq 0, \qquad i = m_e + 1, \dots, m \tag{4.49}$$

$$||d||_\infty \leq \Delta_k. \tag{4.50}$$

More details can be found in Yuan (1992,1993).

There are trust region algorithms specially for linear constrained problems. Most of these algorithms use projected gradients. And once the active set is identified (Burke and Moré, 1988), the problem is equivalent to an unconstrained optimization in the null space of the active constraints. Thus, the results for unconstrained optimization can be easily extended to linearly constrained optimization. For more details, please see Burke, Moré and Toraldo (1990), Conn, Gould and Toint (1988), Gay (1988), Moré (1988) and Toint (1988).

Algorithm 4.1 can be generated to solve general constrained problems that have inequality constraints (4.3), one possible way to modify subproblem (4.11)-(4.13) is to replace (4.12) by

$$||(c_k + A_k d)^-||_2^2 \leq \xi_k \ , \tag{4.51}$$

where the superscript "-" has the same meaning as in (4.43)-(4.44). The feasible region of (4.51) is still a convex set. The difficulty is that the constraint function in (4.51) is a piecewise quadratic function, which has discontinuities in second order derivative. Recently, Dennis and Williamson (1991) has extended the Celis-Dennis-Tapia algorithm to general constrained problem. Another recent algorithm given by Burke (1990) also handles general constraints.

5. Nonsmooth optimization

In this section, we consider trust region algorithms for the following "composite nonsmooth optimization" problem:

$$\min_{x \in \Re^n} h(F(x)), \tag{5.1}$$

where $F(x) = (f_1(x), f_2(x), \dots , f_m(x))^T$, $h(\cdot)$ is a convex function defined in $\Re^m$ and $f_i(x)(i = 1, \dots , m)$ are m continuous differentiable functions defined in $\Re^n$. This form of the objective function in (5.1) occurs frequently in discrete approximation and data fitting calculations. One application of (5.1) is to solve a constrained problem by minimizing the L_1 exact penalty function (4.9). Another special subclass of (5.1) is the minimization of some norm of a set of nonlinear equations (see, Duff, Nocedal and Reid, 1987; El-Hallabi and Tapia, 1987; and Powell, 1984b). We focus our attentions to problem (5.1) because algorithms for (5.1) can be extended to general nonsmooth optimization (for example, see Dennis, Li and Tapia, 1989).

Similar to (3.2)-(3.3), a trust region subproblem can be given as follows:

$$\min_{d \in \Re^n} h(F(x_k) + A(x_k)^T d) + \frac{1}{2} d^T B_k d = \phi_k(d) \tag{5.2}$$

$$\text{subject to} \qquad ||d||_2 \leq \Delta_k \tag{5.3}$$

where $A(x_k) = \nabla F(x_k)^T$ is the Jacobi matrix. Assume s_k is the solution, it is easy to see that there exists $\mu_k \geq 0$ such that

$$0 \in A(x_k)\partial h(F(x_k) + A(x_k)^T s_k) + (B_k + \mu_k I)d \tag{5.4}$$

where $\partial h(\cdot)$ indicates the subgradient of $h(\cdot)$ (see Rockafellar, 1970). Using the following function:

$$\Psi_t(x) = \max_{d \in \Re^n} \{h(F(x)) - h(F(x) + A(x)^T d) | \quad ||d||_2 \leq t\} \quad t \geq 0, \tag{5.5}$$

Yuan (1985a) proves that

LEMMA 5.1. *Let d_k be a solution of* (5.2)-(5.3), *and let $\Psi_t(x)$ be defined by* (5.4). *Then, we have*

$$h(F(x_k)) - \phi_k(d_k) \geq \frac{1}{2}\Psi_{\Delta_k}(x_k)\min\{1, \quad \Psi_{\Delta_k}(x_k)/||B_k||_2\Delta_k^2\}\ . \tag{5.6}$$

The above lemma is a generalization of Lemma 3.3. And indeed, it reduces to Lemma 3.3 if $m = 1$ and $h(F) = F$ in which case $\Psi_t(x) = t||\nabla F(x)||_2$.

A model trust region algorithm for nonsmooth optimization problem (5.1) is given by Fletcher (1982a), which is also a special case of the following general algorithm:

ALGORITHM 5.2.

Step 1 *Given* $x_1 \in \Re^n$, $\Delta_1 > 0$,
$0 < c_3 < c_4 < 1 < c_1$, $0 \leq c_0 \leq c_2 < 1$, $c_2 > 0$, $k := 1$.

Step 2 *If x_k is acceptable as a solution then stop;*
Solve (5.2)-(5.3) *giving* s_k.

Step 3 *Compute*

$$r_k = [h(F(x_k)) - h(F(x_k + s_k))]/[\phi_k(0) - \phi_k(s_k)] \tag{5.7}$$

$$x_{k+1} = \begin{cases} x_k & \textit{if } r_k \leq c_0 \\ x_k + s_k & \textit{otherwise} \end{cases} \tag{5.8}$$

Choose Δ_{k+1} that satisfies

$$\Delta_{k+1} \in \begin{cases} [c_3||d_k||_2, \ c_4\Delta_k] & \textit{if } r_k < c_2 \\ [\Delta_k, \ c_1\Delta_k] & \textit{otherwise;} \end{cases} \tag{5.9}$$

Step 4 *update B_{k+1};*
$k := k + 1$; go to Step 2.

The convergence of Algorithm 5.2 is first given by Fletcher (1981), for a specific set of $c_i (i = 0, 1, \ldots, 4)$. Fletcher's proof remains valid for general parameters $c_i (i = 0, 1, \ldots, 4)$ as long as they satisfy the conditions in Step 1 of Algorithm 5.2. Thus we have the following global convergence result:

THEOREM 5.3. *If the function $h(F(x))$ is bounded below, and if $||B_k||_2$ is uniformly bounded, the sequence $\{x_k\}$ generated by Algorithm 5.2 is not bounded away from stationary points of problem (5.1).*

The above theorem is still true if we relax the uniformly boundedness of $||B_k||_2$ by condition (3.22) or (3.28). The proofs for these results can be found in Yuan (1985a).

The rate of convergence of Algorithm 5.2 has been studied by many authors, including Powell and Yuan (1984) and Womersley (1985). Most of the local superlinear convergence analyses require the following assumptions.

ASSUMPTION 5.4.

1) *x_k converges to x^*;*

2) *The second order sufficiency condition holds at x^*, that is*

$$d^T W^* d > 0 \tag{5.10}$$

for all d satisfying $P^* d = 0$, where P^* is a projector from $\Re^n$ to the null space of $A(x^*)^T$ and where

$$W^* = \sum_{i=1}^{m} \lambda_i^* \nabla^2 f_i(x^*) \tag{5.11}$$

and $\lambda_i^* (i = 1, \ldots, m)$ are Lagrange multipliers that satisfy $\sum_{i=1}^m \lambda_i^* \nabla f_i(x^*) = 0$.

Under Assumption 5.4, one can prove that (for example, see Powell and Yuan, 1984; Womersley, 1985):

THEOREM 5.5. *If conditions in Theorem 5.3 and Assumption 5.4 are satisfied, the sequence $\{x_k\}$ converges to x^* Q-superlinearly if and only if*

$$\lim_{k \to \infty} ||P^*(W^* - B_k)s_k||_2 / ||s_k||_2 = 0 \tag{5.12}$$

and the trust region bound is inactive for all large k.

Unfortunately, it is not always true for the trust region constraint to be inactive for large k. Actually, by constructing a minimax problem in $\Re^2$, Yuan (1984) shows that it is possible for the trust region bound to be active at every iteration, which could give only linearly convergence even though all other conditions in Theorem 5.5, including (5.12), are satisfied. This phenomenon, caused by the nonsmoothness of the objective function, is very similar to the Marotos effect. For a smooth function $f(x)$, the condition (3.25) implies that

$$f(x_k + s_k) = f(x_k) + s_k^T g_k + \frac{1}{2} s_k^T B_k s_k + o(||s_k||_2^2) \tag{5.13}$$

which gives that $\lim_{k \to \infty} r_k = 1$, and consequently superlinear convergence follows. However, in the nonsmooth case the relation

$$h(F(x_k + s_k)) = h(F(x_k) + A(x_k)^T s_k) + \frac{1}{2} s_k^T B_k s_k + o(||s_k||_2^2) \tag{5.14}$$

does not generally hold, even though $B_k = W^*$ for all k. Therefore when x_k converges to x^* along a path such that $h(F(x)) - h(F(x^*))$ is $O(||x_k - x^*||_2^2)$, the error in (5.14) may cause the ratio r_k not to indicate that x_k is converging, which leads to a reduction in the trust region bound, and consequently linear convergence follows.

One way of improving the rate of convergence of Algorithm 5.2 is due to Fletcher (1982b). Whenever the ratio r_k based on the trial step s_k indicates that a reduction in trust region radius is necessary, a second order correction step $\bar{s}_k$ is computed by solving the "second order correction" subproblem:

$$\min_{d\in\Re^n} h(F(x_k+s_k)+A(x_k)^Td)+\frac{1}{2}(s_k+d)^TB_k(s_k+d)=\phi_k(d) \quad (5.15)$$

$$\text{subject to} \qquad ||s_k+d||_2 \le \Delta_k. \quad (5.16)$$

Details of the second order correction algorithm can be found in Fletcher (1982b). Under certain conditions, Yuan (1985b) obtains some properties of the second order correction step $\bar{s}_k$:

$$||\bar{s}_k||_2 = O(||s_k||_2^2) \quad (5.17)$$

$$||x_k+s_k+\bar{s}_k-x^*||_2 = o(||x_k-x_*||_2) \quad (5.18)$$

$$h(F(x_k+s_k+\bar{s}_k)) < h(F(x_k)) \quad (5.19)$$

holds for all sufficiently large k. Thus the superlinear convergence of the second order correction algorithm can be established.

THEOREM 5.6. *(Yuan, 1985b) If the function $h(F)$ is a polyhedral convex function, if the gradients of $f_i(x)(i=1,\dots,m)$ are linearly independent, if conditions of Assumption* 5.4 *are satisfied, and if B_k tends to the Hessian of the Lagrangian at the solution x^*, the sequence $\{x_k\}$ generated by the second order correction algorithm converges to x^* Q-superlinearly.*

Recently, Terpolili (1991) studies a trust region algorithm for problem (5.1) when the exact Jacobi matrix $A(x_k)$ is not available, and an approximate Jacobian G_k is used every iteration.

REFERENCES

1. J.V. Burke, *A robust trust region method for constrained nonlinear programming problems*, Report MCS-P131-0190, Mathematics and Computer Science Division, Argonne National Laboratory, Argonne, Illinois, USA, 1990.
2. J.V. Burke and J.J. Moré, *On the identification of active constraints*, SIAM J. Numer. Anal., **25**(1988), 1197–1211.
3. J.V. Burke, J.J. Moré and G. Toraldo, *Convergence properties of trust region methods for linear and convex constraints*, Mathematical Programming, **47**(1990), 305–336
4. R. Byrd, R.B. Schnabel and G.A. Shultz, *A trust region algorithm for nonlinearly constrained optimization*, SIAM Journal of Numerical Analysis, **24**(1987), 1152–1170.

5. M.R. Celis, *A Trust Region Strategy for Nonlinear Equality Constrained Optimization*, Ph. D. thesis, Dept of mathematical Sciences, Rice University, Texas, USA, 1985.
6. M.R. Celis, J.E. Dennis and R.A. Tapia, *A trust region algorithm for nonlinear equality constrained optimization*, in: P.T. Boggs, R.H. Byrd and R.B. Schnabel, eds., Numerical Optimization, SIAM, Philadelphia, 1985, 71–82.
7. T.F. Coleman and A.R. Conn, *Nonlinear programming via an exact penalty function: asymptotic analysis*, Mathematical Programming, **24**(1982), 123–136.
8. A.R. Conn, *Large-scale nonlinear constrainted optimization*, presented at the ICIAM'91, Washington DC, USA, 1991.
9. A.R. Conn, N.I.M. Gould and Ph.L. Toint, *Global convergence of a class of trust region algorithms for optimization with simple bounds*, SIAM J. Numer. Anal., **25**(1988), 433–460.
10. J.E. Dennis, *Trust region methods for constrained optimization*, presented at the 14th Intern. Symp. on Math. Prog., Amsterdam, The Netherlands, 1991.
11. J.E. Dennis, S.B. Li and R.A. Tapia, *A unified approach to global convergence of trust-region methods for nonsmooth optimization*, Technical Report, 89-5, Math. Sciences Dept. Rice University, Texas, USA, 1989.
12. J.E. Dennis and H.H.W. Mei, *Two new unconstrained optimization algorithms which use function and gradient values*, J. Opt. Theory and Applns., **28**(1979), 453–482.
13. J.E. Dennis and J.J. Moré. *A characterization of superlinear convergence and its application to quasi-Newton methods*, Math. Comp., **28**(1974), 549–560.
14. J.E. Dennis and K.A. Williamson, *A robust trust region algorithm for nonlinear programming*, Technical Report, Math. Sciences Dept., Rice University, Texas, USA. presented at the ICIAM 91 Conference, Washington DC, 1991.
15. I.S. Duff, J. Nocedal, and J.K. Reid, *The use of linear programming for the solution of sparse sets of nonlinear equations*, SIAM J. Sci. Stat. Comput., **8**(1987), 99–108.
16. M. El-Alem, *A global convergence theory for the Celis-Dennis-Tapia trust region algorithm for constrained optimization*, Technical Report, 88-10, Math. Sciences Dept., Rice University, Texas, USA, 1988.
17. M. El-Hallabi and R.A. Tapia, *A global convergence theory for arbitrary norm trust-region methods for nonlinear equations*, Technical Report, 87-25, Math. Sciences Dept., Rice University, Texas, USA, 1987.
18. R. Fletcher, *Practical Methods of Optimization, Vol. 1, Unconstrained Optimization*, John Wiley and Sons, Chichester, 1980.
19. R. Fletcher, *Practical Methods of Optimization, Vol. 2, Constrained Optimization*, John Wiley and Sons, Chichester, 1981.
20. R. Fletcher, *A model algorithm for composite NDO problem*, Math. Prog. Study, **17**(1982), 67–76. (1982a)
21. R. Fletcher, *Second order correction for nondifferentiable optimization*, in: G.A. Watson, ed., Numerical Analysis, Springer-Verlag, Berlin, 1982, 85–115. (1982b)
22. R. Fletcher, *Practical Methods of Optimization* (second edition), John Wiley and Sons, Chichester, 1987.
23. D.M. Gay, *Computing optimal local constrained step*, SIAM J. Sci. Stat. Comp., **2**(1981), 186–197.
24. D.M. Gay, *A trust region approach to linearly constrained optimization*, in: D.F. Griffiths, ed., Lecture Notes in Mathematics 1066: Numerical Analysis, Springer-Verlag, Berlin, 1978, 72–105.
25. H.F.H. Khalfan, *Topics in quasi-Newton methods for unconstrained optimization*, PhD thesis, University of Colorado, 1989.
26. K. Levenberg, *A method for the solution of certain nonlinear problems in least squares*, Qart. Appl. Math., **2**(1944), 164–166.
27. N. Maratos, *Exact Penalty Function Algorithms for Finite Dimensional and Control Optimization Problems*, Ph. D. thesis, Imperial College Sci. Tech., University of London, 1978.
28. D.W. Marquardt, *An algorithm for least-squares estimation of nonlinear inequalities*,

SIAM J. Appl. Math., **11**(1963), 431–441.
29. D.Q. Mayne and E. Polak, *A superlinearly convergent algorithm for constrained optimization problems*, Math. Prog. Study, **16**(1982), 45–61.
30. J.J. Moré, *The Levenberg-Marquardt algorithm: implementation and theory*, in: G.A. Watson, ed., Lecture Notes in Mathematics 630: Numerical Analysis, Springer-Verlag, Berlin, 1978, 105–116.
31. J.J. Moré, *Recent developments in algorithms and software for trust region methods*, in: A. Bachem, M. Grötschel and B. Korte, eds., Mathematical Programming: The State of the Art, Springer-Verlag, Berlin, 1983, 258–287.
32. J.J. Moré, *Trust region and projected gradients*, Report ANL/MCS-TM-107, Mathematics and Computer Science Division, Argonne National Laboratory, Argonne, Illinois, USA, 1988.
33. J.J. Moré and D.C. Sorensen, *Computing a trust region step*, SIAM J. Sci. Stat. Comp., **4**(1983), 553–572.
34. J. Nocedal and Y. Yuan, *Combining trust region and line search techniques*, Technical Report, NAM06, Dept of Computer Science, Northwestern University, Illinois, USA, 1991.
35. E. O. Omojokun, Trust Region Algorithms for Optimization with Nonlinear Equality and Inequality Constraints, Ph. D. Thesis, University of Colorado at Boulder, 1989.
36. M.J.D. Powell, *A new algorithm for unconstrained optimization*, in: J.B. Rosen, O.L. Mangasarian and K. Ritter, eds., Nonlinear Programming, Academic Press, New York, 1970, 31–66. (1970a)
37. M.J.D. Powell, *A hybrid method for nonlinear equations*, in: P. Robinowitz, ed., Numerical Methods for Nonlinear Algebraic Equations, Gordon and Breach Science, London, 1970, 87–144. (1970b)
38. M.J.D. Powell, *Convergence properties of a class of minimization algorithms*, in: O.L. Mangasarian, R.R. Meyer and S.M. Robinson, eds., Nonlinear Programming 2, Academic Press, New York, 1975, 1–27.
39. M.J.D. Powell, *On the global convergence of trust region algorithms for unconstrained optimization*, Math. Prog., **29**(1984), 297–303. (1984a)
40. M.J.D. Powell, *Genera algorithms for discrete nonlinear approximation calculations*, in: L.L. Schumarker, ed., Approximation Theory IV, Academic Press, New York, 1984, 187–218. (1984b)
41. M.J.D. Powell and Y. Yuan, *Conditions for superlinear convergence in l_1 and l_∞ solutions of overdetermined nonlinear equations*, IMA J. Numerical Analysis, **4**(1984), 241–251.
42. M.J.D. Powell and Y. Yuan, *A trust region algorithm for equality constrained optimization*, Mathematical Programming, **49**(1991), 189–211.
43. G.A. Shultz, R.B. Schnabel and R.H. Byrd, *A family of trust-region-based algorithms for unconstrained minimization with strong global convergence*, SIAM J. Numer. Anal., **22**(1985), 47–67.
44. D.C. Sorensen, *Newton's method with a model trust region modification*, SIAM J. Numer. Anal., **20**(1982), 409–426. (1982a)
45. D.C. Sorensen, *Trust region methods for unconstrained optimization*, in: M.J.D. Powell, ed., Nonlinear Optimization 1981, Academic Press, London, 1982, 29–38. (1982b)
46. T. Steihaug, *The conjugate gradient method and trust regions in large scale optimization*, SIAM J. Numer. Anal., **20**(1983), 626–637.
47. P. Terpolili, *Trust region method in non-differentiable optimization: the use of exact and inexact local method*, presented at 14th Intern. Symp. on Math. Prog., Amsterdam, The Netherlands, 1991.
48. S.W. Thomas, *Sequential estimation techniques for quasi-Newton algorithms*, Technical Report, 75-227, Dept of Computer Science, Cornell University, Ithaca, NY, USA, 1975.
49. Ph.L. Toint, *Global convergence of a class of trust region methods for nonconvex minimization in Hilbert space*, IMA J. Numer. Anal., **8**(1988), 231–252.
50. A. Vardi, *A trust region algorithm for equality constrained minimization: convergence properties and implementation*, SIAM J. Numer. Anal., **22**(1985), 575–591.
51. R. S. Womersley, *Local properties of algorithms for minimizing nonsmooth composite func-*

tions, Mathematical Programming, **32**(1985), 69–89.

52. Y. Yuan, *An example of only linearly convergence of trust region algorithms for nonsmooth optimization*, IMA J. Numerical Analysis, **4**(1984), 327–335.
53. Y. Yuan, *Conditions for convergence of trust region algorithms for nonsmooth optimization*, Mathematical Programming, **31**(1985), 220–228. (1985a)
54. Y. Yuan, *On the superlinear convergence of a trust region algorithm for nonsmooth optimization*, Mathematical Programming, **31**(1985), 269–285. (1985b)
55. Y. Yuan, *A dual algorithm for minimizing a quadratic function with two quadratic constraints*, Report DAMTP 1988/NA3, University of Cambridge. (to appear in J. Computational Mathematics)
56. Y. Yuan, *On a subproblem of trust region algorithms for constrained optimization*, Mathematical Programming, **47**(1990), 53–63.
57. Y. Yuan, *A new trust region algorithm for nonlinear optimization*, Report No. 394, Schwerpunktprogramm der Deutschen Forsch, Anwendungsbezogene Optimierung und Steuerung, Inst. Angew. Math., Uni. Würzburg, Germany, 1992.
58. Y. Yuan, *Local convergence and numerical results of a trust region algorithm*, Report, Computing Center, Academia Sinica, Beijing, China 1993

Computing Center, Academia Sinica, P.O. Box 2719, Beijing 100080, China

Recent Titles in This Series

(Continued from the front of this publication)

132 **Mark Gotay, Jerrold Marsden, and Vincent Moncrief, Editors,** Mathematical aspects of classical field theory, 1992

131 **L. A. Bokut′, Yu. L. Ershov, and A. I. Kostrikin, Editors,** Proceedings of the International Conference on Algebra Dedicated to the Memory of A. I. Mal′cev, Parts 1, 2, and 3, 1992

130 **L. Fuchs, K. R. Goodearl, J. T. Stafford, and C. Vinsonhaler, Editors,** Abelian groups and noncommutative rings, 1992

129 **John R. Graef and Jack K. Hale, Editors,** Oscillation and dynamics in delay equations, 1992

128 **Ridgley Lange and Shengwang Wang,** New approaches in spectral decomposition, 1992

127 **Vladimir Oliker and Andrejs Treibergs, Editors,** Geometry and nonlinear partial differential equations, 1992

126 **R. Keith Dennis, Claudio Pedrini, and Michael R. Stein, Editors,** Algebraic K-theory, commutative algebra, and algebraic geometry, 1992

125 **F. Thomas Bruss, Thomas S. Ferguson, and Stephen M. Samuels, Editors,** Strategies for sequential search and selection in real time, 1992

124 **Darrell Haile and James Osterburg, Editors,** Azumaya algebras, actions, and modules, 1992

123 **Steven L. Kleiman and Anders Thorup, Editors,** Enumerative algebraic geometry, 1991

122 **D. H. Sattinger, C. A. Tracy, and S. Venakides, Editors,** Inverse scattering and applications, 1991

121 **Alex J. Feingold, Igor B. Frenkel, and John F. X. Ries,** Spinor construction of vertex operator algebras, triality, and $E_8^{(1)}$, 1991

120 **Robert S. Doran, Editor,** Selfadjoint and nonselfadjoint operator algebras and operator theory, 1991

119 **Robert A. Melter, Azriel Rosenfeld, and Prabir Bhattacharya, Editors,** Vision geometry, 1991

118 **Yan Shi-Jian, Wang Jiagang, and Yang Chung-chun, Editors,** Probability theory and its applications in China, 1991

117 **Morton Brown, Editor,** Continuum theory and dynamical systems, 1991

116 **Brian Harbourne and Robert Speiser, Editors,** Algebraic geometry: Sundance 1988, 1991

115 **Nancy Flournoy and Robert K. Tsutakawa, Editors,** Statistical multiple integration, 1991

114 **Jeffrey C. Lagarias and Michael J. Todd, Editors,** Mathematical developments arising from linear programming, 1990

113 **Eric Grinberg and Eric Todd Quinto, Editors,** Integral geometry and tomography, 1990

112 **Philip J. Brown and Wayne A. Fuller, Editors,** Statistical analysis of measurement error models and applications, 1990

111 **Earl S. Kramer and Spyros S. Magliveras, Editors,** Finite geometries and combinatorial designs, 1990

110 **Georgia Benkart and J. Marshall Osborn, Editors,** Lie algebras and related topics, 1990

109 **Benjamin Fine, Anthony Gaglione, and Francis C. Y. Tang, Editors,** Combinatorial group theory, 1990

108 **Melvyn S. Berger, Editor,** Mathematics of nonlinear science, 1990

107 **Mario Milman and Tomas Schonbek, Editors,** Harmonic analysis and partial differential equations, 1990

106 **Wilfried Sieg, Editor,** Logic and computation, 1990

105 **Jerome Kaminker, Editor,** Geometric and topological invariants of elliptic operators, 1990

(See the AMS catalog for earlier titles)

MICROECONOMICS

MICROECONOMICS

TENTH

Richard G. Lipsey

Simon Fraser University

CANADIAN

Christopher T. S. Ragan

McGill University

EDITION

Toronto

Canadian Cataloguing in Publication Data

Lipsey, Richard G., 1928-
Microeconomics

10th Canadian ed.
ISBN 0-201-66469-0

1. Micoeconomics. I. Ragan, Christopher. II. Title.

HB172.L56 2001 338.5 C00-930276-X

0-201-66469-0

Vice President, Editorial Director: Michael Young
Acquisitions Editor: Dave Ward
Marketing Manager: James Buchanan
Developmental Editor: Maurice Esses
Production Editor: Mary Ann McCutcheon
Copy Editor: Edie Franks
Production Coordinator: Deborah Starks
Page Layout: Carol Magee
Permissions and Photo Research: Susan Wallace-Cox
Art Director: Mary Opper
Interior Design: Anthony Leung
Cover Design: Artville LLC
Cover Image: Artville LLC

2 3 4 5 05 04 03 02 01

Printed and bound in the United States